Biodiversity and Ecosystem Functioning

Biodiversity and Ecosystem Functioning

Synthesis and Perspectives

EDITED BY

Michel Loreau
Laboratoire d'Ecologie, Ecole Normale Supérieure, Paris, France

Shahid Naeem
Department of Zoology, University of Washington, USA

Pablo Inchausti
Laboratoire d'Ecologie, Ecole Normale Supérieure, Paris, France

OXFORD
UNIVERSITY PRESS

This book has been printed digitally and produced in a standard specification
in order to ensure its continuing availability

OXFORD
UNIVERSITY PRESS

Great Clarendon Street, Oxford OX2 6DP
Oxford University Press is a department of the University of Oxford.
It furthers the University's objective of excellence in research, scholarship,
and education by publishing worldwide in
Oxford New York

Auckland Cape Town Dar es Salaam Hong Kong Karachi
Kuala Lumpur Madrid Melbourne Mexico City Nairobi
New Delhi Shanghai Taipei Toronto
With offices in
Argentina Austria Brazil Chile Czech Republic France Greece
Guatemala Hungary Italy Japan South Korea Poland Portugal
Singapore Switzerland Thailand Turkey Ukraine Vietnam

Oxford is a registered trade mark of Oxford University Press
in the UK and in certain other countries

Published in the United States
by Oxford University Press Inc., New York

© Oxford University Press 2002

The moral rights of the author have been asserted

Database right Oxford University Press (maker)

Reprinted 2009

ISBN 978-0-19-851571-5

Preface

The study of biodiversity and ecosystem functioning has followed a pattern that often characterizes history in science. This pattern is best described as periods of empirical and theoretical development bracketed by periods of synthesis (Kuhn 1962; Kingsolver and Paine 1991). This is not an even course; new developments are often accompanied by debate or controversy (Dunwoody 1999).

A conference, entitled *Biodiversity and ecosystem functioning: synthesis and perspectives*, was held in Paris, France, on 6–9 December 2000 under the auspices of the International Geosphere–Biosphere Programme—Global Change and Terrestrial Ecosystems (IGBP–GCTE) and DIVERSITAS, international programmes that foster communication among scientists involved in global change and biodiversity research. The conference was designed to facilitate synthesis of nearly a decade of observation, theory, and experiment in biodiversity and ecosystem functioning research. Its goals were to identify central principles, certainties, uncertainties, future directions, and policy implications in this area. A brief report of the conference was published in *Trends in Ecology and Evolution* (Hughes and Petchey 2001), and a summary of its main findings was published in *Science* (Loreau *et al.* 2001). This volume provides overviews, position papers, and reports from the synthesis workshops of the conference, which together give a synthetic and balanced account of the current knowledge and future challenges in the fast growing area of biodiversity and ecosystem functioning.

The conference was a delight. Virtually every invitation was accepted (indeed, many could not be invited or were turned away to keep the workshops of manageable size) in the interest of resolving the issues. The distribution of participants was broad, most importantly being weighted towards junior and emerging researchers. The presentations, workshops, and panel discussions were extraordinarily cordial, friendly, and interactive. Not unexpectedly, some left with as strong an opinion as they arrived with, but all were encouraged to explore the issues in greater depth and all had a greater appreciation of the perspectives and the fascinating science behind the varied perspectives.

The conference was made possible by the financial support provided by the European Science Foundation LINKECOL programme, the Centre National de la Recherche Scientifique (France), and the US National Science Foundation (DEB NSF DEB 973343). Some who attended contributed to the workshops and panel discussions although they could not contribute to the chapters. In addition, we wish to acknowledge the help of many anonymous individuals who provided critical reviews of the chapters, and Paola Paradisi, Régine Mfoumou, Christelle Blée, Marie-Bernadette Tesson and Susie Dennison who helped with logistics. And to all those that space does not provide for a proper acknowledgment, we thank for help in making the conference the success that it was.

Michel Loreau, Shahid Naeem and Pablo Inchausti
14 January 2002

Contents

Contributors

Richard D. Bardgett, Department of Biological Sciences, Institute of Environmental and Natural Sciences, University of Lancaster, Lancaster LA1 4YQ, UK. e-mail r.bardgett@lancaster.ac.uk

Jan Bengtsson, Department of Ecology and Crop Production Science, Swedish University of Agricultural Sciences, SLU, BOX 7043, S-750 07 Uppsala, Sweden. e-mail Jan.Bengtsson@evp.slu.se

Nina Buchmann, Max-Planck-Institute for Biogeochemistry, P.O. Box 10 01 64, 07701 Jena, Germany. e-mail buchmann@bgc-jena.mpg.de

Johannes H. C. Cornelissen, Department of Systems Ecology, Free University, De Boelelaan 1087, 1081 HV Amsterdam, The Netherlands. e-mail hansco@bio.vu.nl

Valérie Degrange, UMR CNRS 5557—Ecologie Microbienne du Sol, UFR de Biologie—Université Lyon I, 43 Bd. du 11 Novembre 1918, 69622 Villeurbanne cedex, France. e-mail Valerie.Degrange@univ-lyon1.fr

Sandra Díaz, Instituto Multidisciplinario de Biología Vegetal (IMBIV), Universidad Nacional de Córdoba—CONICET, Casilla de Correo 495, Vélez Sarsfield 299, 5000 Córdoba, Argentina. e-mail sdiaz@com.uncor.edu

Amy Downing, Department of Ecology and Evolution, The University of Chicago, 1101 E. 57th Street, Chicago, IL 60637, USA. e-mail adowning@midway.uchicago.edu

Mark Emmerson, Department of Biology, University of York, P.O. Box 373, York YO10 5YW, UK. e-mail mce1@york.ac.uk

Katia Engelhardt, University of Maryland, Center for Environmental Science, Appalachian Lab, 301 Braddock Road, Frostburg, MD 21532-2307, USA. e-mail engelhardt@al.umces.edu

Mark Gessner, Department of Limnology, Limnological Research Center, 6047 Kastanienbaum, Switzerland. e-mail mark.gessner@eawag.ch

Paul Giller, University College Cork, Department of Zoology, Lee Maltings Prospect Row, Cork, Ireland. e-mail deanofscience@ucc.ie

Andrew Gonzalez, Laboratoire d'Ecologie, Ecole Normale Supérieure, 46 rue d'Ulm, Paris 75005, France. e-mail gonzalez@wotan.ens.fr

Bryan Griffiths, Scottish Crop Research Institute, Invergowrie, Dundee DD2 5DA, UK. e-mail B.Griffiths@scri.sari.ac.uk

Philip Grime, Unit of Comparative Plant Ecology, Department Animal & Plant Sciences, University of Sheffield, Sheffield S10 2TN, UK. e-mail j.p.grime@sheffield.ac.uk

Andy Hector, NERC Centre for Population Biology, Imperial College at Silwood Park, Ascot, Berkshire GB-SL5 7PY, UK. e-mail a.hector01@ic.ac.uk

Katarina Hedlund, Department of Ecology, Lund University, Sölvegatan 37, S-22362 Lund, Sweden. e-mail Katarina.Hedlund@zooekol.lu.se

Sarah E. Hobbie, Department of Ecology, Evolution, and Behavior, University of Minnesota, 1987 Upper Buford Circle, St. Paul MN 55108, USA. e-mail shobbie@umn.edu

David Hooper, Department of Biology, Western Washington University, Bellingham, WA 98225-9160, USA. e-mail hooper@biol.wwu.edu

Jennifer B. Hughes, Department of Ecology and Evolutionary Biology, Brown University, Providence, Rhode Island, USA. e-mail Jennifer_Hughes@brown.edu

Florence Hulot, University of Amsterdam, Faculty of Science, Aquatic Microbiology—IBED, Nieuwe Achtergracht 127, NL-1018 WS Amsterdam, The Netherlands. e-mail fhulot@science.uva.nl

Michael Huston, Environmental Sciences Division, Oak Ridge National Laboratory, Oak Ridge, TN 37831-6355, USA. e-mail hustonma@ornl.gov

Mark Huxham, Napier University, School of Life Sciences, 10 Colinton Road, Edinburgh, Scotland EH10 4NY, UK. e-mail m.huxham@napier.ac.uk

Pablo Inchausti, Laboratoire d'Ecologie, Ecole Normale Supérieure, 46 rue d'Ulm, Paris 75005, France. e-mail inchausti@biologie.ens.fr

Anthony R. Ives, Department of Zoology, University of Wisconsin-Madison, Madison, WI 53706, USA. e-mail arives@facstaff.wisc.edu

Jasmin Joshi, Institut für Umweltwissenschaften, Universität Zürich, Winterthurerstrasse 190, Zürich CH-8057, Switzerland. e-mail joshi@uwinst.unizh.ch

Theodore Kennedy, Department of Ecology, Evolution, and Behavior, University of Minnesota, 1987 Upper Buford Cr. St. Paul, MN 55108, USA. e-mail kenn0148@tc.umn.edu

Jean Knops, School of Biological Sciences, University of Nebraska, 348 Manter Hall, Lincoln, NE 68588-0118, USA. e-mail Jknops@unlnotes.unl.edu

Julia Koricheva, Section of Ecology, Department of Biology, University of Turku, FIN-20014, Turku, Finland. e-mail julkoric@utu.fi

Deborah Lawrence, Department of Environmental Sciences, P.O. Box 400123, University of Virginia, Charlottesville, VA 22904-4123, USA. e-mail dl3c@virginia.edu

Paul Leadley, Université Paris-Sud XI, Ecologie des Populations et Communautés, Bâtiment 362, F-91405 Orsay Cedex, France. e-mail paul.leadley@epc.u-psud.fr

Jonathan M. Levine, NERC Centre for Population Biology, Imperial College at Silwood Park, Ascot, Berkshire SL5 7PY, UK. e-mail j.levine@ic.ac.uk

Michel Loreau, Laboratoire d'Ecologie, UMR 7625, Ecole Normale Supérieure, 46 rue d'Ulm, F-75230 Paris Cedex 05, France. e-mail loreau@ens.fr

Allen McBride, Class of 2004, Swarthmore College, 500 College Avenue, Swarthmore, PA 19081, USA. e-mail Amcbrid1@swarthmore.edu

Jill McGrady-Steed, Department of Ecology, Evolution, & Natural Resources, 14 College Farm Road, Cook College, Rutgers University, New Brunswick, NJ 08901, USA. e-mail jkm46@eden.rutgers.edu

Florian Mermillod, UMR CNRS 5023, Ecologie des Hydrosystème Fluviaux, Université Lyon 1, 69622 Villeurbanne Cedex, France. e-mail Florian.mermillod@univ-lyon1.fr

Juha Mikola, Department of Biological and Environmental Science, University of Jyväskylä, P.O. Box 35 (YAC), 40351 Jyväskylä, Finland. e-mail jmikola@cc.jyu.fi

John C. Moore, Department of Biology, University of Northern Colorado, Greeley, 80523 CO, USA. e-mail johnm@NREL.ColoState.EDU

H. A. Mooney, Department of Biological Sciences, Stanford University, Stanford, CA, 94305, USA. e-mail hmooney@jasper.stanford.edu

Peter J. Morin, Department of Ecology, Evolution, & Natural Resources, 14 College Farm Road, Cook College, Rutgers University, New Brunswick, NJ 08901, USA. e-mail pjmorin@rci.rutgers.edu

Shahid Naeem, Department of Zoology, University of Washington, 24 Kincaid Hall, BOX 351800, Seattle, WA 98195-1800, USA. e-mail naeems@u.washington.edu

Ivan Nijs, Department of Biology, University of Antwerp (UIA), Universiteitsplein 1, B-2610 Wilrijk, Belgium. e-mail inijs@uia.ua.ac.be

Jon Norberg, Department of Systems Ecology, 106 91 Stockholm, Stockholm University, Sweden. e-mail jon.norberg@ecology.su.se

Lennart Persson, Department of Ecology and Environmental Science, Animal Ecology, Umeå University, 901 87 Umeå, Sweden. e-mail Lennart.Persson@eg.umu.se

Owen L. Petchey, Department of Animal and Plant Sciences, University of Sheffield, Alfred Denny Building, Sheffield, S10 2TN, UK. e-mail o.petchey@sheffield.ac.uk

Dave Raffaelli, Environment Department, University of York, Heslington, York YO10 5DD, UK. e-mail drs3@york.ac.uk

Peter Reich, Department of Forest Resources, University of Minnesota, 1530 No. Cleveland Avenue, St. Paul, MN 55108, USA. e-mail preich@forestry.umn.edu

Jacques Roy, Centre d'Ecologie Fonctionnelle et Evolutive, GDR 1936 DIV-ECO, CNRS 34293 Montpellier Cedex 5, France. e-mail roy@cefe.cnrs-mop.fr

Peter C. de Ruiter, Department of Environmental Sciences, University Utrecht, P.O. Box 80115, 3508 TC Utrecht, The Netherlands. e-mail p.deruiter@geog.uu.nl

Osvaldo E. Sala, University of Buenos Aires, Department of Ecology, Facultad de Agronomia, Av San Martin 4453, Buenos Aires 1417, Argentina. e-mail sala@ifeva.edu.ar

Bernhard Schmid, Institut für Umweltwissenschaften, Universität Zürich, Winterthurerstrasse 190, Zürich CH-8057, Switzerland. e-mail bschmid@uwinst.unizh.ch

Martin Solan, Ocean Laboratory, Department of Zoology, University of Aberdeen, Newburgh, Aberdeenshire, AB41 6AA, UK. e-mail m.solan@abdn.ac.uk

Eva Spehn, Institut der Botanisches, University Basel, Schoenbeinstr.6, 4056, Basel, Switzerland. e-mail Eva.Spehn@unibas.ch

Amy Symstad, Illinois Natural History Survey, Lost Mound Field Station, 3159 Crim Dr, Savanna, IL 61074, USA. e-mail asymstad@inhs.uiuc.edu

David Tilman, Department of Ecology, Evolution and Behavior, University of Minnesota, 1987 Upper Buford Circle, St. Paul, MN 55108-6097, USA.
e-mail tilman@umn.edu

Marcel van der Heijden, Department of Systems Ecology, Free University, De Boelelaan 1087, 1081 HV Amersterdam, The Netherlands.
e-mail heijden@bio.vu.nl

Wim van der Putten, Department of Multitrophic Interactions, NIOO-CTO, P.O. Box 40, 6666 ZG Heteren, The Netherlands.
e-mail putten@cto.nioo.knaw.nl

Liesbeth van Peer, Research Group of Plant and Vegetation Ecology, Department of Biology, University of Antwerp, Universiteitsplein 1, 2610 Wilrijk, Belgium. e-mail lvanpeer@uia.ua.ac.be

John Vandermeer, Department of Ecology and Evolutionary Biology, 830 North University, Natural Science Building (Kraus), University of Michigan, Ann Arbor, MI 48109-1048, USA. e-mail jvander@umich.edu

Montserrat Vilà, Centre de Recerca Ecològica i Aplicacions Forestals, Universitat Autònoma de Barcelona, 08193 Barcelona, Spain.
e-mail vila@cc.uab.es

David A. Wardle, Department of Forest Vegetation Ecology, Faculty of Forestry, Swedish University of Agricultural Sciences, SE901-83 Umea, Sweden.
e-mail david.wardle@svek.slu.se

David Wedin, School of Natural Resource Sciences, University of Nebraska, 104 Plant Industry, Lincoln, NE 68583-0814, USA.
e-mail dwedin1@unl.edu

Volkmar Wolters, Department of Animal Ecology, Justus-Liebig-University, Heinrich-Buff-Ring 26-32 (IFZ), 35392 Giessen, Germany.
e-mail Volkmar.Wolters@allzool.bio.uni-giessen.de

PART I

Introduction

Biodiversity and ecosystem functioning: the emergence of a synthetic ecological framework

S. Naeem, M. Loreau, and P. Inchausti

1.1 Understanding the significance of biodiversity

Earth's biota is not a passive epiphenomenon of Earth's physical conditions and geochemical processes. Through the collective metabolic and growth activities of its trillions of organisms, Earth's biota[1] moves hundreds of thousands of tons of elements and compounds between the hydrosphere, atmosphere, and lithosphere every year. It is this biogeochemical activity that determines soil fertility, air and water quality, and the habitability of ecosystems, biomes, and ultimately the Earth itself (Lovelock 1979; Butcher *et al.* 1992; Schlesinger 1997). Indeed, biogeochemistry makes Earth a unique planet in the solar system (Ernst 2000).

While the functional[2] significance of Earth's biota to ecosystem or Earth-system functioning is well established, the significance of Earth's *biodiversity*[3] has remained unknown until recently. For example, we have a well-developed understanding of photosynthetic production at minute (e.g. subcellular) scales (Hall *et al.* 1993) and a well-developed understanding of primary productivity at global scales (Roy *et al.* 2001), yet we have, by comparison, little understanding of how plant diversity in a grassy meadow, desert, or forest affects production at the ecosystem, biome, or global scale.

At a time when biodiversity is undergoing dramatic changes in distribution and abundance (Ehrlich 1988; Wilson 1988; Soulé 1991; Reaka-Kudla *et al.* 1997; Stork 1997), predicting the ecosystem or Earth-system consequences of such change is a critical issue (Ehrlich and Wilson 1991; Chapin III *et al.* 2000). At a conference in Bayreuth, Germany, organized by E.-D. Schulze and H. A. Mooney in 1991, the proceedings of which were published in 1993 (Schulze and Mooney 1993), ecologists formerly reviewed what was known about the relationship between biodiversity and Earth-system and ecosystem functioning (henceforth, 'biodiversity-functioning' research). Since then, this focus has become a major thrust in contemporary ecology, reflecting a modern synthesis in which the study of biodiversity (e.g. distribution and abundance) is merged with the study of ecosystem functioning (e.g. biogeochemical processes). H. A. Mooney (Chapter 2) traces the events that led to this 1991 symposium and the explosion of research that followed shortly after its publication.

Although the studies that have contributed to this discipline represent a broad array of individuals and their collective expertise, a considerable amount of debate has arisen concerning its findings (André *et al.* 1994; Givnish 1994; Aarssen 1997; Garnier *et al.* 1997; Grime 1997; Huston 1997; Tilman *et al.* 1997a; Wardle *et al.* 1997c, 2000b; Hector 1998; Hodgson *et al.* 1998; Lawton *et al.* 1998; Loreau 1998b;

---

[1] By 'biota' we mean all biological entities in a habitat, ecosystem, or larger region, independent of its diversity.

[2] By 'functional' or 'functioning', we mean the activities, processes, or properties of ecosystems that are influenced by its biota. In no case is 'purpose' inferred in our usage of these terms.

[3] 'Biodiversity' refers to the extent of genetic, taxonomic, and ecological diversity over all spatial and temporal scales (Harper and Hawksworth 1995).

Wardle 1998, 1999; Naeem 1999; van der Heijden *et al.* 1999; Hector *et al.* 2000b; Huston *et al.* 2000; Naeem 2000; Tilman 2000) and its presentation in the press has been negative (Guterman 2000; Kaiser 2000).

A conference held in Paris, France, in December 2000, entitled *Biodiversity and ecosystem functioning: synthesis and perspectives* (henceforth, the Synthesis Conference), brought together researchers representing the full gamut of expertise and opinion on the empirical and theoretical foundations of biodiversity-functioning research. This volume is the outcome of that conference (see Preface), and this chapter provides a brief review of the topic and the content of this volume.

1.2 A brief history of biodiversity and ecosystem functioning

1.2.1 Early history

Initially, there may have been little question concerning the relationship between biodiversity and ecosystem functioning, though the perspective was predominantly one of metaphysical harmony among species and their environment (Egerton 2001). For example, Aristotle (384–322 BC) considered all entities to be made up of five elements (earth, fire, water, air and a fifth element known as the ether or the *quinta essencia*). Thus organisms, habitat, and environment were seen as one and it would likely have been an uninteresting question to ask if biodiversity and ecosystem functioning were related. This powerful construct endured nearly 2000 years until the scientific revolution of seventeenth-century Europe. While the positive aspects of the revolution are well documented, the abandonment of Aristotelian thinking fractionated the sciences, and natural history in particular underwent considerable transformation (Henry 1997). This hindered progress in biodiversity-functioning research, which requires multi-disciplinary approaches that integrate across such fields as botany, zoology, microbiology, chemistry, physics, and geology, to name just a few.

Today, ecologists and environmentalists understand that environment and habitat are the endpoints of the collective activities of abiotic and biotic processes shaped by history. The biosphere[4] is recognized as a vast, staggeringly complex, highly dynamic system made up of some 10–100 million species that share over 3.5 billion years of history and currently occupies virtually all $5.10 \times 10^{14}\,\mathrm{m^2}$ of the Earth's terrestrial and aquatic surfaces. Clearly, to understand the functioning of Earth systems requires not only understanding biogeochemistry, but also the role that biodiversity plays in this complex system.

In spite of its rapid growth, however, the inclusion of biodiversity in Earth-system and ecosystem science has only recently become a growing part of ecological research (Mooney, Chapter 2). This lack of inclusion most likely stems from the fact that ecology has been historically divided primarily into two disciplines: community ecology and ecosystem ecology (Likens 1992; Grimm 1995; Loreau 2000b; Loreau *et al.* 2001). Community ecology focused on how extrinsic factors[5] such as climate, disturbance, or site fertility affect biodiversity and how intrinsic factors[6] affect biodiversity dynamics. In contrast, ecosystem ecologists have focused on the rates, dynamics, and stability of energy flow and nutrient cycling within ecosystems. Over the last decade, however, synthetic studies that consider both biodiversity and ecosystem functioning have grown to become an integral part of the ecological literature (Fig. 1.1).

Like much of science, however, if one searches, earlier works that predate current activities are often found. For example, the importance of atmospheric greenhouse gasses in climate was recognized by Jean-Baptiste Fourier in 1827 (Houghton 1997) and the logistic model was first described by Pierre-François Verhulst in the 1830s. Similarly, Darwin himself (McNaughton 1993) and the ecological

---

[4] 'Biosphere' is the global domain within which biodiversity is found. This domain is located between the Earth's lithosphere and atmosphere, occupying a layer that includes parts of the atmosphere, hydrosphere, and lithosphere.

[5] By 'extrinsic' we mean primarily abiotic processes such as disturbance and climate.

[6] By 'intrinsic' we mean biotic factors such as biotic interactions (e.g. competition, predation, mutualism) or community structure (e.g. the type, strength, and number of biotic interactions among species in a community that often describe webs or networks of material and energy flow among species).

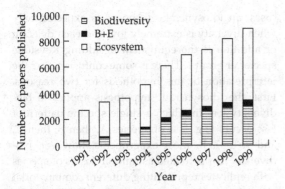

Figure 1.1 The emerging synthesis of biodiversity and ecosystem research. Results represent publications that included either 'biodiversity', 'ecosystem', or both (black fill in centre of bars). Note the dramatic increase in biodiversity research and the steady rise in papers that use both terms in their titles or abstracts. BIOSIS was the literature data base used for this figure.

experiments he cited (Hector and Hooper, 2002) predate current biodiversity-functioning research by 150 years. Perhaps the most prominent research, including that which inspired Darwin, has centred on agro-ecological efforts to improve yield through intercropping (Trenbath 1974; Vandermeer 1989; Swift and Anderson 1993). Although this research is distinct in its motivation and intent, the recent explosion of research concerning biodiversity and ecosystem functioning has venerable roots.

1.2.2 The central hypothesis of biodiversity and ecosystem functioning

The Bayreuth conference formally identified the central idea that was graphically portrayed by Vitousek and Hooper (1993) in their contribution to the symposium and has since expanded to an extensive list of hypotheses (Schläpfer and Schmid 1999). On the surface, it is a relatively simple idea with only two points in a bivariate plane that describe the central idea (Fig. 1.2). The axes defining the plane are biodiversity on the x-axis as the independent variable and some ecosystem process on the y-axis as the dependent variable. The first point of interest is the point at or near zero biodiversity. If there is no biodiversity (e.g. no plants) there is no ecosystem functioning (e.g. no production). The second point is the natural level of biodiversity where there is a highly predictable amount of

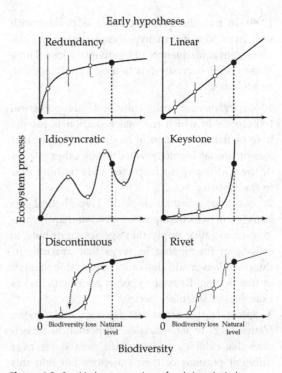

Figure 1.2 Graphical representations of early hypothetical relationships between biodiversity and ecosystem processes. These were meant primarily as heuristic devices or graphical representations of testable hypotheses representing a variety of potential mechanisms. Beyond the point at the origin (where there is no diversity and therefore no measurable processes) and the often highly predictable second point at natural levels of diversity, there was insufficient empirical and theoretical information to know under which circumstances which of the above possible relationships applied to ecosystems. Contemporary research rarely refers to these early hypotheses, although the terminology is still often in use when referring to different classes of associations.

functioning. The central question concerns what the trajectory might look like for a particular function for a given variation in biodiversity away from the second point. For example, what are the ecosystem consequences (the shape of the trajectory in the bivariate space) of local extinction (biodiversity loss, moving to the left of the second point) or invasion (biodiversity augmentation, moving to the right of the second point)?

When ecologists were asked what the shape of the trajectory might be (though the emphasis has traditionally been on biodiversity loss) a wonderful breadth of ideas emerged (Schläpfer and Schmid

1999). In fact, this bivariate space is packed with well over 50 different hypotheses concerning the ecosystem consequences of biodiversity loss. Three classes of biodiversity-functioning hypotheses can be identified.

1. *Species are primarily redundant.*[7] Hypothetical trajectories in which the major portion is insensitive or flat to variation in biodiversity imply that loss of species is compensated for by other species or the addition of such species adds nothing new to the system.

2. *Species are primarily singular.*[8] Hypothetical trajectories in which slopes are predominantly positive or negative imply that species contribute to ecosystem functioning in ways that are unique, thus their loss or addition causes detectable changes in functioning. Keystone species are often cited as examples of singular species.

3. *Species impacts are context-dependent and therefore idiosyncratic*[9] *or unpredictable.* Hypothetical trajectories that exhibit a variety of different slopes over different portions of their trajectory fall into this category. In such cases, the impact of loss or addition of a species depends on conditions (e.g. community composition, site fertility, disturbance regime) under which the local extinction or addition occurs.

Biodiversity-functioning research frequently uses terminology associated with these classes of hypotheses. The 'Rivet' hypothesis, accredited to Paul and Anne Ehrlich, reflects the notion that redundancy is important to a point where once so many species are lost, the system fails, much the way an engineered system fails when it looses too many rivets. The 'Redundancy' hypothesis refers to an asymptotic trajectory that asymptotes at extremely low levels of diversity. The 'Keystone' species refers to a trajectory in which functioning plummets as soon as biodiversity declines from its natural levels. The 'Idiosyncratic' hypothesis is the label ascribed to a trajectory that exhibits no clear trend. In many

---

[7] 'Redundant' implies that species are at least partially substitutable.

[8] 'Singular' implies that species make unique contributions to ecosystem functioning.

[9] 'Idiosyncratic' implies that a species makes different contributions to ecosystems depending on extrinsic and intrinsic factors.

cases, an idiosyncratic response is expected only when diversity is extremely low and each deletion or addition is the equivalent of adding Keystone species or groups. There is some confusion over the interpretation of this hypothesis for two reasons. First, the Idiosyncratic hypothesis applies to biodiversity loss in which a single specific pattern of loss occurs (e.g. loss of species A, then B, then C, and so on). In most experiments, however, biodiversity loss is treated as random losses of species with replicates representing different combinatorial permutations (e.g. loss of A or B or C followed by loss of A and B or B and C or C and A, and so on) and the trajectory plots the average change in functioning in response to the random biodiversity loss. The second common error is to associate the Idiosyncratic hypothesis with an inability to reject the null hypothesis or the hypothesis that the slope of the relationship is indistinguishable from zero (the absence of any evidence that biodiversity drives ecosystem functioning). Idiosyncratic does not mean that there is no effect of variation in biodiversity nor does it necessarily imply that response to variation in diversity is unpredictable. It merely implies that the slope of the relationship is not monotonic.

1.3 Articulating the hypothesis

The hypotheses outlined in Fig. 1.1 represent a heuristic framework that serves to organize our thoughts about the relationship between biodiversity and ecosystem functioning. It also provides a clear picture of what needs to be tested—reject the null hypothesis of no relationship between biodiversity as an independent variable and ecosystem functioning as a dependent variable.

New ideas are tested primarily by 'articulating the hypothesis' (Kuhn 1962) and experiments that followed the Bayreuth conference did exactly that. They created a gradient in biodiversity under homogeneous extrinsic conditions (e.g. fertility, climate, space, history) and monitored a variety of functions as response variables. Subsequent experiments, all of which are reviewed in this volume, added additional dimensions or searched for mechanisms.

The first such experiment designed to articulate the hypothesis was conducted in the Ecotron (Naeem *et al.* 1994a, 1995), a controlled environmental facility designed for ecological research (Lawton *et al.* 1993). It established replicates of high-, intermediate-, and low-diversity, terrestrial ecosystems. Thus, each level represented increasingly depauperate versions of the highest diversity level. It also simultaneously established 164 combinations of the 16 species of plants used in the main experiment at constant densities (16 individuals per pot) following a combinatorial design (Naeem *et al.* 1996). Its initial report (Naeem *et al.* 1994a) concluded, 'Our study demonstrates for the first time under controlled environmental conditions, that loss of biodiversity, in addition to loss of genetic resources, loss of productivity, loss of ecosystem buffering against ecological perturbation, and loss of aesthetic and commercially valuable resources, may also alter or impair the services that ecosystems provide. However, different ecosystem processes respond differently to loss of biodiversity providing some support for several current hypotheses. To the extent that loss of plant biodiversity in the real world means a reduction in the ability of ecosystems to fix CO_2, we also tentatively conclude that the loss of diversity may reduce the ability of terrestrial ecosystems to absorb anthropogenic CO_2.'

The Ecotron study effectively demonstrated that the relationship between diversity and ecosystem functioning was not flat. It hypothesized that niche complementarity,[10] or a greater efficiency of light utilization in more diverse communities due to differences in growth forms among species, was the mechanism responsible for the observed positive relationship between diversity and production. It also emphasized that processes varied in response and that other factors in nature contribute to biodiversity-functioning relationships and the value of biodiversity.

Early experiments attract considerable attention in science and the Ecotron study was no exception

(Moore 1996). While it successfully articulated the hypotheses, however, its design in which lower levels of diversity were nested sets of higher levels, could not address many issues that would emerge as biodiversity-functioning research evolved (Lawton *et al.* 1998; Allison 1999).

Two influential early studies by Tilman and colleagues would focus biodiversity-functioning on plant diversity and plant production and this focus would dramatically increase the visibility of biodiversity and ecosystem functioning research. They would also catalyse emerging debates over the interpretation of biodiversity-functioning studies. The first paper, by Tilman and Downing (1994), reported that nitrogen-induced reductions in plant diversity lowered ecosystem resistance and resilience. Later, Tilman *et al.* (1996) reported that controlled variation in plant diversity was positively associated with production in experimental prairie grassland plots. The larger spatial and temporal scales of these experiments and their outdoor or field nature provided greater assurance that the findings were more applicable to natural systems than the short-term, growth-chamber studies of the Ecotron. The proposed mechanism was again niche complementarity, either for nitrogen or water use, rather than light as was proposed in the Ecotron study.

From these experimental beginnings, only six years ago, an explosion of research ensued that, in each case, added important dimensions to the expanding field of biodiversity-functioning research. McGrady-Steed *et al.* (1997) manipulated diversity across a range of autotrophic and heterotrophic species using microbial microcosms. This study would also be the first to examine invasibility as an ecosystem property associated with biodiversity. Naeem and Li (1997) would similarly use microbial microcosms, this time manipulating the number of species per functional group.[11] Hopper and Vitousek (1997) would manipulate functional diversity while Tilman *et al.* (1997a) would manipulate both species and functional group richness.

Many experimental studies followed, in each case adding more dimensions. Some hallmark

---

[10] 'Niche complementarity' refers to the greater efficiency (in space or over time) of resource use by a community of species whose niches are complementary (i.e. non-overlapping). For example, a community of shallow-rooting and deep-rooting plant species mines mineral nutrients in a fixed volume of soil better than either group does by itself.

[11] 'Functional group' refers to a set of species that are similar, and at least partially substitutable in their contribution to a specific ecosystem process (see Hooper *et al.*, Chapter 17).

contributions include Van der Heijden *et al.* (1998a,b) who examined mycorrhizal fungal diversity, and Hector *et al.* (1999) who examined within-site and across-site biodiversity-functioning relationships, and Sankaran and McNaughton (1999) who examined extrinsic factors. The explosion of research has gone on to tackle other dimensions such as different systems, like wetlands (Engelhardt and Ritchie 2001) or marine systems (Emmerson *et al.* 2001), or the role of producer–decomposer interactions (Naeem *et al.* 2000a), nutrients (Hulot *et al.* 2000), invasive species (Knops *et al.* 1999; Levine 2000b; Naeem *et al.* 2000b), and plant pathogens and insect diversity (Siemann 1998; Knops *et al.* 1999). Many other studies are reviewed in the following chapters.

1.3.1 Theoretical developments

Theory lagged slightly behind the experimental work in biodiversity-functioning research. While the earliest ideas can be traced to McNaughton (1977) and intercropping theory (Vandermeer 1989), models developed explicitly to address biodiversity-functioning really did not emerge until recently (Loreau *et al.*, Chapter 7). Tilman *et al.* (1997a) and Loreau (1998a–c) contributed early models exploring how competition for resources and niche complementarity could explain the relationship between plant species richness and plant production based on plant resource use models. Doak *et al.* (1998) made a significant contribution by arguing that statistical averaging could account for apparent stabilization of aggregate community properties (i.e. ecosystem processes) without invoking niche complementarity. Yachi and Loreau (1999) would introduce a mathematical formalization of the concept of biological insurance. Naeem (1998) would introduce the concept of ecosystem reliability. Hughes and Roughgarden (1998) would examine how interaction strengths affected biodiversity-functioning relationships. As in the experimental studies, theoretical studies began focusing on the initial issues raised by the experiments that articulated the central biodiversity-functioning hypothesis, but theory has grown to cover increasingly sophisticated topics.

1.3.2 How best to interpret the findings?

Two fundamental issues concerning the interpretation of biodiversity-functioning arose shortly after the publication of the early studies. First, given the weight of correlational studies that suggested contrary patterns of association between plant diversity and production or other ecosystem processes, it seems unlikely that patterns observed in biodiversity-functioning experiments are relevant in the face of overwhelming influences of extrinsic factors (Grime 1997; Wardle *et al.* 1997b). Second, it is difficult to separate effects due to the increasing probability that species with major impacts on whatever process is being measured are present in higher diversity experiments (the sampling effect) from effects due to niche complementarity (Aarssen 1997; Huston 1997; Wardle 2001). Additional issues such as the role of the below-ground community (Wardle *et al.* 2000b), the possibility that functional diversity is far more critical than species diversity (Hooper and Vitousek 1997, 1998), and the fact that higher diversity replicates are more similar to one another in composition than lower diversity communities were also raised (Wardle 1998; Fukami *et al.* 2001).

Although each query was met with a rebuttal by the authors of the biodiversity-functioning studies, the replies were deemed inadequate by those who disagreed with the interpretation of the findings. This is often the case in science—interpretation of results can vary dramatically among researchers. Often an experiment is deemed 'flawed' by those who disagree with the original author's interpretation, while the original authors see their rebuttals as satisfying the concerns raised and continue to perceive the original studies to be correct. Such exchange is commonplace in science although the tone of the debate in this instance was regrettable and led to negative effects when reported in the press (Mooney, Chapter 2).

1.3.3 The consequences of debate among researchers: the science/public disconnect

Given the excitement generated by the early findings, the ensuing explosion of research by a wide array of researchers, the fact that only a fraction of the studies were under criticism, the fact that the criticisms were being incorporated into the discipline, and the importance of the issues, it may seem surprising that the emerging debate wound up being reported by the press to represent the signs of a questionable science. A couple of selected quotes from the press convey this.

A long-simmering debate among ecologists over the importance of biodiversity to the health of ecosystems has erupted into a full-blown war. Opposing camps are duelling over the quality of key experiments, and some are flinging barbs at meetings and in journals.

                                                    Kaiser (2000)

The altercation went public when, in a letter in the July issue of the *Bulletin of the Ecological Society of America*, eight ecologists bluntly charged that the report was 'biased' and 'little more than a propaganda document'; made 'indefensible statements'; and set a 'dangerous precedent' for scientific societies by presenting only one side of the debate, even though the report seemed to represent the entire 7,600-member society.

                                                  Gutterman (2000)

In reality, there is no scientific discipline that is without its debate unless it is a stagnant discipline. The more dynamic, the more rapidly a discipline is evolving, the more it is surrounded by vigorous debate. The greatest advances in science are often surrounded by the most vigorous debates (Hellman 1998).

The direction science takes follows the weight of the evidence without waiting for debate to be resolved. Public debate, however, does not operate in this fashion. In a public debate, evidence is presented by both sides and juries, judges, or oracles decide which side is right and which wrong (Franklin 2001). In a court of law, in the press, or in a public debate, each party is accorded equal privileges. Each side may identify witnesses, assemble evidence, or elect representatives to present their case. Judgment is based on the persuasiveness of the different arguments to juries, tribunals, judges, other empowered individuals, or by the public who pass judgment individually or collectively by vote.

Scientists, if they turn to the public, are often not familiar with the way it handles evidence (Mooney, Chapter 2). As the science of ecology increasingly enters the public and policy arena, it may increasingly find that scientific debate may be

misunderstood much the way Creationists in the United States continue to use the debate in evolutionary biology to dismiss its findings. Synthetic approaches, however, can allow progress in scientific research while minimizing debate.

1.4 The Synthesis Conference: a critical phase in biodiversity-functioning research

The debate helped to crystallize several issues that served as guideposts for future directions in biodiversity-functioning research and motivated the Synthesis Conference. The first set of chapters (Chapters 3–6) highlight the core issues in the debate over the appropriate interpretation of biodiversity and ecosystem functioning research focusing primarily on the Cedar Creek and BIO-DEPTH grassland experiments. With these issues in mind, the remaining chapters explore how biodiversity and ecosystem functioning touches upon ecological stability, trophic levels and structure, and other dimensions in ecology.

In the first section, Tilman *et al.* (Chapter 3), using the Cedar Creek studies, and Hector *et al.* (Chapter 4), using the BIODEPTH studies, extensively analyse data from these experimental grassland systems to explore the relative contributions of different mechanisms to the biodiversity-functioning relationships. They interpret their evidence as a strong support to their original conclusions about the importance of niche complementarity and plant diversity in governing the positive plant biodiversity-production relations observed in these studies, but their more modern synthetic approach treats these effects as the result of multiple causes that include sampling, facilitation, and perhaps other causes. Schmid *et al.* (Chapter 6) detail the astonishing complexity involved in analysing such data, focusing on the dominant parametric statistical approaches employed in these studies. In contrast, Huston and McBride (Chapter 5), while allowing for the importance of diversity to ecosystems in general, nevertheless criticize the biodiversity-functioning programme. They argue that a variety of statistical problems limit the interpretation of current biodiversity-functioning studies and provide alternative interpretations of current findings.

While these first chapters underscore the complexities one encounters in biodiversity-functioning research, the remaining sections consider the broader scope of biodiversity's potential role in a variety of ecological processes. The second section revisits the stability-diversity debate beginning with Loreau *et al.*'s Chapter 7 that contrasts and compares traditional research in ecological stability with the emerging biodiversity-stability research. This new stability research, unlike its predecessor, explicitly addresses the links between species variability and the variability of aggregate ecosystem properties and shows how population responses to environmental fluctuations and evolutionary modifications provide new insights into this venerable issue. Hughes *et al.* (Chapter 8) derive the core theoretical foundations behind recent theoretical approaches that attempt to understand the relationship between biodiversity and variance in ecosystem properties. De Ruiter *et al.* (Chapter 9), focusing on the enormously complex belowground systems, emphasize that biodiversity may play important roles in a variety of unexpected ways, in system stability. Finally, Levine *et al.* (Chapter 10) examine how biodiversity contributes to invasibility, an aspect of stability first addressed by Elton (1958). Levine *et al.* contrast how biodiversity and the covarying extrinsic determinants of biodiversity determine ecosystem invasibility. These chapters clear up the misleading sense of *déjà vu* that some may have felt in the face of what appeared to be a re-emergence of the old stability-diversity debate.

The next section explores systems other than grassland plant communities, with an emphasis on the trophic dimension of biodiversity-functioning research. The role of trophic groups, trophic structure, food chains and food webs remains among one the largest issues in ecology, yet the role of the trophic dimension in biodiversity-functioning research has only recently begun to catch up with the progress made in plant studies. Petchey *et al.* (Chapter 11) explore the utility of microcosm research and review key microcosm experimental studies in this context, pointing to their value as proving grounds for otherwise empirically intractable theory that often calls for high levels of replication and many generations. Emmerson *et al.* (Chapter 12) shed what light they can on the role

of biodiversity in marine ecosystems given the paucity of marine biodiversity-functioning studies, emphasizing the importance of trophic groups and heterotrophic processes in these systems. Raffaelli *et al.* (Chapter 13) address the fact that linkages among species created by common energy and nutrient pathways mean that changes in one species invariably, either directly or indirectly, have impacts on others. These linkages generate patterns in distribution and abundance, feedback in population cycles, and determine the fate of energy and nutrient flow in ecosystems.

Nowhere else is the importance of these trophic linkages more clear than in below-ground or the decomposer subsystems where >90% of the energy that flows through an ecosystem ultimately passes. In fact, including de Ruiter *et al.* (Chapter 9), four chapters in this volume examine this subsystem. Wardle and van der Putten (Chapter 14) document the lack of evidence for biodiversity-functioning relationships in decomposer systems. A key issue raised in this chapter concerns the sensitivity of ecosystems to extrinsic factors, suggesting that extrinsic factors, not intrinsic factors (i.e. biodiversity), regulate functioning. Mikola *et al.* (Chapter 15) argue, based on many experimental and observational studies, that biodiversity is hardly likely to provide the kinds of relationships one has observed in contemporary biodiversity-functioning studies of grassland plots and microcosm experiments when it comes to below-ground communities. Thus, they second the cautionary message delivered by Wardle and van der Putten. Finally, van der Heijden and Cornelissen (Chapter 16) focus on the often-neglected symbiotic microorganisms that are common (up to 80%) associates of terrestrial plants emphasizing that diversity of these organisms may play important roles in governing above-ground production and carbon cycling.

The trophic dimension of ecosystems, however, is only one of several ecological dimensions that biodiversity-functioning research touches upon. The challenge of addressing the taxonomic–functional diversity dimension in biodiversity-functioning research is addressed by Hooper *et al.* (Chapter 17). In their chapter, they demonstrate that the distinction between taxonomic and functional diversity is critical not only to resolving debates, but to making progress in effective experimental design and policy development. The spatial and temporal dimensions are addressed by Bengtsson et al. (Chapter 18) who remind us that the power of any ecological science is its ability to provide scale-invariant principles, but current biodiversity and ecosystem functioning research is, to be blunt, pathetically limited in scale. Finally, the human dimension is addressed by Vandermeer et al. (Chapter 19), who provide several insights into what managed ecosystems are in comparison to unmanaged systems, how the biodiversity-functioning debate has long been a part of management issues, and provide a number of valuable ways to begin the badly needed dialog between managers and researchers.

1.5 Concluding comments

We have tried, in a limited space, to provide a brief synopsis of the biodiversity-functioning research programme, its central ideas, its terminology, and the issues it contends with to facilitate the reading of this volume for the uninitiated. We have also indicated, though it might be surprising to some, that the ecological consequences of changing patterns in biodiversity, either through extinction or addition, was poorly known until recently, but that over a brief span of time (<10 years) some insights have been derived from empirical and theoretical studies. Not surprisingly, early studies that articulated the hypotheses introduced more questions

than they addressed and the ensuing debate that surrounded interpretations of the rapidly accumulating findings generated a sense that it was possible that biodiversity really did not matter.

Few, however, if any would claim that there is no role for biodiversity in ecosystem processes or ecosystem functioning and that the ecosystem services humans derive from them are affected by the nature of the biota that govern these processes. There is still a debate over the relative or specific role of extrinsic factors, genetic, taxonomic, or functional diversity in ecosystem processes, but the scientific community should not ignore the issue because it is complex, confusing, or unclear. Rather, this should be seen as a challenge to be met. Although the relative contribution of multiple causes to ecological effects, such as the relative importance of soil fertility and plant species richness to production, are important and unresolved, the scientific exchange and debates should stimulate scientists towards resolving the issue, not deter further investigation. A fortunate outcome of the debate is that it has stimulated the quest for a better understanding of mechanisms.

One thing this volume makes clear is that ecological truth lies at the confluence of observation, theory, and experiment. It is through discourse among empiricists and theorists that findings and theory are sorted and matched and where there is a lack of correspondence, new challenges identified. This volume represents a critical step in this direction.

CHAPTER 2

The debate on the role of biodiversity in ecosystem functioning

H. A. Mooney

2.1 The issue

During the past decade there has been an impressive development of research inquiry into the issue of the role of biodiversity in the functioning of ecosystems. Large numbers of research scientists have been attracted to this area, rooms in research meetings have been filled with people presenting and listening to the results of new inquiries into this topic. The literature is burgeoning with new findings and insights. However, recently there has been an unusual discord within this research area that is having some unfortunate consequences in both the larger realm of science and policy but also at the fundamental level of a negative feedback to further inquiry. Young scientists are expressing some fear of stepping into an area where the community is divided and where chances of approval of proposed work might be thwarted by partisan review. This is certainly unfortunate. Of course, vigorous debate and falsification of hypotheses are the basic stuff of scientific inquiry, but it would seem that the present situation has escalated beyond the norms for science and that we are in danger of losing credibility for ecosystem ecology as a discipline because of a small technical battle over an ephemeral issue. Headlines by science journalists such as 'Have ecologists oversold biodiversity?' (Guterman 2000) and 'Rift over biodiversity divides ecologists' (Kaiser 2000) give the impression that ecologists have circled the wagons and are shooting inward as one noted ecologist has observed. The language of doubt used in the debate has escalated from

science words such 'hidden treatments' and 'sampling effects' to *ad hominen* attacks such as 'selective citation of literature' and 'stating opinions as facts'. How did we get to this state of affairs, and more importantly how do we get out of it?

2.1.1 One road to here

Before looking to the future I look backward to how we got into this issue in the first place. There of course is a long history of interest by ecologists in the role of biodiversity in ecological processes. The recent interest in the ecosystem functioning aspect of biodiversity specifically can be traced to the results of a meeting that was held in Mitwitz, Germany in 1991 and published in 1994 (Schulze and Mooney 1994). What sparked my own personal interest in this area and what stimulated the organization of that meeting along with Detlef Schulze was the development of the international research programme of global change (International Geosphere Biosphere Programme). I noted elsewhere (Mooney 1990, 1991, 1999) that the development of Earth system science was progressing in such a way that the role, in any meaningful way, of the biota in regulating the Earth's biogeochemical and energy cycles was being ignored. The physical scientists who were leading the development of global models could not accommodate biotic complexity and were quite satisfied with a simple biosphere that was essentially a green slime over the terrestrial parts of the biosphere and with the water cycle represented, in part, by a bucket rather than by vegetation. The early Earth system models had no biotic texture of any sort and it looked unlikely that those putting together earth system science would incorporate such complexity.

In the past decade, the ecological community did move from their plot-size based science to the incorporation of the larger scales needed for Earth system models. However, even the best of these models incorporates little consideration of diversity, although they do consider ecosystem-averaged biological processes such as canopy roughness, rooting depth, stomatal behaviour, allocation, and even phenology. The scientific triumph of the past decade has certainly been the linking of earth surface biological processes with atmospheric phenomena. Still, however, the actual biotic richness of the Earth surface has been poorly included in these models.

2.1.2 Mitwitz—laying the foundation

The Scientific Committee on the Problems of the Environment (SCOPE) initiated a research programme on the ecosystem functioning on biodiversity to help close the gap between knowledge on biological diversity and research on the functioning of ecosystems. The starting point of this research effort was the Mitwitz meeting noted above. Two things happened at the meeting. One, a series of contributed papers provided the state of the knowledge at that time on this issue and, two, on the basis of these contributions the goals of the remaining SCOPE programme were set. The stated goals of the programme were:

1. Does biodiversity 'count' in system processes (e.g. nutrient retention, decomposition, production etc.), including atmospheric feedbacks, over short- and long-term time spans, and in face of global change (climate change, land-use, and invasions)?
2. How is system stability and resistance affected by species diversity, and how will global change affect these relationships?

It is important to note the breadth of the first statement. First it is couched in terms of atmospheric feedbacks and global changes reflecting its origins from the Earth system science perspective. Secondly, the term utilized is biodiversity and not species

diversity and in fact diversity was defined at this meeting in terms of populations, species, functional groups, systems and landscapes, keeping with the approach defined in the Convention on Biological Diversity noted below.

For some reason, this mandate for the study of the role of biodiversity in ecosystem functioning, in its broad dimensions, has not been pursued to any great extent. Rather, in what Jean-Pierre LaDanff calls 'the species vortex', virtually all efforts have been devoted to the study of species biological diversity, and even here, until recently studies have been centred on species richness. This focus is important even though it ignores the assessment of more immediate and massive changes in biodiversity that are occurring as a result of land use and even atmospheric change, nitrogen deposition in particular (Wedin and Tilman 1996; Egerton-Warburton and Allen 2000).

The tools that were to be utilized in the SCOPE programme were those readily at hand, that is, analyses of data mostly gathered for other purposes. Specifically noted were experiments or manipulations involving altering biodiversity by using additions, subtractions, fragmentation, and disturbance. At the same time, it was hoped that the programme would stimulate explicit experiments and long-term observations. The call for explicit experiments was certainly followed with extraordinary results that are reported in detail in this volume.

2.1.3 The Global Biodiversity Assessment—community 'buy-in'

As the SCOPE process was underway, a call for an assessment of the status of biodiversity was initiated (Global Biodiversity Assessment or GBA) and the SCOPE programme effort was expanded somewhat from its original study plan (it became more comprehensive in terms of the biome types it would consider). Further, and importantly, it provided an opportunity to get comprehensive input from the global scientific community on these issues. On publication (UNEP 1995) the GBA contained two chapters on ecosystem functioning and included 38 international authors on one and 66 lead authors on another with dozens of additional contributors. There were over 100 international reviewers on these two chapters. The final SCOPE synthesis volume (Mooney *et al.* 1996) on the project had 59 authors representing many different nations.

Curiously, even though the GBA process was inclusive and consensus driven it was flawed in one way, according to certain constituencies, that is policy makers. They had not asked for the document and thus questioned its legitimacy, at least in the Convention on Biological Diversity process. In essence, the GBA represented the scientists' view of what they thought the important problems were based on literature review. It did not, evidently, take into account issues that were directly being encountered by policy makers. I will return to this issue below.

2.1.4 Further review and the roles of SCOPE and GCTE

After the conclusion of the SCOPE initial programme and GBA, SCOPE decided to take on a next phase of the effort. They reviewed the state of our knowledge on below-ground biological diversity and its relationship to ecosystem functioning, an area that did not receive sufficient attention in the GBA. Already, the new SCOPE focus on below-ground biological diversity has resulted in tremendous amount of new and exciting work (see e.g. Freckman *et al.* (1997) and special issue of BioScience, February, 1999).

Similarly, in another integrated international effort, the Global Change and Terrestrial Ecosystems (GCTE) programme of the International Geosphere Biosphere Programme initiated a programme for new experimentation in this area. The present volume is a product of that programme. Both the SCOPE assessment process and the GCTE research effort are co-ordinated through DIVERSITAS, another international programme that is concerned with biodiversity science as a whole.

2.1.5 Nature's services

An important parallel development with the SCOPE programme was an effort to try to quantify services provided by natural systems (Daily *et al.*

1997). Whereas the SCOPE programme looked at ecosystem functional traits independent of value to humans, the Nature's Services approach was societally oriented. Since there is a great amount of new research that is needed to fully understand diversity/functioning relationships, it follows that there is an equally daunting challenge ahead of us in quantifying ecosystem services. There are many non-ecological scholars, particularly economists, very interested in this approach, as is the policy community, for obvious reasons. Already, innovative approaches are being taken to establish marketplaces for ecosystem services (Daily and Ellison, 2002). Further, many see this issue as a way to connect the general public to some understanding of ecological principles. A recent article entitled 'Crossing the moat: using ecosystem services to communicate ecological ideas beyond the ivory tower' (Kranz 2000) conveys the idea that scientists must think in different ways to accomplish these goals.

I would now like to shift focus here for a while in order to build back to the opening issue of how we as a community can provide information to society that is helpful while maintaining the strength of the scientific process. There has been a dramatic shift in the past couple of decades in the interaction between science and policy in environmental sciences. There probably are a number of reasons for this, but one for sure is that environmental problems are becoming more acute and widespread. Mitigation or remedy of these problems may call for strong action that will not be accepted unless there is widespread scientific support for that action. In some cases new science is being done in real time with policy formulation, as was the case with the Montreal Protocol on stratospheric ozone beginning with its adoption in 1987. Policy is being made under uncertainty regarding the basis of the phenomenon under question. In such cases, policy decisions are flexible and allow for revisiting decisions as new information becomes available. As this information was being gathered, a mechanism for evaluating it in an impartial way by the scientific community was put in place—the so-called 'science assessment'. The Montreal Protocol was a dramatic success of interaction between scientists and policy makers. The relative simplicity of the issue (a single class of compounds causing a single chain of reactions) and a relatively easy technical fix (in retrospect) to the problem certainly led to this success.

The Intergovernmental Panel on Climate Change (IPCC) has proven to be an excellent mechanism for providing quality and consensus advice to governments about the options regarding controlling or mitigating climate change beginning with the first report in 1990. There are a number of lessons learnt from this assessment that are of value for the policy/science interface. For one, the IPCC serves an international convention, the Framework Convention on Climate Change, which is science based. The IPCC, funded by governments, yet is operated independently. The very best scientists provide input into the process of assessments. These assessments utilize published information only, and go through a very rigorous peer review process that involves literally thousands of scientists and a consensus is reached on the reliability of the information that is available. The process is transparent and inclusive.

The structure of the IPCC is in part the reason for its success. Another reason may be the strategy utilized for assessing much of the data. A scenario of doubling CO_2 has been the standard under which most of the models, both climate and biology have been evaluated. Thus, the complex interactions that were put in play when modelling the future were all tested against a single metric. More recent assessments have more complicated scenario building.

2.1.6 The Millennium Assessment—ecosystem services at the foundation

We are now about to embark on a new assessment that will be directed at the status of the world's ecosystems (Ayensu *et al.* 1999) and how it is changing. This important assessment will differ from those described above in that it is designed to serve the needs of a number of conventions including the Convention on Desertification, The Ramsar Convention (wetlands) as well as the Convention on Biological Diversity. It will both examine status and trends of ecosystems in local regions as well

as globally. Importantly, it will be designed specifically to look at the capacity of natural and managed systems to provide the goods and services upon which society depends now, and into the future. Thus, the programme will call upon the kind of information that ecologists can provide on how ecosystems operate and how human activities are disrupting the natural processes that deliver goods and services (http://www.millenniumassessment. org/en/index.htm). Of course, many disciplines, such as forestry, agriculture, and animal husbandry have worked toward delivering certain goods from natural and managed systems such as food and fibre. Hydrologists and atmospheric scientists have intensively studied the delivery of clean air and water and the control of floods and storm surges. There has not been an equal effort on managing biogeochemical cycles, except perhaps for sewage, although we are certainly heading that way in view of the massive disruption of the global nitrogen cycle and the off-site consequences of these disruptions, such as the 'dead zones' in the Gulf of Mexico (Ferber 2001). Such studies will have to be broader in geographical extent than is normally the case. The Millennium Assessment will utilize heavily the kind of information being provided by those working on the role of biodiversity in ecosystem functioning. It is important that this research area continues its strong advancement.

2.2 How did we get to this state of affairs?

Returning to the main theme of this essay, it would appear that the current disharmony among those working on the ecosystem functioning and biodiversity issue is due to the relative newness of the research area, and the development of a greater demand, and even urgency for good science in our educational and policy process. This has put a strain on our traditional science/society interaction, where the scientists provide information to technical journals and leave the translation to society to other constituencies in unspecified time frames. In the particular case at issue, some have contended that the movement from new science to policy implications has been too fast and has not been done in a consensus manner. There is certainly merit in this argument and the issue here is how to move on in a more positive manner and in a way that gives confidence to those outside our field that we are not too busy trying to wound ourselves to give advice that is needed.

2.2.1 How to move on

At the same time that we are being asked to bring the best of science to the attention of the society at large, there are evolving mechanisms where we can do so and do so in a manner that is consensus driven. Some of these have been discussed above. Of course, there will always be dissent on any scientific issue and this dissent is taken into account in the normal process of doing science—that is by testing new hypotheses. The issue is though that policy makers often ask, 'what does your community think on this matter?'. What we need to do and have done in the IPCC-type process, is to give our best analysis of what the prevailing data shows and with what degree of certainty. In the ecosystem functioning and biodiversity issue, at present our state of knowledge is well encompassed in the 'uncertainty principle' that is the basis of much of the discussions on biodiversity. The uncertainty principle is under attack, however, and politicians want more guidance than this statement of ignorance. The GCTE programme, under whose auspices we are meeting here is a great mechanism for bringing scientists together to work toward reducing uncertainty, as well as reducing confusion among those who depend on the advice that we give. The Millennium Assessment will provide another mechanism for developing confidence in what the literature tells us at a given time, since the merits of each study will be debated among a large international group and confidences in what it says will be explicitly noted. We should work to use these evolving mechanisms and not try to end run the community process through the popular press that in the long run does not serve us well. The popular press is of course generally dedicated to providing both sides on any issue, which is certainly an admirable feature. However, it can be misleading to the public as seen in the climate change debate. In spite of the unusually comprehensive and

scrupulously adhered to open process of assessment of data, the press gives equal weight to the voice of 98% of the community and the 2% dissenters. However, in the end the voice of the assessment process has the greater credibility in influencing policy.

2.2.2 Back to the science

The past few years has seen very rapid progress in the study of species and functional-type richness in relation to ecosystem functioning. In particular, there have been some very major experiments on the issue that have led to new insights. These experiments are beginning to give us a crucial time dimension on diversity/function. Attention is now being directed to experiments with greater ecosystem richness, utilizing trophic level complexity. Landscape diversity is being considered. New models are being developed that are able to build understanding beyond our immediate present experimental capabilities. Many of these new directions are included in this volume.

We need to move back to the more difficult study of natural ecosystems. Such studies could bolster modelling and experimental study approaches. In doing so we need to pay more attention to the ecosystem functioning part of the equation since only a few parameters have been included in most studies to date. Shaver *et al.* (2000) have recently reviewed how 'changes in species composition and abundance affect ecosystem processes (net primary production, carbon and nutrient cycling, energy fluxes) through changes in tissue chemistry, demography (both parts and individuals), turnover rates, and vegetation structure (physiognomy, leaf area index)'. The interactions of the many dimensions of community and ecosystem structure and functioning, as related to diversity, need a much fuller exploration.

Further, we need those who study, in quantitative terms, the interactions among natural and managed ecosystems in terms of pollination and pest control services. The study of agroforestry, and integrated pest management has certainly given us insights into these processes. We need to integrate them more formally into the study of ecosystem functioning of biodiversity. Field studies are difficult because of confounding variables. However, we are developing an experimental ecosystem science and approaches for manipulating natural systems should be utilized—such as the species removal experiments underway by the GCTE programme (http://www.gcte.org/).

It is my prediction that because of the richness of ideas in this area that the functioning of ecologists will continue to be stimulated to produce exciting results of importance to a wide range of information users. Let us use the available mechanisms to work toward not only getting this new information but also in providing the results to users in a way that is helpful. We have the means to do this.

2.3 Summary remarks

The research area on the role of biodiversity and ecosystem functioning is an exciting, important and rapidly developing field. Those involved in this area will certainly continue to make great research progress, however, for maximum impact of their findings they should do so in a manner that is constructive and builds on the efforts of all involved. The recent paper by Loreau *et al.* (2001) serves as an excellent model of how this research community can come together to bring light, rather than heat only, to this important field.

There are many new frontiers that need to be addressed in order to have the findings of this field have a larger utility. In particular, attention should shift away a bit from the small-scale single functional group (e.g. herbs) species richness/ecosystem functioning focus, which has dominated this research area, to examining processes involving different trophic levels, as well as phenomena at the landscape level. Also, and this is true for other areas of ecological study, we need to explore the time dimensions of responses of ecosystems to environmental and biotic changes brought on by land use and other global changes in a diversity context.

PART II

Core areas of debate: biodiversity and ecosystem processes in grassland ecosystems

CHAPTER 3

Plant diversity and composition: effects on productivity and nutrient dynamics of experimental grasslands

D. Tilman, J. Knops, D. Wedin, and P. Reich

3.1 Introduction

There are few times in science when as many novel hypotheses are developed as when new results challenge an existing paradigm. Such was the case when, by 1997, recent studies suggested links between diversity and ecosystem processes (e.g. Frank and McNaughton 1991; McNaughton 1993; Dodd *et al*. 1994; Naeem *et al*. 1994a, 1995; Tilman and Downing 1994; Tilman *et al*. 1996, 1997a; Hooper and Vitousek 1997; Naeem and Li 1997). Some of the major conclusions of the preceding two decades—its paradigms—had been that diversity was controlled by disturbance and productivity, and that ecosystem functioning was controlled by the traits of the dominant resident species. The new work asserted that diversity was as important as composition in determining ecosystem functioning. The gauntlet had been dropped, and the debate, ripe with creativity and fraught with all the problems that occur when well-worn words are used in new ways, had began (e.g. Aarssen 1997; Huston 1997; Garnier *et al*. 1997; Grime 1997; Tilman *et al*. 1997a,b; Wardle *et al*. 1997b).

Unlike some earlier ecological debates that faded away unresolved, the one on biodiversity heralds an era in which large scale, long-term, well-replicated field studies combined with rigorous mathematical theory offer hope of a resolution. The resolution of the biodiversity debate will require clarification of terminology (see other chapters in this book), synthesis of existing theory and results (e.g. Loreau *et al*. 2001), and new experimental evidence capable of testing alternative hypotheses. Here, we present results of a biodiversity field experiment (Tilman *et al*. 1997a) that has high replication and has run for a sufficiently long time, seven years, to allow interspecific interactions to equilibrate. It has the potential to test many of the alternative hypotheses that have been proposed.

It is useful to divide the central hypotheses or models underlying the biodiversity debate into two classes. One set of models can be called simple sampling effect models (Aarssen 1997; Huston 1997; Tilman *et al*. 1997b). These models are named after the sampling process, which is the greater chance, for randomly assembled communities, of any given species being present at higher diversity (Huston 1997). Simple sampling effect models combine this sampling process with the simple (and often reasonable) assumption that interactions among species cause a single species to become highly (or completely) dominant. The combination of the sampling process and simple interspecific interactions causes the functioning of ecosystems initiated with high diversity to be determined by the traits of whatever species became dominant, whereas the functioning of low-diversity ecosystems would merely reflect the average traits of the entire species pool. For instance, Huston (1997) suggested that the apparent effects of plant diversity on total community plant abundances (Tilman *et al*. 1996) might be a short-term transient caused by species having different maximal rates of vegetative growth. Fast-growing species would dominate initially, leading to higher biomass, an effect that would be more likely to occur at higher diversity because of the greater chance that a fast-growing species would present at higher diversity. However, if species were identical in all other ways, the greater productivity of higher-diversity plots would be transient. Plots would not differ in productivity once species had grown to equilibrial abundances (Huston 1997; Pacala and Tilman 2002).

Another simple sampling effect model suggests that the effect of diversity on productivity might result from a species pool that contained some poorly performing species (Huston 1997). Such species would lower the average performance of low-diversity plots, but higher-diversity plots would have greater productivity because of the greater chance of a highly productive species being present (Aarssen 1997; Huston 1997). One way to formally explore this mechanism is to construct a simple analytical model in which species compete for a single resource, species attain superior competitive ability because of greater resource use efficiency, and this greater efficiency causes better competitors to be more productive (Tilman et al. 1997b). In this model, on an average, the effect of diversity on productivity is proportional to the extent of interspecific differences in resource use efficiency, both diversity and species composition have approximately equal effects on productivity, and more diverse communities consume and retain more of the limiting resource.

Huston (1997) and Aarssen (1997) both suggested that the reported effects of diversity on productivity (Naeem et al. 1994a, 1995; Tilman et al. 1996, 1997a) might have been the result of the sampling process combined with simple interspecific interactions. They considered such effects to be artifacts rather than 'real' effects of diversity. We will not argue the semantics of causation, but consider such simple sampling effects to be fundamental and ecologically important effects of diversity.

Niche differentiation and facilitation are the major alternatives to simple sampling effect models. The concept of niche differentiation and coexistence is venerable (e.g. MacArthur 1968, 1970, 1972) as is that of increased community biomass because of facilitation (e.g. Clements 1916). When extended to diversity and ecosystem functioning, niche differentiation and facilitation models also assume that a sampling process occurs, i.e. that communities are assembled at random and thus that any given species or combination of species has a greater chance of occurring at higher diversity (Tilman et al. 1997b; Loreau 2000a; Loreau et al. 2001). In models of niche differentiation and/or facilitation (henceforth called niche models), productivity and nutrient use are greater at higher diversity (e.g. Tilman et al. 1997b; Loreau 1998a,b, 2000a; Tilman 1999a; Lehman and Tilman 2000) because of interactions among combinations of species. Such combinations are unimportant in simple sampling effect models. For instance, if a cool season and a warm season species were to grow together, their complementary growth patterns could lead to greater total primary productivity than possible if either were growing alone. Similarly, if plant species differed in their rooting depths, in their requirements for soil resources versus light, in their ability to fix nitrogen versus to use nitrogen efficiently, or in other such ways, combinations of species with such differences should be able to coexist, more fully exploit resources, and attain greater productivity. Indeed, several different analytical niche models that encompass the essence of these alternative mechanisms predict that productivity and resource use should be asymptotically increasing functions of diversity (e.g. chapters in this book and Tilman et al. 1997b; Tilman 1999a; Lehman and Tilman 2000).

Niche models have characteristics that distinguish them from simple sampling effect models. First, niche models predict multi-species coexistence whereas sampling effect models have, as their logical outcome, dominance by a single species. The long-term persistence of many species would contradict simple sampling effect models. Second, these models predict different patterns of plot-to-plot variation in community biomass and nutrient levels. Specifically, simple sampling effect models predict that if all species are grown in monoculture, the most diverse plots would never be more productive than the best monocultures (e.g. Huston 1997; Tilman et al. 1997b). This means that the upper bound in variation in productivity would be flat, independent of diversity. In contrast, niche models predict that this upper bound would, itself, be an increasing function of diversity because some combinations of

$N+1$ species should be more productive than any combination of N species (e.g. Tilman *et al.* 1997b; Lehman and Tilman 2000). Third, niche models predict that combinations of species with complementary traits should be more productive than plots of equal species number that are composed of functionally similar species. Thus, species number and functional group composition should be important co-determinants of productivity and resource use in niche models. Fourth, simple sampling effect models predict that high productivity in high diversity plots should be caused by the presence, and dominance, of those species that are most productive in monoculture. In contrast, niche models predict that the high productivity of high-diversity plots should be caused by complementarity in the traits of particular species combinations. Both simple sampling effect models and niche models predict that diversity matters precisely because species differ in their traits, causing composition to be as important a determinant of productivity and nutrient use as diversity (e.g. Aarssen 1997; Huston 1997; Tilman *et al.* 1997b; Tilman 1999a).

We use the results of our biodiversity experiment to test among these alternative predictions. Results of an adjacent smaller experiment (Biodiversity I) have been reported elsewhere (Tilman *et al.* 1996, 2001, 2002; Knops *et al.* 1999; Naeem *et al.* 2000b). Here, we report results from the larger experiment (Biodiversity II; Tilman *et al.* 1997a), focusing on the last three years (1998, 1999, and 2000). We expand on the analyses in Tilman *et al.* (2001) by presenting additional information on species dynamics, dominance patterns, and responses of species in monocultures versus high-diversity treatments, and by evaluating results in light of the consensus presented by Loreau *et al.* (2001).

3.2 Methods

We briefly summarize methods presented in Tilman *et al.* (1997a, 2001). After herbiciding vegetation in July 1993 and removing the upper 6–8 cm of soil to reduce the seed bank, a 9 ha field was plowed, disked, and divided into 342 13 m × 13 m plots. In 1998, the plot size was reduced to 9 m × 9 m by

mown walkways. The 168 randomly located plots of Biodiversity II were seeded in May 1994 to perennials consisting of four species in each of the four functional groups [C_4 grasses (warm season); C_3 grasses (cool season); herbaceous legumes; nonlegume forbs] and two woody species (Table 3.1). The woody plants are sufficiently slow growing that they have contributed little to results.

Diversity treatments consist of 1, 2, 4, 8, or 16 species planted into a plot with 39, 35, 29, 30, and 35 replicates, respectively, of each diversity level. The species composition of each plot was chosen from this pool of 18 species by a separate random draw. By having many replicates of each level of diversity, but with species compositions randomly chosen, it is possible to determine the effects of differences in diversity while averaging across differences in composition. Because all species were planted in monoculture, with all but three in at least two monoculture plots, it is possible to compare the responses of each species in monoculture to the responses of higher-diversity combinations of these

Table 3.1 Species used in Biodiversity II. All species are perennials either native or naturalized to the prairie and savannah grasslands of east central Minnesota. Abbreviations are used in Fig. 1

Species	Abbreviation	Functional Group
Andropogon gerardi	Ag	C_4 Grass
Panicum virgatum	Pv	C_4 Grass
Schizachyrium scoparium	Ss	C_4 Grass
Sorghastrum nutans	Sn	C_4 Grass
Agropyron smithii	As	C_3 Grass
Elymus canadensis	Ec	C_3 Grass
Koeleria cristata	Kc	C_3 Grass
Poa pratensis	Pp	C_3 Grass
Achillea millefolium	Am	Forb
Asclepias Tuberose	At	Forb
Liatris aspera	La	Forb
Monarda fistulosa	Mf	Forb
(*Solidago rigida*)	Sr	Forb
Amorpha canescens	Ac	Legume
Lespedeza capitata	Lc	Legume
Lupinus perennis	Lp	Legume
Petalostemum purpureum	Ppu	Legume
Quercus ellipsoidalis	Qe	Woody
Quercus macrocarpa	Qm	Woody

same species. Three species (*Elymus canadensis, Poa pratensis, Panicum virgatum*) did not establish in one of two original monocultures, even after re-seeding, and these three plots were abandoned after 1996.

In May 1994, each plot received $10\,g\,m^{-2}$ of seed, in total, with equal masses of each species being added. Plots were not watered in 1994 and hand-weeded throughout the experiment. Selective herbicides were used on some plots from 1994 to 1997. All plots were burned each spring in late April or early May, before plant growth began. Burning consumed almost all litter in most plots.

In our initial analyses (Tilman *et al.* 1997a), we reported results for 76 additional plots that had functional group compositions drawn from a species pool augmented with 16 additional species. We do not report results for these plots nor for the 46 additional plots planted with 32 species because combining results from different species pools could introduce bias. Moreover, the 16 additional species were not grown in monoculture, making it impossible to compare results for these plots to monocultures of each species.

A dry, windy spring in 1994 caused soil and seed erosion and poor germination. To correct this, all plots were re-seeded in May 1995 at half the 1994 rate and watered (about $2.5\,cm^3\,wk^{-1}$) during the growing seasons of 1995, 1996, and 1997 to aid germination and establishment. Because virtually no *Solidago rigida*, a non-legume forb, germinated during 1994, in 1995 we re-seeded all plots planted originally to *S. rigida* with appropriate densities of a replacement non-legume forb, *Monarda fistulosa*. Some *S. rigida* became evident by 1996, and were allowed to remain. Second, in 1996 we discovered that the legume *Petalostemum villosum* had been seeded into some plots instead of *Amorpha canescens*, one of the four legume species. *P. villosum* and a seed contaminant, its congener *P. candidum* were allowed to remain. *P. villosum* was not planted into any 16-species plots. None of these species (*S. rigida, M. fistulosa, A. canescens, P. villosum, P. candidum*) were among the 8 most abundant species of 16-species plots (Fig. 3.1(a)).

To determine the responses attributable to functional group composition, we classified each plot by the functional groups to which it was planted.

With five functional groups (C_4 grasses, C_3 grasses, legumes, other forbs, and woody plants), there are 31 different possible functional group compositions ($2^5 - 1$), of which 28 occurred in the 168 plots.

Plant community above-ground biomass, measured around the time of peak standing crop, is a correlate of primary productivity in herbaceous communities because all above-ground living biomass is produced during the growing season. Total community biomass (above-ground plus below-ground plant biomass) measures carbon stored in living biomass. Total plant per cent cover (visually estimated, to species in four $0.5\,m \times 1.0\,m$ subplots per plot) provides a non-destructive estimate of above-ground abundances of species and calculation of H', the Shannon diversity index. We directly measured total community above-ground biomass by clipping four (1997–99) or eight (2000) $0.1\,m \times 3.0\,m$ strips within each plot in August, the time of peak above-ground biomass in native savannah vegetation. Clipped vegetation was sorted to living vascular plants and litter, dried and weighed. Root mass was determined from three $5\,cm$ diameter by $30\,cm$ depth soil cores per clip strip immediately after vegetation was clipped. During the middle of the growing season from 1996 through 1999, we collected four $2.5\,cm$ diameter by $20\,cm$ depth soil cores per plot. These were mixed and then sub-sampled for NO_3 extractable with $0.01\,M$ KCl. During 1997, we similarly sampled soils from the $40–60\,cm$ depth for NO_3 concentration below the rooting zone.

All data were analysed using the software JMP (SAS 2000). In general linear model (GLM) analyses of covariance using type I sums of squares, plant species number was entered first as a continuous variable and functional group composition was entered second as a categorical variable. Such analyses determine the effects attributable to the directly manipulated variable, species number, and test for the dependence of residual variation on functional group composition. In some cases, type III sums of squares were used to test for simultaneous effects of two or more variables. Such tests are highly conservative in ascribing effects to correlated variables. Results of statistical tests are termed insignificant if $P > 0.05$, significant if $P < 0.05$, and highly significant if $P < 0.001$.

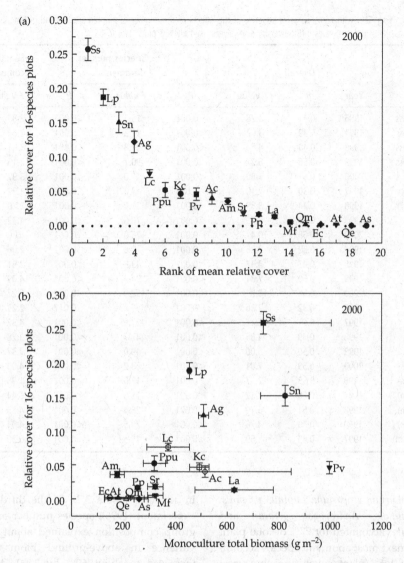

Figure 3.1 (a) Dominance-diversity curve for average per cent relative cover in 2000 of all species planted in the 16-species treatment plots. Relative cover is the per cent cover of a species out of the total plant cover of a plot. (b) Relationship between relative cover of species in the 16-species plots in 2000 and their total biomass in monoculture.

3.3 Results

All 18 species grew and persisted in monoculture and all but one, *Agropyron smithii*, were present in at least some of the 16-species plots in which they were planted. In 2000, the four cover subplots within an average 16-species treatment plot contained 11.6 planted species, which comprised an average of 97% of total plant cover, and an additional

6.4 non-planted (weedy) species, which comprised 3% of total plant cover. Similarly, the cover subplots in the average 2-species treatment plot contained 1.97 planted species (comprising 87% of total cover) and an additional 9.8 non-planted species (comprising 13% of total cover). The non-planted species were mainly agricultural annual weeds. By 2000, the 16-species plots had a dominance-diversity curve in which the five most abundant

Table 3.2 Analyses, using GLM with type I sums of squares, of the effects of the number of planted species and of functional group composition on various measured responses in Biodiversity II. Analyses used all 168 plots

Variable analysed	Year	Overall			Species number treatment		Functional group composition	
		R^2	F value	P	F value	P	F value	P
Above-ground biomass	1996	0.41	3.28	<0.001	15.5	<0.001	2.83	<0.001
Above-ground biomass	1997	0.39	3.02	<0.001	12.3	<0.001	2.60	<0.001
Above-ground biomass	1998	0.49	4.80	<0.001	31.8	<0.001	3.81	<0.001
Above-ground biomass	1999	0.56	6.27	<0.001	90.7	<0.001	3.15	<0.001
Above-ground biomass	2000	0.61	7.80	<0.001	111.0	<0.001	3.97	<0.001
Root biomass	1997	0.30	2.10	0.003	7.07	0.009	1.92	0.008
Root biomass	1998	0.44	3.89	<0.001	35.9	<0.001	2.70	<0.001
Root biomass	1999	0.57	6.69	<0.001	70.6	<0.001	4.32	<0.001
Root biomass	2000	0.68	10.4	<0.001	128.0	<0.001	6.00	<0.001
Total biomass	1997	0.32	2.26	0.001	9.8	0.002	2.06	0.004
Total biomass	1998	0.47	4.37	<0.001	43.8	<0.001	2.91	<0.001
Total biomass	1999	0.60	7.31	<0.001	94.2	<0.001	4.09	<0.001
Total biomass	2000	0.68	10.5	<0.001	152.0	<0.001	5.27	<0.001
Total plant cover	1996	0.32	2.36	0.001	5.87	0.017	2.23	0.001
Total plant cover	1997	0.51	5.17	<0.001	27.9	<0.001	4.32	<0.001
Total plant cover	1998	0.49	4.86	<0.001	47.7	<0.001	3.28	<0.001
Total plant cover	1999	0.50	5.00	<0.001	49.0	<0.001	3.37	<0.001
Total plant cover	2000	0.59	7.01	<0.001	88.2	<0.001	4.00	<0.001
Soil nitrate (0–20 cm)	1996	0.33	2.37	0.001	13.0	0.001	1.98	0.006
Soil nitrate (0–20 cm)	1997	0.60	7.12	<0.001	25.5	<0.001	6.44	<0.001
Soil nitrate (0–20 cm)	1998	0.51	5.19	<0.001	28.4	<0.001	4.33	<0.001
Soil nitrate (0–20 cm)	1999	0.49	4.76	<0.001	24.5	<0.001	4.04	<0.001
Soil nitrate (40–60 cm)	1997	0.45	5.59	<0.001	15.59	0.001	5.26	<0.001

species (*Schizachyrium scoparium, Lupinus perennis, Sorghastrum nutans, Andropogon gerardi*, and *Lespedeza Capitata*) accounted for 77% of total plant cover. The five next most abundant species accounted for 20% of total plant cover, and the rarest 9 species accounted for 3% of total cover (Fig. 3.1(a)). There was a significant, though weak, correlation between the year 2000 abundances of species in monoculture plots and in the 16-species plots ($r = 0.49$, $N = 19$, $P = 0.03$; Fig. 3.1(b)).

Above-ground biomass. Based on GLM analyses that used type I sums of squares, above-ground plant biomass increased highly significantly with species number (entered first, as a continuous variable) in all years (1996–2000), and a significant amount of the residual variation in above-ground biomass was explained by functional group composition (entered second, as a categorical variable)

in all years (Table 3.2). In the third year of this experiment (1996), species number and functional group composition explained about 40% of total variance in above-ground biomass, and this increased to about 60% by 2000. Above-ground biomass was an approximately saturating function of species number in 1998, was more of a linearly increasing function in 1999, and was a clearly linear function of species number, for species number from 2 to 16, in 2000 (Fig. 3.2(a)).

Root biomass. The total root mass of plots was a highly significant increasing function of species number in all years when measured (1997–2000), and functional group composition explained a highly significant amount of residual variance in all years (Table 3.2). These two variables explained progressively more of the variance in root mass as the plots matured (Table 3.2). In 1998, 1999, and

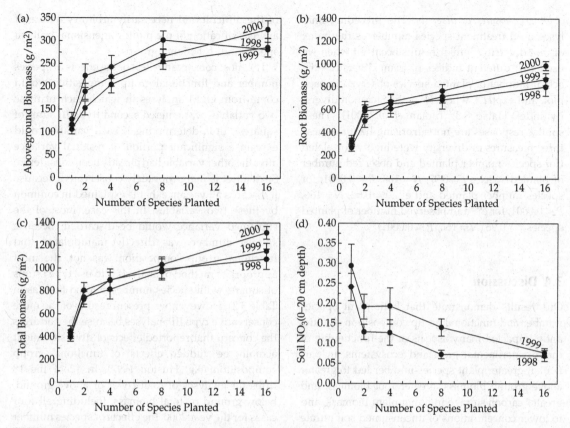

Figure 3.2 Dependence in 1998, 1999, and 2000, of above-ground biomass (a), root biomass (b), total biomass (c), and soil nitrate at 0–20 cm depth (d) (no nitrate data collected in 2000) on the number of species planted. Means and standard errors are shown for each treatment.

2000, there was a sharp increase in root mass from the average monoculture to the average 2-species plot, then root mass increased approximately linearly as species number increased from 2 to 16 (Fig. 3.2(b)).

Total biomass. Total biomass, which is the sum of above-ground living plant biomass and root biomass, was significantly dependent on the species number treatment and on functional group composition, much as for above-ground biomass and total biomass (Table 3.2). Species number and functional group composition explained about 1/3 of the variance in total biomass in 1997 and about 2/3 by 2000. By 1999 and 2000, total biomass was an approximately linearly increasing function of species number for species number from 2 to 16 (Fig. 3.2(c)).

Total plant cover. Total plant cover (the summed per cent covers for all plant species in a subplot) increased significantly with species number, and functional group composition explained a significant amount of residual variation in all years (1996–2000; Table 3.2). The total variance explained by species number and composition doubled from 1996 to 2000 (Table 3.2).

Extractable soil NO₃. Soil NO₃ at 0–20 cm depth was a highly significant decreasing function of plant species number for all years (1997–99; Table 3.2; Fig. 3.2(d)). In GLM analyses with type I sums of squares, functional group composition (entered second) explained highly significant amounts of residual variation in soil nitrate in all years (Table 3.2). Soil nitrate at 40–60 cm depth showed similar patterns when measured (1997; Table 3.2).

Other measures of diversity. The preceding analyses used treatment species number as the index of plot diversity. Similar results occurred when we used two different indices of plant diversity—the actual number of planted species observed in each plot, and exp[H'], where H' is the Shannon diversity index (Table 3.3; Tilman *et al.* 2001). These similar responses are not surprising because these three measures of diversity were highly correlated (for species number planted and observed number of planted species $r = 0.97$, $N = 168$, $P < 0.0001$; for species number planted and $e^{H'}$, $r = 0.92$, $N = 166$, $P < 0.0001$; for $e^{H'}$ and observed number of planted species $r = 0.96$, $N = 166$, $P < 0.0001$).

3.4 Discussion

Our results demonstrate that both plant species number and functional group composition significantly impacted many aspects of the functioning of these experimental grassland ecosystems. In particular, greater plant species number led to greater above-ground biomass, greater root biomass, and greater carbon storage in living plant biomass, and to lower concentrations of unconsumed soil nitrate both in the rooting zone and below it. Indeed, GLM analyses showed that plant species number and functional group composition, combined, had highly significant overall effects all 21 times that various responses have been measured to date. When entered first, species number was significant in all 21 cases, as was functional group composition when entered second in these analyses.

One of the points of debate has been whether it is plant species number per se, or differences in plant functional traits, that causes ecosystem functioning to depend on diversity (e.g. Chapin *et al.* 1996b, 2000; Diaz and Cabido 2001; Huston 1997; Grime 1997). Clearly, these are not mutually exclusive possibilities. The number of functional traits and the number of species are always positively associated in nature, and in our experiments. We consider species number—our preferred measure of diversity—to be a convenient and practical measure of the range of species traits in a system and how well such traits 'cover' the temporal, spatial, and biotic heterogeneity of a habitat. We do this because

of the difficult and necessarily arbitrary nature of any classification of the multi-dimensional traits of species into functional groups.

The most conservative test of the effects of species number and functional group composition would come from GLM analyses in which each of these two variables was entered second (type III sums of squares), thus determining if each variable could explain a significant portion of residual variance after the other variable had already been considered. Such a test is inappropriately conservative because it discards all variance that is explained in common by these two variables. In this case, most of the explained variance would be discarded. Because species number was directly manipulated and functional group composition was not, the most appropriate method of analysis is to use type I sums of squares with species number entered first (as in Table 3.2). However, we present results of the more conservative type III analyses because they address the concern that reported effects of diversity might actually be 'hidden effects' of functional group composition (e.g. Huston 1997). In 12 of the 18 cases for analyses of total cover or above-ground, below-ground or total biomass, including all four cases for the year 2000, the effects of species number remained significant ($P < 0.05$) even after first controlling for functional group composition. Thus, these highly conservative tests (Tilman *et al.* 2001; see also Table 3.3) demonstrate that the observed effects of species number and other measures of diversity on plant community abundance were not solely the result of a 'hidden composition treatment' but were, indeed, real effects of plant species number.

A different result occurred, though, when such analyses were performed on soil nitrate. When species number was entered first in GLM type I analyses, there were highly significant effects of species number on rooting zone soil NO_3 in 1996, 1997, 1998, and 1999 and on deep soil NO_3 in 1997 (Table 3.2). However, none of these relationships remained significant ($P > 0.05$ in all cases) when species number was entered second, after functional group composition. This occurred mainly because of the marked differences between the effects of C_4 grasses and legumes on soil NO_3. In all years C_4 grasses greatly reduced soil NO_3. In some

Table 3.3 Effects of two different diversity indices (observed species number or e^H) on total and above-ground biomass in 2000. Observed species number is the number of planted species observed in one or more of four $0.5\,m^2$ quadrats within each plot. Effective species richness is $e^{H'}$, where H' is the Shannon diversity index based on observed abundances (relative per cent cover) of planted species. GLM analyses determined effects of one or the other diversity index (continuous variable; entered first using type I SS) and of functional group composition (categorical variable; entered second) on total biomass and on above-ground biomass. $N = 168$. Overall model d.f. $= 28$ and error d.f. $= 139$, with plant diversity d.f. $= 1$ and composition d.f. $= 27$. The last columns show similar analyses, but with plant diversity entered second, after composition (i.e. type III sums of squares)

Variable analysed	Year	Overall			Diversity index (entered first)		Functional group comp. (entered second)		Diversity index (entered second)	
		R^2	F value	P	F value	P	F value	P	F value	P
Observed species number										
Total Biomass	2000	0.68	10.6	<0.001	157	<0.001	5.20	<0.001	13.3	<0.001
Above-ground Biomass	2000	0.63	8.40	<0.001	133	<0.001	3.78	<0.001	17.7	<0.001
Effective species richness ($e^{H'}$)										
Total Biomass	2000	0.66	9.53	<0.001	136	<0.001	4.84	<0.001	7.82	0.006
Above-ground Biomass	2000	0.61	7.73	<0.001	132	<0.001	3.11	<0.001	14.0	<0.001

years, the presence of legumes led to higher NO_3 levels (Fig. 3.3). In the case of soil NO_3, then, there were great differences among functional groups, and these differences swamped out the effects of species number. This suggests that there is likely to be no simple resolution to the debate over whether it is species number or functional traits that determine ecosystem functioning. Clearly, it is both and one or the other may seem more important depending on the ecosystem processes of interest and the traits of the species in the species pool. This being so, we believe that this debate should be put aside and replaced with the realization that species number, functional group number, and functional group composition are just different ways to measure the range of species traits in an ecosystem. All are valid, and highly correlated, indices of diversity.

3.4.1 The predominance of niche effects

Our results allow us to test between simple sampling effect models and niche models as potential causes of the greater biomass observed at greater species number. One simple sampling effect model predicts that positive effects of diversity on biomass would be solely a short-lived, transient effect caused by interspecific differences in maximal growth rates. Throughout the seven years of this experiment, the dependence of above-ground biomass, of

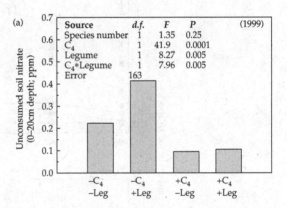

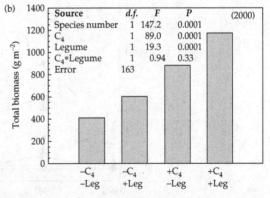

Figure 3.3 (a) Dependence of levels of unconsumed soil nitrate (in 1999, 0–20 cm depth) and of (b) total biomass (in 2000) when results were analysed with the presence/absence of C_4 grasses, legumes, the legume $\times$ C_4 grass interaction, and planted species number as independent variables.

root biomass, and of total biomass on species number became increasingly stronger and more linear, and the proportion of total variance explained by diversity and functional group composition almost doubled. This refutes this sampling effect hypothesis, but it is plausible that some of the early effects we observed were caused by differences in maximal growth rates, as suggested by Huston (1997).

All sampling effect models assume that a single species would dominate (or even displace all other species) and thus impose its traits on the community (Aarssen 1997; Huston 1997; Tilman *et al.* 1997b). We did not observe such competitive displacement. Rather, we observed long-term persistence in high-diversity plots of all but one of the planted species (Fig. 3.4), with high-diversity treatments having from 8 to 15 species coexisting within a $0.5\,\text{m}^2$

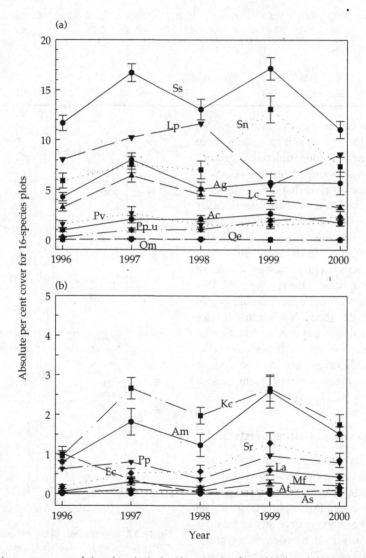

Figure 3.4 Mean absolute per cent cover of planted species in the 16-species plots from 1996 through 2000. Note that these are not relative covers (i.e. have not been normalized to have total plant cover be 100%), and thus are not comparable to the data of Fig. 1. Rather, these data are the per cent of ground covered by each species each year.

subplot, and with all the planted species except *A. smithii* persisting at some locations within each plot. Indeed, excluding *A. smithii*, the five rarest species of the 16-species plots occurred in all the 16-species plots in which they were planted. Such multi-species coexistence is similar to the coexistence of similar numbers of species in the native grasslands of Cedar Creek (Tilman *et al.* 1996). The species that became dominant in the biodiversity experiments, mainly C_4 grasses such as *S. scoparium*, are also dominants in nearby native grasslands (Tilman 1987, 1988), and the rarer species of the biodiversity experiment tend to be rarer species in native grasslands. The dominance-diversity distributions observed in the biodiversity plots (Fig. 3.1(a)) are similar to those of prairie-grasslands in the area, further demonstrating the coexistence of a large number of plant species in the plots.

Simple sampling effect models (e.g. Huston 1997; Tilman *et al.* 1997b) predict that the upper bound of the variation in total biomass or productivity should be a flat line, i.e. higher-diversity plots should never be more productive than the most productive monoculture. In niche models, interspecific complementarity and facilitation cause some higher diversity plots to have greater productivity than any monocultures. For above-ground biomass, root biomass, and total biomass in 1999 and 2000, the upper bound of variation was a clearly increasing function of diversity (Fig. 3.5). This strongly supports niche models. Indeed, on an average for the years 1999 (Tilman *et al.* 2001) and 2000 (Fig. 3.5), about half of all 16-species plots had greater total biomass or above-ground biomass than the single most productive monoculture plot (Fig. 3.6), and the average 16-species plot had slightly greater total biomass than the single best monoculture plot (Fig. 3.5). Similarly, about 30% of the 8-species plots had greater total or above-ground biomass than the single best monoculture for the years 1999 and 2000. These patterns are a strong demonstration of over-yielding, showing the clear signature of niche models and demonstrating that simple sampling effect models were not the major cause of the greater biomass of higher diversity plots.

Although strong over-yielding occurred for many measured variables in 1998, and for all measured variables in 1999 and 2000, earlier results

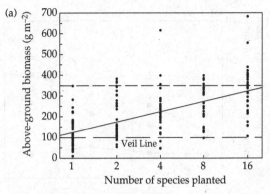

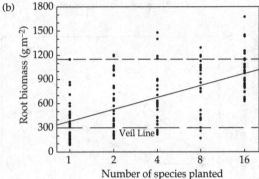

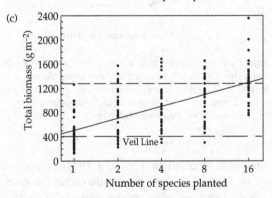

Figure 3.5 Dependence of above-ground biomass (a), of root biomass (b), and of total biomass (c) on planted species number, shown for each of the 168 plots in 2000. The solid line is a regression of a variable on the log of species number. The lower dashed line is a veil line showing which data points were excluded in one set of analyses. The upper dashed line shows the biomass of the single best species in monoculture. Any points above this line meet the strictest criterion of exhibiting over-yielding.

did not provide such clear support for niche models (Fig. 3.6). In 1997, the per cent of plots with greater total biomass than the best monoculture was not an increasing function of species number.

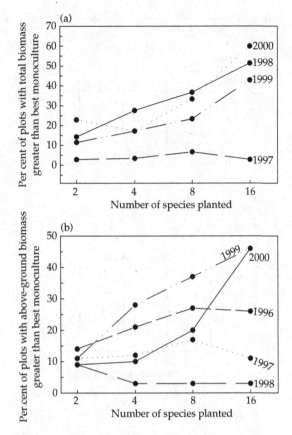

Figure 3.6 The per cent of plots of a given diversity treatment that had greater total biomass (a) or greater above-ground biomass (b) than the single monoculture plot with the highest total or above-ground biomass in each of the years shown. Results shown for all years for which relevant data were collected.

This also occurred in 1997 and 1998 for above-ground biomass. These results are consistent with simple sampling effect models but not with the niche models. Although other analyses have suggested that niche complementarity was occurring in those early years (Tilman *et al.* 1996, 1997a), such complementarity was not strong enough to erase the signature of sampling. This suggests that sampling effects were stronger during the early years of the experiment, and that niche effects came to dominate as the communities matured.

Another simple sampling effect model predicts that high productivity in high diversity plots should be caused by the presence, and dominance, of those species that attained the greatest productivity in monoculture (Aarssen 1997; Huston 1997; Tilman *et al.* 1997b). The two species with the greatest average total biomass in monoculture were *P. virgatum* and *S. nutans* (Fig. 3.1), but these were, respectively, the eighth and the third most abundant species, on average, in the 16-species plots. Thus, they neither dominated plots when present, nor imposed their high productivity on high-diversity plots. Moreover, the most abundant species of the high-diversity plots, *S. scoparium*, comprised only about 1/4 of total plant cover. There was a weak but significant correlation between species abundances in monoculture and their abundances in high diversity plots (Fig. 3.1(b)). The existence of this correlation means that the processes assumed in this sampling effect model likely contributed to some of the observed effect of diversity on biomass. However, the lack of strong dominance by one species and the increasing upper bounds in total and above-ground biomass (Figs 3.5 and 3.6) show that niche effects were predominant. Other analyses showed that the presence of one or two higher-productivity species did not eliminate the effect of diversity on productivity (Tilman *et al.* 2001), further demonstrating that the higher productivity of the high-diversity plots was not caused just by the greater probability that higher-productivity species were present.

It had been suggested that the presence of a few low-viability species could have caused the diversity effects we reported (Huston 1997). To test this hypothesis, we identified the five species with the lowest total biomass in monoculture for the year 2000 and then excluded all plots containing any combinations of just these species from analyses. In the remaining 131 plots, total biomass was highly significantly dependent on species number and functional group composition in both type I and type III analyses (Tilman *et al.* 2001). We found a similar result for above-ground biomass in 2000. In another set of analyses designed to test the low-viability hypothesis, we excluded from analysis of year 2000 results the 30 plots with the lowest total biomass (those points below the 'veil line' of Fig. 3.5(c)). We found that the remaining plots showed highly significant effects of both species number and functional group composition on total biomass, using both type I and type III GLM analyses (Tilman *et al.* 2001). We did similar analyses for above-ground biomass and for root biomass, and observed similar results. Thus the presence of poorly performing species in the species pool was not the sole explanation for the strong effects of species number on biomass.

Although our long-term results strongly support the predictions of niche models, this does not mean that the sampling process is unimportant. Both niche and sampling effect models assume that a sampling process occurs, i.e. that any given species or combination of species is more likely to occur at higher diversity (Loreau 2000a; Loreau *et al.* 2001). The failure of our results to support simple sampling effect models means that the sampling process impacts ecosystem functioning not via the greater likelihood that a particular single species is present in higher-diversity plots but rather via the greater likelihood that particular *combinations* of species are present at higher diversity. In niche models, combinations of species matter because a species that is better at handling one condition is worse at others. This makes species have complementary traits, causing species combinations to better use a spatially or temporally heterogeneous habitat than could any single species (e.g. McNaughton 1993; Vitousek and Hooper 1993; Tilman *et al.* 1997b, Loreau 2000a). If species compositions are determined randomly, complementarity increases as diversity increases because the chance of any given species combination being sampled is greater at higher diversity.

3.4.2 Why diversity mattered

Why did greater diversity lead to greater productivity (above-ground biomass) and to more complete use of the limiting soil nutrient, nitrate? Although we have not found a single simple explanation for these effects, evidence supports the importance of interspecific differences in resource requirements and in temporal and spatial patterns of resource utilization, and of facilitation. Using results for the year 2000, ANOVAs' found significant positive effects on above-ground biomass of the presence of legumes ($P < 0.001$), of C_4 grasses ($P < 0.01$), and of forbs ($P < 0.05$), but not of either C_3 grasses ($P > 0.05$) or woody species ($P > 0.05$). Similarly,

total biomass in the year 2000 was significantly greater because of the presence of C_4 grasses ($P < 0.001$) and legumes ($P < 0.001$), but not other functional groups. The presence of C_4 grasses led to the greatest increase in total biomass (Fig. 3.3(b)), consistent with their high nitrogen use efficiency and the low level to which C_4 grasses reduced the level of soil nitrate (Fig. 3.3(a)). The presence of legumes also increased total biomass (Fig. 3.3(b)), an effect that led to higher levels of soil nitrate (Fig. 3.3(a)), likely because of decomposition of the high-N litter of these nitrogen-fixing plant species. The presence of both C_4 grasses and legumes led to the greatest total biomass (Fig. 3.3(b)) and to levels of unconsumed soil nitrate comparable to those when just C_4 grasses were present (Fig. 3.3(a)). Total biomass when both C_4 grasses and legumes were present was somewhat more than additive indicating that the C_4 grasses consumed, and produced biomass with, the nitrate mineralized because of legumes. This demonstrates that a significant cause of greater biomass at greater diversity is the greater chance of occurrence of a high nitrogen use efficiency warm-season species with a nitrogen-fixing cool-season species. However, GLM analyses showed that total biomass was still a highly significant and increasing function of species number even after controlling for the presence/absence of C_4 grasses, legumes, and the legume $\times C_4$ grass interaction (Fig. 3.3(b)). Thus, the number of species mattered in addition to the presence/absence of two major functional groups.

Another insight comes from considering how many species might be responsible for the observed effects of diversity on total biomass (carbon stored in living tissue) and on above-ground biomass (productivity). We explored this question in two ways. We used ANOVA to determine the effects of the presence or absence of each species, when entered as main effects, on total or above-ground biomass for each year. We found significant effects of five of the six most abundant species of the high-diversity mixtures (Fig. 3.1), suggesting that about five species might account for much of the observed effects of diversity. Because these five species (or any five species, for that matter) were only likely to co-occur in the 16-species plots, their joint importance could explain why the 16-species plots had significantly greater total and above-ground biomass in 2000 than the 8-species plots (Tilman *et al.* 2001). However, our experiment was not designed to test for the effects of the presence/absence of particular species, and, indeed, has relatively little power to do so. Thus, it is possible that other rarer species contributed to the observed effects of diversity on biomass, but did so with effects that are not detectable with this approach.

A different analysis tested this possibility. We ranked species based on per cent cover (abundance) in 2000 in the 16-species plots. We used this to create 17 new diversity indices. For each plot, we determined how many of the N most abundant species had been planted in it, where $N = 2, 3, 4, \ldots, 18$. Above-ground biomass in 2000 was most dependent on how many of the 9 to 13 most abundant species had been planted in each plot (Fig. 3.7; Tilman *et al.* 2001). This suggests that, in addition to the effects of five dominant species, the

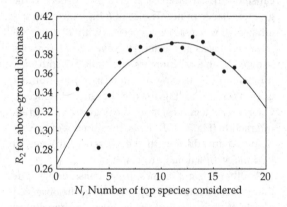

Figure 3.7 The dependence of a measure of goodness of fit on species number. Goodness of fit, as estimated by R^2 values, for regressions of above-ground biomass in 2000 on the log of the number of the N top species that were present in each plot. For instance, when $N = 4$, the four species with the greatest per cent cover in 2000 in the 16-species plots were considered. These four species were *S. scoparium, L. perennis, S. nutans* and *A. gerardi.* For each plot, we determined how many of these four species had been planted in the plot, and used this number as a diversity index. Log [diversity index +1] was then used as the independent variable in a regression in which total biomass in 2000 was the dependent variable. Such regressions were repeated for the other values of N, number of top species considered, ranging from 2 to 18. The best fit occurred for $9 \leq N \leq 13$. Similar analyses, but using total biomass in 2000, had highest R^2 values for $N = 4$. See Tilman *et al.* (2001) for further details.

presence of from 4 to 8 rarer species also contributed to the greater above-ground biomass of more diverse treatments. In contrast, total biomass was most dependent on how many of the four most abundant species were present (Tilman *et al.* 2001).

3.4.3 Diversity and ecosystem functioning on larger spatial scales

Diversity experiments have manipulated diversity within relatively small field plots (9 m × 9 m in our case) or even in pots in a greenhouse (e.g. Naeem *et al.* 1996; Symstad *et al.* 1998). These spatial scales are mechanistically appropriate because the effects of diversity should come from local interactions within the neighbourhoods of individual plants (e.g. Pacala and Silander 1985, 1990). Looked at this way, our results demonstrate that 0.5 m^2 neighbourhoods containing about 11 species (16-species plots) are significantly more productive than those containing 7 species (8-species plots). How might this local diversity correspond to diversity on a larger scale?

At Cedar Creek, a 0.5 hectare field has to contain about 72 plant species for an average 0.5 m^2 plot within the field to contain about 11 species (Tilman 1999b). Our observed dependence of neighbourhood diversity on field diversity fit the classical species-area relationship, $S = c A^z$, in which diversity, S, scales as area, A, raised to the power, z, with $z = 0.21$. Thus, for a local area (of size A_L) imbedded in a region (of size A_R) to have a local number of S_L species, the larger region would have to have a species number of S_R, where

$$S_R = S_L \left(A_R / A_L \right)^Z$$

Given our observed z value of 0.21, the species-area relationship suggests that an area of 1 km^2 would need to contain 232 species for the average 0.5 m^2 within it to contain 11 species. Thus, this would suggest that a 1 km^2 area containing 232 species (comparable, via species-area scaling, to the neighbourhood diversity of our 16-species treatment) would be significantly more productive than a 1 km^2 area containing 147 species (comparable to the neighbourhood diversity of our 8-species treatment). Clearly, this intriguing possibility has never been tested experimentally, but it suggests that many more species should be required to assure functioning on larger landscape units than indicated by small-scale diversity experiments.

3.5 Conclusions

Both productivity and nutrient use increased as species number increased in our experimental prairie grasslands. Our long-term results lend much greater support to niche differentiation processes than to sampling effect models. Greater productivity was not a transient and did not result, on an average, solely from the presence of particular single species, but rather from the presence of particular combinations of species, especially of species in different functional groups. The joint presence of legumes and of C_4 grasses and a higher number of species simultaneously contributed to the higher productivity of higher-diversity treatments. The concordance of our results with earlier results (e.g. Naeem *et al.* 1995; Tilman *et al.* 1996; Hector *et al.*

1999) and with predictions of theory suggests that higher species number may lead to greater productivity, in general, in exploitative competitive communities. Both species number and functional group composition—which we consider to be two different measures of diversity—impacted many aspects of ecosystem functioning in our experiment, and their relative strengths depended on the response being measured. In combination with other results (cited earlier in this paper and presented in the other chapters in this book), our results suggest that species number and composition are, along with disturbance, climate, and soil fertility, major determinants of productivity and nutrient dynamics in terrestrial ecosystems.

We thank the National Science Foundation (NSF/DEB 9411972, NSF/DEB 0080382, NSF/DEB 9629566) for support, T. Mielke, N. Larson, L. Johnson, and Cedar Creek Summer Research Interns for assistance, and M. Loreau, P. Inchausti, and S. Naeem for comments.

CHAPTER 4

Biodiversity manipulation experiments: studies replicated at multiple sites

A. Hector, M. Loreau, B. Schmid, and the BIODEPTH project[1]

4.1 Introduction

Biodiversity is usually defined in a general sense as a collective term for all biological differences at scales ranging from genes to ecosystems (e.g. Harper and Hawksworth 1994). This breadth of meaning is both a strength and weakness of the term, which poses many challenges in its study. In this chapter, we focus on the BIODEPTH project, presenting an overview of published analyses of the combined site datasets from Hector *et al.* (1999) and Loreau and Hector (2001), and re-examining the debate caused by the early results (Hector *et al.* 1999, 2000b; Huston *et al.* 2000). We present some additional material and new methods and analysis that we hope will clarify some points of debate and draw comparisons with the small number of other multi-site biodiversity experiments that have been conducted so far. Many of the issues we discuss are not specific to biodiversity studies but apply widely in ecology and should therefore be of broad interest. These can be classified into two broad groups:

1. The importance of specifying precise, testable hypotheses and identifying causality when examining complex explanatory and response variables,

in this case biodiversity and ecosystem processes. This is particularly important where explanatory variables have multiple co-varying components, as for the many compositional and richness aspects of diversity.

2. Identifying the effects of extrinsic factors (e.g. climate) that vary and act at larger scales, and between systems, from intrinsic factors (including community processes) that act at a local scale within systems.

4.2 An overview of the BIODEPTH experiment and results

We present only a brief overview of the combined site analyses here as the BIODEPTH project is described extensively in many other publications: the overall project and analysis is described in Hector *et al.* (1999) and the details of the individual site designs can be found in the following papers for Switzerland (Diemer *et al.* 1997; Joshi *et al.* 2000; Koricheva *et al.* 2000; Spehn *et al.* 2000a,b; Stephan *et al.* 2000; Diemer and Schmid 2001), Greece (Troumbis *et al.* 2000), Sweden (Mulder *et al.* 1999; Koricheva *et al.* 2000), Silwood (Hector *et al.* 2000a, 2001a,b), Portugal (Caldeira *et al.* 2001), Germany (Scherer-Lorenzen 1999) and for the theoretical modelling component (Loreau 1998a, 2000a; Loreau and Hector 2001). While our analysis of combined datasets from all sites has concentrated on above-ground biomass production, these papers examine the effects of plant diversity on other processes at individual sites including: above-ground

[1] C. Beierkuhnlein, M. C. Caldeira, M. Diemer, P. G. Dimitrakopoulos, J. A. Finn, H. Freitas, P. S. Giller, J. Good, R. Harris, P. Högberg, K. Huss-Danell, J. Joshi, A. Jumpponen, C. Körner, P. W. Leadley, A. Minns, C. P. H. Mulder, G. O'Donovan, S. J. Otway, J. S. Pereira, A. Prinz, D. J. Read, M. Scherer-Lorenzen, E-D. Schulze, A-S. D. Siamantziouras, E. M. Spehn, A. C. Terry, A. Y. Troumbis, F. I. Woodward, S. Yachi, and J. H. Lawton.

space-filling and canopy structure; root production; decomposition and nutrient cycling; soil biodiversity, activity and other below-ground processes; community water use efficiency, invasion resistance and the diversity and abundance of insects and other invertebrates.

A major aim of the BIODEPTH project was to explicitly test for statistically significant differences between locations by combining data from all sites in single analyses. To date, our published multisite analyses have largely examined above-ground biomass production (but see Joshi *et al.* 2001; Spehn *et al.* 2002). To summarize, the BIODEPTH project replicated the same basic experiment at eight different locations around Europe. At each site, we established experimental plant communities from seed so that we could control numbers and types of species and functional groups (species predicted to have similar effects on processes). The primary aim was to generate a gradient of species richness at each location with five levels ranging from monocultures to numbers of species estimated to be found in 4 m²

plots in unmanipulated grasslands at each site. Each level of species richness was replicated with several plant communities with different compositions, and each particular species combination was also replicated (see also Bell 1990; McGrady-Steed *et al.* 1997; Petchey *et al.* 1999; McGrady-Steed and Morin 2000) in order to separate these two components of diversity (Givnish 1994). Species were chosen from the pool of co-occurring species at each site at random with certain constraints (detailed in the above publications). Primary amongst these was constraining the functional composition of the communities so that we could also examine this aspect of diversity, at least in part. We then used standardized protocols to examine responses in a suite of ecosystem and community processes, including above-ground biomass production on which we concentrate here.

As we examined the productivity-diversity relationships emerging at individual sites, it was clear that there were differences between sites (Fig. 4.1, Table 4.1; Loreau *et al.* 2001); we return to these

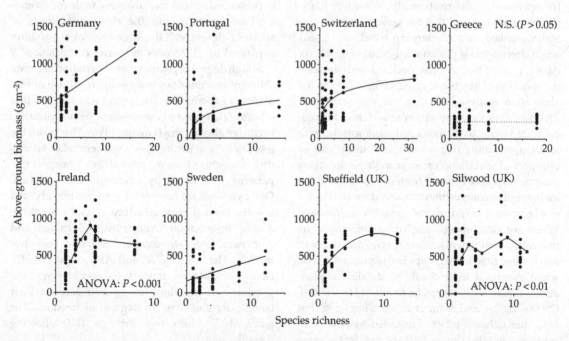

Figure 4.1 The different relationships between biodiversity and above-ground biomass production found in the analyses of individual site patterns when the highest adjusted R^2 criterion was used to select chosen models. Points are individual plot values and lines are slopes from linear (Germany, Sweden), log-linear (Portugal, Switzerland) or quadratic (Sheffield) models, join richness level means (solids squares with SEMs) for significant ANOVA (Ireland, Silwood), or show non-significant relationships (Greece, broken line).

Table 4.1 Summary of individual site analyses comparing major diversity-function hypotheses. Table entries are adjusted R^2 of different models scaled relative to the grand mean fitted in model 1 (e.g. adjusted R^2 of a given model, N, = (rms[model 1] − rms[model N])/ rms[model 1], where rms = residual mean square), which we used to compare the five models to take account of both their explanatory power (proportion of total variance) and complexity (degrees of freedom). The selected best model for each site when analysed individually is highlighted in bold. Model 1: Productivity = Grand mean; Model 2: Productivity = Constant + b (Species number); Model 3: Productivity = Constant + b log (Species number); Model 4: Productivity = Constant + b (Species number) + c (Species number)^2; Model 5: Productivity = Grand mean + a_i (where a_i is the effect of species level i)

Location	Model 1 null	Model 2 linear	Model 3 log-linear	Model 4 quadratic	Model 5 ANOVA
Germany	0	**0.3031**	0.2652	0.2919	0.2845
Ireland	0	0.0422	0.1312	0.2072	**0.3039**
Silwood	0	0.0461	0.0846	0.1031	**0.1661**
Sheffield	0	0.4161	0.4672	**0.4895**	0.4784
Switzerland	0	0.1145	**0.1726**	0.1347	0.1541
Portugal	0	0.1605	**0.1755**	0.1593	0.1588
Sweden	0	**0.1165**	0.0840	0.1058	0.1053
Greece	0	−0.0032	−0.0046	−0.0236	−0.0662

differences below. However, when combined in an overall analysis we found that these differences in the species richness relationships were not statistically significant and that the general pattern was well described by a relationship which was linear when diversity was put on a $\log_2$ scale (Fig. 4.2(a)). Above-ground biomass also declined with decreasing numbers of functional groups (Fig. 4.2(b)) in the alternative analysis focusing on this aspect of diversity. Species composition was clearly biologically important as it interacted significantly with location, revealing that where the same species or mixtures of species occurred at multiple locations their performance differed from site-to-site, and the main effect of composition accounted for the largest single portion (39%) of the variation in biomass. When we decomposed the composition into its individual components, many species and the herb and legume functional groups had significant effects when examined individually. A detailed explanation of these analyses can be found in Hector *et al.* (2000b, 2002a) and Schmid *et al.*, Chapter 6, but note that richness effects are tested against compositional effects (Tilman 1997a,b) and, by the same logic, interactions between richness and other terms against the corresponding interaction with composition.

The results reported above pose an apparent problem: how can we reconcile the differences between individual site analyses with the overall log-linear pattern, and the significant effects of species richness with the large amount of variation explained by differences in species composition?

A high degree of compositional variation between different communities is a common feature of biodiversity experiments. One recent realization is that effects of richness and composition are not mutually exclusive alternatives (Lawton 1998). This raises the issue of how to distinguish consistent relationships due to richness from potentially 'idiosyncratic' patterns dominated by compositional variation. One approach we have tried is to use the adjusted R^2—the normal error structure special case of the Akaiki Information Criteria (e.g. Burnham and Anderson 1998)—to compare different statistical models. The adjusted R^2 and AIC assess the efficiency of different models by considering their goodness of fit while taking into account their complexity and 'cost' in degrees of freedom. The adjusted R^2 does this through the following formulation:

$$1 - \frac{n-1}{n-p-1}(1-R^2)$$

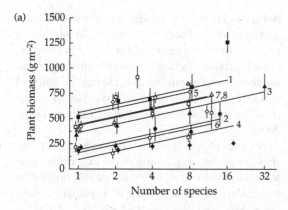

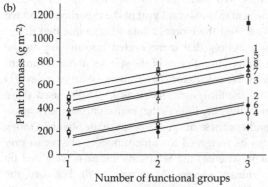

Figure 4.2 Above-ground biomass production increases with higher plant species richness (a) and functional group richness (b). Lines are regression slopes from the linear regression analyses presented in Hector *et al.* (1999) and symbols (staggered for clarity) are richness level means and standard errors for: closed squares = Germany, line 1; closed circles = Portugal, line 2; closed triangles = Switzerland, line 3; solid diamonds = Greece, line 4; open squares = Ireland, line 5; open circles = Sweden, line 6; open diamonds = Sheffield (UK), line 7; open diamonds = Silwood Park (UK), line 8.

where n is the sample size, p is the number of explanatory variables and R^2 is the proportion of the total sums of squares explained by the model. We used this procedure for the eight individual site patterns of above-ground biomass presented in Hector *et al.* (1999). Various schemes of alternative hypotheses have been proposed as possible diversity-function relationships (Vitousek and Hooper 1993; Lawton 1994; Naeem 1998; Schläpfer and Schmid 1999). We compared the following five statistical models: a grand mean; a linear effect of species richness; a linear effect of species richness when diversity is put on a $\log_2$ scale (curvilinear on untransformed axes); a quadratic

effect of species richness (maximum or minimum productivity at intermediate levels of species richness); and when significant differences between factor levels in an analysis of variance are more efficient in explaining variation than simple linear or curvilinear relationships.

Table 4.1 shows that each of these five relationships was the most efficient model, using this criterion, for at least one site of the BIODEPTH experiment. How can we reconcile this picture of differences in species richness responses, presented in Fig. 2 of Hector *et al.* (1999), with the picture of consistency presented in Fig. 1 of the same paper (compare Figs 4.1 and 4.2 of this chapter)? The differences stem in part from the different aims of hypothesis testing—which aims for a model with the minimal number of significant terms—versus model selection procedures that focus on finding the model that explains the most variation, or which is most efficient in explaining the most variation for the least cost in degrees of freedom (the adjusted R^2 and AIC approaches; note that there are other methods for doing this which differ in how they assess the most efficient model). For our data, the approach presented here in Table 4.1 focuses on analysing each site *individually* and with the aim of finding the most efficient model at each location. However, this approach sometimes includes terms in models which are not statistically significant and which are therefore eliminated with procedures that aim to find the model with the minimal number of significant terms, hence the two approaches sometimes differ in the models they select. Note also that although we were able to select one best model at each site using the adjusted R^2, in many cases there is little to choose between alternative models. A second major point, is that even if a relationship is non-significant at one (or even all) of the individual sites tested, it does not follow that the more powerful combined test will also fail to show significance. We feel that progress in identifying general phenomena in ecology on objective grounds will require developing methods for testing of statistical consistency or differences in effects at different locations. Clearly, the challenge in doing this will be in identifying analysis strategies that can identify general patterns while also taking into account detailed differences between locations. While there is clearly some way to go in this process, it is nonetheless underway in other areas of ecology (e.g. Reader *et al.* 1994; Zak *et al.* 1994; Gough *et al.* 2000; Gross *et al.* 2000). Results for multisite biodiversity experiments show a mixture of significant differences between locations (e.g. Van der Putten *et al.* 2000; Leps *et al.* 2001) and consistent general patterns (e.g. Bullock *et al.* 2001).

One of the particular complexities that arises when multiple locations are compared (Hector *et al.* 1999; Van der Putten *et al.* 2000; Bullock *et al.* 2001; Emmerson *et al.* 2001) is the additional between-sites components for species richness effects. In addition, while different sites are probably unlikely to be identical in their species lists, they may not overlap in species composition at all (some of the sites in Emmerson *et al.* 2001) or they may partially overlap with some species and mixtures re-occurring at two or more locations. For BIO-DEPTH, reoccurrence of the same species mixtures at different locations was simply a product of our within-site species selection. An alternative, if more complex approach, would be to deliberately control how mixtures are common or unique to sites in the design. In either case, the advantage is that it is possible to test for different effects of diversity at different locations, by testing the location-by-richness terms against the location-by-composition interaction for example (the analysis of BIODEPTH and biodiversity experiments in general are discussed in Schmid *et al.*, Chapter 6, and Hector *et al.* (2002a)). In BIODEPTH, the advantage of this was that it allowed us to determine that although species richness effects varied in individual location analyses, these differences were non-significant compared to the compositional variation of the same species assemblages grown at different sites (Hector *et al.* 1999, 2000b, 2002).

4.2.1 Experimental planned levels of diversity versus observed values

One of the problems that arises from the multi-faceted nature of biodiversity is that it is difficult, indeed impossible, to control all aspects at once in a single experimental design. For example, in the BIODEPTH experiments, we controlled the initial numbers of species sown into our plots but we did not try to control the relative abundances of these species after the initial seed sowing. In addition, a small number of species did not persist in our plots reducing the numbers present relative to the planned numbers. Might our results change if different measures of diversity are used instead of the values from the initial experimental design (Huston *et al.* 2000)?

We re-analysed the data on above-ground biomass from the second year of the experiment presented in Hector *et al.* (1999) to test this possibility. We calculated three alternative explanatory variables. We used the observed number of species present in the second year of the experiment and we also used the biomass data to calculate two diversity indices that incorporated information on the relative abundances of the species in our communities. We calculated the Shannon–Weaver index (H) and the Simpsons' index as $1 - D$. We used these two diversity indices that both combine the richness and evenness of species because the Shannon index can be weighted towards numbers of species present, whereas the Simpsons' index is weighted by dominant species (Magurran 1988). We used the indices on their original scales for comparison with $\log_2$(species richness), which we found described our data well, but the antilog of H' is a further possibility. Diversity indices can have properties, such as non-normal distributions, that can be problematical for analysis; we use them in these exploratory analyses as in practice they produced reasonable residual plots and ANOVA is a relatively robust technique.

We found, that just as for sown species richness, all three alternative explanatory variables had highly significant positive relationships with biomass (observed species richness: $F_{1,196} = 48.92$, $P < 0.001$; Shannon index: $F_{1,196} = 26.4$, $P < 0.001$; Simpsons' index: $F_{1,196} = 27.49$, $P < 0.001$) and that these relationships did not differ significantly between locations (all location-by-diversity interaction terms $F < 2$ and $P > 0.05$). Surprisingly, even after controlling for the Shannon index values there was a highly significant residual effect of the sown number of species ($F_{1,195} = 28.17$, $P < 0.001$). There are a number of possible explanations for this unexpected result. Setting aside the trivial point that it is likely that any survey of diversity at a particular place and time will underestimate the true diversity, more importantly, a process measured at a certain place and point in time may be influenced not only by the diversity of organisms currently present at that location but also by neighbouring individuals and by species that were present but have now disappeared. This second point is not specific to biodiversity experiments but may be true more generally: the results of an experiment may reflect its history and not just the explanatory variables at the time of measurement (Harper 1977). Interestingly, a third possibility concerns the size of the pool and hence the diversity of functional traits from which a number of species is derived; a 'multi-species sampling effect' (Loreau et al. 2001). For example, consider two sets of four species, one of which is derived from an initial (sown) set of eight species and the other from a group of six. It may be that the subset derived from the set of eight could be more productive than that derived from the set of six if the set of eight species had a greater initial range of functional traits, some of which was passed on to the subset of four species, thereby providing a greater potential for complementary and positive interactions. Clearly, there is much to be gained from additional analyses examining different aspects of diversity but it appears that the planned species richness is not only the 'correct' explanatory variable in terms of the experimental design and hypotheses but that it is also a good reflection of longer-term effects and can have greater explanatory power than measurements of diversity at one point in space and time during the experiment.

4.2.2 Partitioning biodiversity effects

The other most controversial aspects of the early BIODEPTH results concerned identification of the mechanisms generating the negative relationship between above-ground biomass and declining diversity. Initial explanations for the effects of biodiversity on ecosystem processes focused on niche complementarity (Naeem et al. 1994a; Tilman et al. 1996) through the partitioning of resources. However, there is a simpler way, which was initially missed, in which diversity can affect ecosystem functioning even in the absence of resource use complementarity. Under the sampling effect hypothesis

(Aarssen 1997; Huston 1997; Tilman et al. 1997a), when communities are assembled at random from a pool of species, more diverse mixtures have a higher probability of containing a species with extreme traits which could become dominant and drive ecosystem functioning. A more general 'selection effect' (Loreau 1998a, 2000a; Loreau and Hector 2001) is obtained if two assumptions of the sampling effect are relaxed. First, a single species need not dominate completely and, second, dominance need not be perfectly positively correlated with increasing monoculture productivity or biomass. Thus, biodiversity effects can be grouped into two classes: 'complementarity effects' (including resource-partitioning complementarity, positive interactions and negative interference) and 'selection effects' that occur through dominance or subordinance of species with particular traits. We have recently presented a new method that performs an additive partitioning of the individual contributions of the two effects in biodiversity experiments.

Our additive partition unifies and relates in a single equation previous measures based on the relative yields and proportional deviation from expected value approaches (Garnier et al. 1997; Hector 1998; Hooper 1998; Loreau 1998b; Emmerson and Raffaelli 2000; Dukes 2001b), with a way of estimating selection in mixed populations analagous to the Price equation from evolutionary genetics (Price 1970, 1995; Frank 1997). The method provides absolute estimates of different biodiversity effects allowing quantitative comparison of their respective contributions. The Price equation (Price 1970, 1995; Frank 1997) is typically used to separate changes in character traits that are due to the direct effects of natural selection in altering the frequencies of different alleles from those due to interactions between alleles in the altered population; the fidelity of transmission. Analogously in our method, 'ecological selection' occurs when changes in the relative yields of species in a mixture are non-randomly related to their traits (e.g. yields) in monoculture causing dominance and subordinance of species. The selection effect is therefore determined by the covariation between monoculture traits and relative abundance in mixtures in the same way as in the Price equation. Positive selection occurs if species with higher-than-average monoculture yields dominate mixtures, and negative selection if the opposite is true. The complementarity effect measures change in the average relative yields of species in mixtures compared to their expected values under the null hypothesis that yields in mixture will equal the monoculture yield times the proportion of the species in mixture. That is, the expected relative yield of species i is

$$RY_{e,i} = P_i M_i$$

The observed relative yield is

$$RY_{o,i} = Y_{o,i}/M_i$$

and the deviation in relative yields equals

$$\Delta RY_i = RY_{o,i} - RY_{e,i}$$

where P_i is the proportion of species i in mixture; M_i is the monoculture yield of species i; and $Y_{o,i}$ is the observed yield of species i in mixture. The method assesses whether increases in some species in mixtures are balanced by declines in others or whether there is evidence for complementary (resource partitioning), positive (facilitation) or negative (physical or chemical interference) interactions that shift the total yield away from the null prediction which assumes none of these additional interactions. A positive complementarity effect (resource partitioning or facilitation) occurs if the average deviations from expected values of the relative yields of the species in a mixture is higher than expected, and a negative complementarity effect (direct interference) if it is lower. The sum of the selection and complementarity effects gives the net biodiversity effect, which is the difference between the observed yield of a mixture and its expected yield under the null hypothesis that there is no selection effect or complementarity effect. This expected null value is the average of the monoculture yields of the component species weighted by their initial relative abundance in mixture, which for BIODEPTH and similar substitutive designs is $1/N$, where N is the number of species in the mixture and is thus a simple averaging of the single-species yields. The selection, complementarity and net biodiversity effects all have the dimension of the ecosystem property in question (such as yield)

and an expected value of zero under the null hypothesis of no biodiversity effect. Full details of the additive partitioning method can be found in Loreau and Hector (2001) but the basic equation expressed in the terms described above is

$$\text{Net biodiversity effect} = N\overline{\Delta RY}\,\overline{M} + N\text{cov}(\Delta RY, M)$$

All three biodiversity effects can be positive or negative, and complementarity and selection effects can therefore fully or partially cancel each other. There are therefore nine possible qualitative outcomes that arise from the combinations of positive, zero or negative complementarity and selection effects.

We provide a worked example of the method in Table 4.2 and Fig. 4.3 which illustrates these nine possible qualitative outcomes (cases (a)–(i) in the following all refer to both Table 4.2 and Fig. 4.3). The null situation refers to no effect of complementarity or selection and therefore no net effect (e). When they arise, biodiversity patterns can be either purely due to selection or complementarity effects with zero values of the other biodiversity effect ((b) and (d); note that no selection effect arises only when the relative effects of complementarity are equally distributed between species), or a combination of reinforcing positive (a) or negative (i) complementarity and selection effects. Negative selection effects occur when dominance is by a species with a lower-than-average monoculture biomass ((c),(f), (i)), which can then hide positive complementarity (c). Positive and negative effects may even cancel each other out exactly producing no net effect (c). Negative complementarity effects ((g)–(i)) indicate direct interference between species.

In general qualitative terms, joint positive biodiversity effects (a) appear to be the situation for the BIODEPTH experiments in Ireland the Sheffield (Loreau and Hector 2001) and for Cedar Creek (see Tilman *et al.* Chapter 3, 2001; Tilman 2001). Counteracting positive complementarity and negative selection (c) appear to produce a zero-to-negative net effects in a Californian Serpentine Grassland (Hooper and Vitousek 1997; Hooper 1998) and a positive net effect at the Portuguese BIODEPTH site. The overall pattern across all eight BIODEPTH

Table 4.2 Worked examples of different scenarios for the additive partitioning of biodiversity effects showing the different qualitative outcomes illustrated in Fig. 4.3

	M_i	$Y_{e,i}$	$Y_{o,i}$	$RY_{e,i}$	$RY_{o,i}$	ΔRY	NE	CE	SE
(a) Positive SE, positive CE (transgressive overyielding)									
Species A	800	400	700	0.5	0.875	0.375	—	—	—
Species B	200	100	150	0.5	0.75	0.25	—	—	—
Total mixture	—	500	850	1	1.625	—	350	312.5	37.5
Mean	500	—	—	—	—	0.3125			
(b) No SE, positive CE (non-transgressive overyielding)									
Species A	800	400	600	0.5	0.75	0.25	—	—	—
Species B	200	100	150	0.5	0.75	0.25	—	—	—
Total mixture	—	500	750	1	1.5	—	250	250	0
Mean	500	—	—	—	—	0.25			
(c) Negative SE, positive CE									
Species A	800	400	350	0.5	0.4375	−0.0625	—	—	—
Species B	200	100	150	0.5	0.75	0.25	—	—	—
Total mixture	—	500	500	1	1.1875	—	0	93.75	−93.75
Mean	500	—	—	—	—	0.09375			
(d) Positive SE, no CE (no overyielding)									
Species A	800	400	600	0.5	0.75	0.25	—	—	—
Species B	200	100	50	0.5	0.25	−0.25	—	—	—
Total mixture	—	500	650	1	1	—	150	0	150
Mean	500	—	—	—	—	0			
(e) No SE, no CE (Null hypothesis)									
Species A	800	400	400	0.5	0.5	0	—	—	—
Species B	200	100	100	0.5	0.5	0	—	—	—
Total mixture	—	500	500	1	1	—	0	0	0
Mean	500	—	—	—	—	0			
(f) Negative SE, no CE									
Species A	800	400	200	0.5	0.25	−0.25	—	—	—
Species B	200	100	150	0.5	0.75	0.25	—	—	—
Total mixture	—	500	350	1	1	—	−150	0	−150
Mean	500	—	—	—	—	0			
(g) Positive SE, negative CE (interference)									
Species A	800	400	500	0.5	0.625	0.125	—	—	—
Species B	200	100	25	0.5	0.125	−0.375	—	—	—
Total mixture	—	500	525	1	0.75	—	25	−125	150
Mean	500	—	—	—	—	−0.125			
(h) No SE, negative CE									
Species A	800	400	400	0.5	0.5	0	—	—	—
Species B	200	100	100	0.5	0.5	0	—	—	—
Total mixture	—	500	500	1	1	—	0	0	0
Mean	500	—	—	—	—	0			
(i) Negative SE, negative CE									
Species A	800	400	200	0.5	0.25	−0.25	—	—	—
Species B	200	100	75	0.5	0.375	−0.125	—	—	—
Total mixture	—	500	275	1	0.625	—	−225	−187.5	−37.5
Mean	500	—	—	—	—	−0.1875			

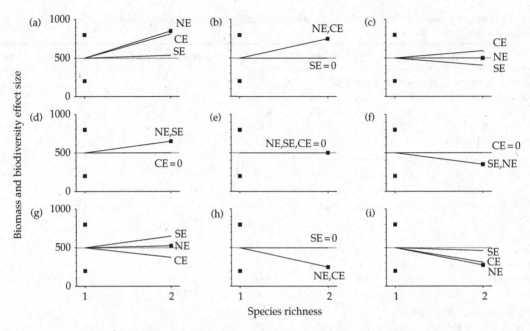

Figure 4.3 Examples of the nine different possible qualitative outcomes of the additive partitioning method presented in Table 4.2: (a) positive selection, complementarity and net effects; (b) zero selection effect, positive complementarity and net effect; (c) equal negative selection effect and positive complementarity effect producing zero net effect (example 1 from Loreau 1998); (d) positive selection effect and zero complementarity effect producing positive net effect; (e) zero selection, complementarity and net effects; (f) negative selection effect and zero complementarity effect producing negative net effect; (g) positive selection effect and negative complementarity effect; (h) zero selection effect and negative complementarity effect producing a negative net effect; (i) negative selection, complementarity and net effects. Symbols are monoculture and two-species mixture total yields, solid lines are biodiversity effects and broken lines show the null scenario of an averaging of monoculture yields (note that biodiversity effects have been scaled relative to this null value of 500 rather than zero so that they can be shown on the same figure as the observed yields).

sites pointed to a positive net effect generated by positive complementarity with a zero selection effect on average (b) as described in Loreau and Hector (2001). Since our approach requires a comparison between the performances of species in mixture and in monoculture, we restricted its application to the subset of experimental mixture plots that contained species for which monoculture yields were available. We discuss only the overall patterns across all sites here but individual site variations are described in Loreau and Hector (2001). The overall log-linear increase in above-ground biomass with species richness for the whole experiment was observed for this subset of the data. The net biodiversity effect was positive (the grand mean was significantly different from zero; Loreau and Hector 2001 Fig. 1(a)) and increased significantly with species richness beyond two species (Fig.4.4(a)). However, the selection effect was variable across

individual localities and overall these variations cancelled out so that the grand mean was not significantly different from zero, and on an average the selection effect was unaffected by changes in species richness. The only factors that influenced the selection effect significantly were locality and species composition (Fig. 4.4(c)). Complementarity effects at individual locations were also variable but the combined analysis revealed significant locality and composition main effects and a significant positive relationship with species richness (Fig. 4.4(b)). The presence of legumes in mixtures had important impacts on their performance; in general, they tended to increase observed yields and the net and complementarity effects, and to generate more extreme selection effects, both positive and negative (Fig. 4.4(c)). However, species richness retained a significant log-linear effect on complementarity even when the presence of legumes was included

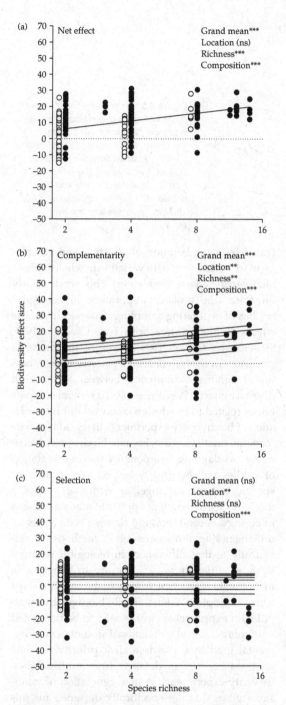

as an additional factor in our across-site analyses (this test is very conservative since part of the species richness effect is absorbed into the legume effect when the latter is fitted first in the analysis).

To summarize, the results of the overall across-site analysis of the complementarity effect showed a much closer match with the log-linear relationship found for the above-ground biomass patterns than did the selection effect. The increased complementarity in species-rich mixtures involved not only complementarity between legumes and other plant types, but also complementarity between species within each of these groups. Therefore, our analysis suggests that the positive relationship between above-ground biomass production and increasing diversity was driven by the complementarity effect and could not be explained by the selection effect. The additive partitioning method has not yet been applied to other multisite studies but it appears that in all cases both sampling and complementarity effects act in combination (Van der Putten *et al.* 2000; Leps *et al.* 2001; Bullock *et al.* 2001; Emmerson *et al.* 2001).

4.2.3 Dominance and above-ground biomass production

A central component of the sampling effect hypothesis concerns the dominance of plant communities by the most productive species in monoculture. Indeed, several researchers have predicted that biomass production in plant communities should be greater for communities that are strongly dominated than for those with a more even distribution of species (e.g. Huston 1997; Grime 1998; Huston *et al.* 2000; Wardle *et al.* 2000b). To test this hypothesis for

from the multiple regression model using species richness on a $\log_2$ scale. Complementarity effect lines from highest elevation to lowest are: Portugal, Switzerland, Silwood, Germany, Sheffield, Ireland, Greece, and Sweden. Selection effect lines from highest elevation to lowest are: Ireland, Germany, Sheffield, Greece, Sweden, Silwood, Switzerland, and Portugal. Values of the biodiversity effect (in g m^{-2}) were square root transformed while preserving the original positive and negative signs to meet the assumptions of analyses. Results are summarized for the grand mean for the three biodiversity effects and for the influence of location, species richness and composition: $* = P < 0.05$, $** = P < 0.01$, $*** = P < 0.001$. Adapted from a figure in *Nature* with permission.

Figure 4.4 Net biodiversity effect ΔY, (a), Complementarity effect (b), $N\overline{\Delta RY M}$, and Selection effect $N\text{cov}(\Delta RY, M)$, (c), as functions of species richness for year two above-ground biomass production. Open circles are plots that do not contain any legume species; filled circles are plots that contain one of more species of legume. Lines are slopes

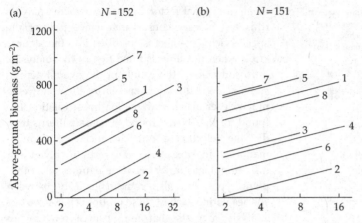

Figure 4.5 The positive effect of species richness on above-ground biomass production is stronger for communities that are less strongly dominated by a single species ((a) most dominant species <70% total biomass; $\log_2$ slope $= 107.6 \pm 16.7\,\mathrm{g\,m^{-2}}$) than for those that are more strongly dominated (b) >70%; $\log_2$ slope $= 67.8 \pm 19.2\,\mathrm{g\,m^{-2}}$). Sample sizes are given as 'N.

the BIODEPTH data, we calculated the proportion of total community biomass of the dominant species in the third year of the experiment and separated communities that were less strongly dominated by a single species (<70% of total above-ground biomass) from those that were more strongly dominated, splitting the dataset approximately in half for a balanced analysis. Counter to predictions, we found a stronger positive effect of increasing species richness on biomass for communities that were less strongly dominated and than for those that were more dominated by a single species (Fig. 4.5). This may have been due to increased dominance in low-productivity sites such as Sweden and Portugal due to the effects of drought at the former and frost and drought in the latter in eliminating individuals of vulnerable species, thereby causing greater relative dominance by the remaining resistant species. These analyses concentrate only on the single most dominant species and we are now examining dominance and evenness over all species. However, this analysis also suggests that dominance by individual species cannot explain productivity patterns in our experiment supporting a role for complementarity interactions.

4.3 Summary

This chapter has focused on the unique aspects of multisite biodiversity experiments. Although only a handful of these experiments exist, all show that changes in biodiversity impact a wide variety of ecosystem processes to some degree. All of these studies also seem to produce results that are gen-

erated by a combination of selection and complementarity effects but with variation in which of these biodiversity effects dominates. This work clearly illustrates the broader importance in ecological studies of separating sampling processes from the effects of biotic interactions (e.g. Oksanen 1996; Stevens and Carson 1999). Some of the multisite studies show overall consistent patterns while others reveal significant differences between locations in diversity effects. We have explored several complex issues related to the design, analysis and interpretation of biodiversity experiments that leads to some recommendations. First, because biodiversity has so many overlapping components the interpretation of studies will be greatly helped by setting very specific hypotheses, together with the proposed mechanisms; a point that applies to many questions in ecology. Second, ecology clearly needs a greater understanding and discussion of methods for statistically testing differences in biological relationships at different sites if we are to progress in distinguishing general from variable relationships across multiple locations. Finally, biodiversity manipulation experiments clearly need to be integrated with more classical correlational studies of environmental gradients; this is a clear priority for new analyses of existing datasets from multisite biodiversity experiments. A new generation of multisite studies that are specifically designed for this purpose could play a key role in achieving this important goal.

Our thanks to A. Kinzig for suggestions on presenting the partitioning method.

CHAPTER 5

Evaluating the relative strengths of biotic versus abiotic controls on ecosystem processes

M. A. Huston and A. C. McBride

5.1 Introduction

There is no doubt that the rates of ecosystem processes and the resulting properties and structure of ecosystems are influenced both by biotic factors, including species properties and interactions among multiple species, and by abiotic factors, such as climate, geology, and soils. The critical issue (and the focus of the debate about diversity and ecosystem function) is the relative contribution of each of these general factors to any particular ecosystem process, under any particular conditions, and at any particular scale.

While much remains to be learnt from future experiments, we can nonetheless evaluate the relative effects of biotic and abiotic factors in the results of current experiments, where they presumably result from processes operating in much the same way they operate under natural conditions. The detection of both types of factors in diversity experiments provides some insights into why these experiments have been so controversial. We will first review the patterns in natural systems that seem to contradict many of the experimental results, and then examine some of the experiments in more detail, in order to distinguish between biotic and abiotic effects.

5.2 Natural patterns of diversity and productivity

The conclusion of virtually all of the 'diversity–productivity' experiments that productivity

increases with increasing species diversity (Naeem et al. 1994a,b, 1995; Tilman 1996; Tilman et al. 1996, 2001; Hector et al. 1999; see however, Hooper and Vitousek 1997; Wardle et al. 1997b; Hooper 1998; Schwartz et al. 2000; Fridley in press) conflicts with the fact that most of the high productivity ecosystems found around the world have strikingly low plant diversity. This phenomenon was well known to early ecologists (Lawes et al. 1882) and was succinctly summarized by Rosenzweig as the 'paradox of enrichment' (1971), referring to the diversity-reducing effects of added nutrients (i.e. eutrophication).

In many regions, plant diversity decreases with increasing productivity over most of the range of productivity (Huston 1980, 1994). Naturally productive ecosystems with low species diversity include phytoplankton and algal blooms, salt marshes, freshwater marshes, riparian forests in the tropics and temperate zones (e.g. *Populus deltoides*, *Prioria copaifera*), bamboo forests, redwood forests, Douglas fir forests, eucalypt forests etc. High diversity plant communities are generally found on relatively unproductive sites, such as chalk grasslands, Mediterranean shrublands, and rainforests on oxisols and ultisols (Dawkins 1959, 1964; Grime 1979, 2001; Mahdi et al. 1989; Berendse 1994a; Huston 1979, 1980, 1993, 1994). If productivity in random-selection diversity experiments increases as a consequence of increasing diversity in experiments, then either (1) the processes involved in the experiments differ from the processes involved in natural patterns of diversity; or (2) the experiments address a different portion of the complete range of productivity, where patterns may differ from those in other portions of the range.

Notwithstanding the inconsistency of the experimental results with high productivity–low diversity and low productivity–high diversity natural ecosystems, the experimental results are compatible with the observation that diversity increases with increasing productivity under low productivity conditions (Grime 1973a,b, 1979; Huston 1979, 1980, 1994; Huston and DeAngelis 1994). When put together, the naturally occurring increase in diversity with increasing productivity at low levels of productivity and the naturally occurring decrease in diversity with increasing productivity at high levels of productivity produce the unimodal 'hump-backed' diversity curve originally described by Grime (1973a,b, 1979; Al-Mufti et al. 1977).

However, mere consistency of the experimental pattern and the natural pattern under low productivity conditions does not mean that they have the same cause. The natural pattern of increasing plant diversity as productivity increases from very low levels can be understood most simply as resource levels rising above the minimum required for survival of various species with different minimum resource requirements (Grime 1973a,b, 1979; Huston 1979; Huston and DeAngelis 1994). This is a phenomenon observed along gradients of increasing resource supply in low resource environments (e.g. water in deserts), or potentially in experiments that use variation in resource availability as a treatment (e.g. Lawes et al. 1882; Huston 1979, 1980, 1994; Tilman 1987, 1988; Fridley 2001).

The increase in average productivity with increasing species richness that has been observed in many of the diversity–productivity experiments occurs under presumably constant and uniform resource conditions (i.e. soil nutrients and water), except to the degree that they are altered by the plants themselves. Thus, any increase in productivity with increasing diversity is expected to result from either (1) complementary resource use among multiple species (with no change in total resources, just a more complete use of them) or (2) facilitative interactions among species, in which one or more species improve conditions (resource availability or a physical condition, such as temperature or pH) for other species, which then grow faster or larger than they would in the absence of the facilitator (Vandermeer 1989; Hooper 1998). The most widespread and important example of facilitation is nitrogen fixation by legumes, which has the effect of increasing nitrogen availability for other plant species that may eventually overgrow and eliminate the legumes (DeWit et al. 1966).

In summary, variation in species richness can play only a minor role in natural patterns of productivity, which are regulated primarily by environmental conditions such as climate and soils. Only when the dominant effects of environmental conditions on productivity are controlled experimentally can the subtle effects of species composition on productivity be detected. Consequently, the failure to adequately control environmental conditions is potentially a problem with any experiment that attempts to evaluate the effect of species richness, or any type of plant interactions, on productivity.

5.3 The influence of soil heterogeneity on diversity–productivity experiments

Small differences in soil conditions can have a large effect on plant growth and species diversity (Fig. 5.1). Natural variation in soil nutrients and water is associated with subtle differences in topography, with even slight depressions generally having higher nutrient and water availability than the surrounding area. Dealing with this natural variability is a major issue in the design of agricultural and ecological field experiments (Trenbath 1974; Vandermeer 1989, see Chapter 19), and has led to such designs as the 'Latin Square', replication by blocks for ANOVA etc.

Separation of 'block effects', some of which could be due to differences in soil conditions, from the main treatment effects requires replication of the full experiment over a set of blocks that may or may not have significant differences in soil conditions. Such replication adds cost and complexity to any experimental design, and may be infeasible for experiments with many different treatment levels. In random-selection diversity experiments, the

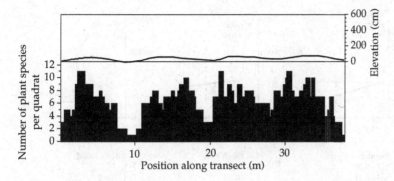

Figure 5.1 Natural variation in plant species richness across the repeating topographic pattern of 'ridge and furrow' grassland in Great Britain. Note that species richness is highest on the ridges, and lowest in the moister, more nutrient-rich furrows, where two clonal species, *Agrostis stolonifera* and *Ranunculus repens*, tend to exclude all the other species. (From unpublished data of G. R. Sagar in Harper 1977).

relatively large number of plots within any particular diversity level has generally precluded any replication by spatial blocks. Consequently, the performance of any treatment plot could potentially be influenced by three factors: (1) the number of species; (2) the particular species that are present; and/or (3) the soil conditions at the location of that plot. Separating the effects of these three factors is the primary challenge in the analysis of the experimental results, as well as the primary cause for disagreement about the interpretation of published results. Unfortunately, this is not a simple task, nor are standard statistical approaches well suited for the task (cf. Schmid *et al.*, Chapter 6).

The effect of soil heterogeneity on experimental results can be seen in the results of the BIODEPTH experiments (Hector *et al.* 1999, see Chapter 4), which were conducted at eight sites in seven European countries. The locations spanned a latitudinal gradient from Sweden to Greece, and included a range of soil conditions and climates. In addition to the variation among the sites at which the experiments were conducted (not replicated, since all sites did not use the same design), there was inevitably soil heterogeneity within the field in which each experiment was conducted.

A striking property of the BIODEPTH results, as well as the results of other experiments of this type, is the high variability in the responses at any given level of species richness (Fig. 5.2). However, this variability is not uniform across all sites, which provides a critical insight into the influence of soil factors on the results of 'pure diversity' treatments. A scattering of very high values, above the level at which most plots are clustered, can be seen at all of the sites but one, the site at Sheffield in UK.

Sheffield was one of three sites that used an experimental design suitable for distinguishing the effects of single species from multiple species using analysis of 'over-yielding' (the other two being Portugal and Sweden). However, Sheffield was the only site in the entire experiment in which the soil in the experimental field was artificially constructed by bringing in truckloads of soil, and mixing and spreading it uniformly over an existing field (A. Hector, personal communication). The exceptional homogeneity of the artificially constructed soil at Sheffield seems to be the explanation for the extremely low variability among the plots at each diversity level, particularly at the highest level where all the plots have identical species composition (Huston *et al.* 2000). This low variance within each treatment level leads to the highest proportion of variance explained by the experimental treatment at any of the eight sites ($r^2 = 0.49$).

The extremely low variance among plots at the same level of species richness on the homogeneous constructed soils of the Sheffield site suggests that the high variance observed at the other sites is at least partially due to variance in soil conditions across the experimental site. At Sheffield and the other two sites designed for over-yielding analysis, the variability at the highest level of species richness can be attributed primarily to soil variability because all of the replicates are identical in species composition. However, the other five sites include species in the high diversity treatments that were not planted in the monocultures. Consequently, some of the variance at these sites may be due to differences in species composition even at the high species richness levels, as well as to the inevitable heterogeneity in soil conditions across the sites, and differences in

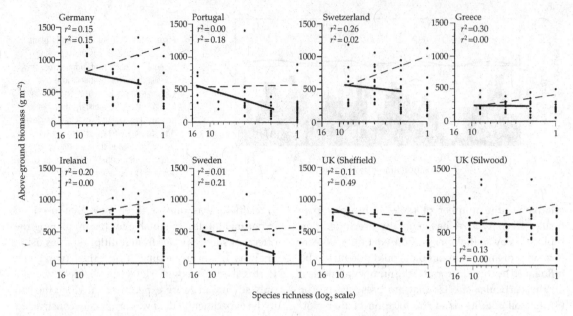

Figure 5.2 Results of the BIODEPTH experiment at eight sites across Europe. Only treatments with species richness less than or equal to the number of species used in monoculture treatment are shown. Solid line is log-linear regression through all plots with two or more species. Dotted line is the log-linear regression through plots expected to contain the species that was most productive in monoculture, for evaluation of the effect of species richness on overyielding. Upper r^2 value refers to the dotted line; lower value refers to solid line (from Huston *et al.* 2000). Note that Sheffield, with artificially constructed soils, has the lowest within-treatment variance at high treatment levels.

species composition at the lower levels of species richness.

In summary, for diversity experiments in which soil conditions are not controlled by site preparation, variability in soil properties across an experimental site can have a strong effect on the magnitude and variability of the measured responses, such as above-ground biomass production. This introduces the possibility that soil heterogeneity can act as a 'hidden treatment' in these experiments, because soil variation can produce the same response, an increase in productivity, that is predicted to occur as a result of the diversity treatments.

A hidden treatment is any factor that is correlated with the treatments in an experiment, but not explicitly controlled, manipulated, measured, or considered in the experimental design and analysis of the experimental results (Huston 1997). If such factors are not recognized or are ignored in the analysis of data, they can lead to misinterpretation of causal relationships because the experimental treatment is assumed (in the context of regression analyses or ANOVA) to cause the observed responses, rather than the hidden treatment factor that actually produced the response.

5.4 Quasi-replication, representation, similarity, and inhomogeneity of variance: a suite of hidden treatments in random-selection experiments

Given that there is inevitably some degree of soil heterogeneity in field experiments, the critical question becomes how inherent soil effects can be distinguished from the effects of the experimental treatment, specifically from the effects of species number or species composition. Obviously, experiments can be established using designs capable of distinguishing field effects from treatment effects. Alternatively, the actual soil properties in each plot can be measured and used as covariates in the analysis. However, neither of these options have been typically used in diversity–ecosystem function experiments, primarily because the logistics and cost of replicating species richness treatments is already quite daunting.

The issue of replication in random-selection diversity experiments is more complex than in standard experiments. Replication (or sample size) is a central issue in the design and analysis of all experiments. In general, higher replication increases the accuracy of estimates of mean response, and also increases the probability that any experimental response that is detected is statistically significant because it allows the response distribution to be determined more accurately. The pattern of variance among responses at different treatment levels is a critical issue in statistical analysis because 'homogeneity of variance' is a basic precondition for the validity of many statistical tests, including ANOVA. The critical assumption is that the variance among replicates in any sample (or treatment level) is the same as the variance in any other sample, i.e. variance is equal (or homogeneous) across all samples that are being compared (Sokal and Rohlf 1981).

Several inherent features in the design of random-selection diversity experiments cause the response variance to differ between treatment levels, compromising the validity of most statistical tests that have been used to analyse these experiments. Furthermore, the pattern of unequal variance is correlated with treatment levels because the cause of variance changes predictably with an increasing number of species. The assumption that variance is equal among treatments is invalid if the causes of variance differ among treatments. This correlation of treatment level with the causes of within-treatment variance increases the probability that the experimental results will be misinterpreted.

Understanding the patterns and causes of within-treatment variance in random-selection diversity experiments requires a careful evaluation of what replication means in these experiments. What is treated as replication in these experiments actually has two very different components, 'representation' and 'true replication', that change in their relative proportions along the treatment gradient from low species richness to high species richness.

Representation is the proportion of the total number of possible species compositions at a specific level of species richness (calculated using combinatorial probabilities) that is actually included in the experiment. The number of unique combinations of

x species drawn from a total species pool of N is calculated as $N!/(x!\ (N-x)!)$. Thus, for an experiment with a total species pool of 18, there are 153 different two-species combinations, and also 153 different 16-species combinations, but only one 18-species combination (Fig. 5.3).

True replication refers to the number of identical plots, i.e. plots with exactly the same set of planted species. For example, in an experiment with a total species pool of 18 that includes a monoculture of each species, representation at the one-species level would be 100%. If there were two monoculture

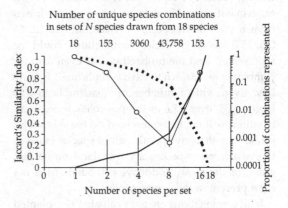

Figure 5.3 Statistical properties related to within-treatment variance that co-vary with treatment level (number of species per set) in random-selection diversity experiments. Jaccard's Similarity Index (left-hand scale and thick line with bars for ±1 SD), illustrates the increasing similarity in species composition among randomly assembled sets of species (quasi-replicates) as the number of species in the set approaches the total number of species, caused by the variance reduction effect (Huston 1997). The proportion of variance (▪▪▪▪) at a specific treatment level due to differences in species composition is the inverse of the similarity among the quasi-replicates at that level. The proportion of combinations represented (right-hand scale and —o—) is the number of randomly drawn quasi-replicates at each treatment level (39, 35, 29, 30, 35 for the five treatment levels) divided by the total number of unique combinations of species at that treatment level (determined by combinatorial probabilities and listed at the top of the figure). The consequences of these co-varying properties are that (1) treatment responses are underestimated at low treatment levels because of under-sampling of the high variability in species composition (and therefore performance) among quasi-replicates, and (2) treatment effects are overestimated at high treatment levels because the low variability in species composition at high treatment levels allows experimental error (such as uncontrolled variability in soil properties) to become the primary source of variability.

plots of each species, the level of true replication would be 2, for a total of 36 plots.

In a 'well-replicated' diversity–productivity experiment, there may be 30 or more plots at each level of species richness (e.g. 1,2,4,8,16). However, these plots are not true replicates but rather 'quasi-replicates' whose statistical properties are correlated (and thus confounded) with the experimental diversity treatments. Note that 'quasi-replicates' represent a different phenomenon than 'pseudo-replicates', *sensu* Hurlbert (1984). The quasi-replicates at each treatment level are composed of a variable proportion of the total possible number of combinations, and, with a very low probability, some true replicates of a few of the combinations. For an experiment with 35 plots of two species drawn from a pool of 18 species, a maximum of 22.8% (35/153) of all possible combinations could be represented, and the probability that even a single combination is drawn twice (i.e. replicated) is quite low. Even with a number of quasi-replicates as large as 35, over 75% of the possible species combinations will not be represented, so there is a high probability that any specific pair of species (such as the two largest species, a particular legume and grass, or the most productive combination) is not even present in the experiment.

If all combinations are not evaluated (i.e. planted at least once in an experiment), then it is impossible to conclude that the observed maximum production is the actual maximum production, especially at low levels of species richness, because there is a high probability that the most productive combination of species was not even planted. Assuming that there is a similar number of plots at each level of species richness, the proportional representation of all possible combinations changes dramatically and non-linearly with increasing species richness in this type of experiment (Fig. 5.3). This pattern is symmetrical, with high representation at low levels of species richness, and also at high levels (near the size of the total pool of species), but with extremely low representation at intermediate levels of richness. For example, at a treatment level of eight species drawn out of a pool of 18, 35 replicates represents 0.08% (35/43,758) of the total number of combinations. Obviously, the results produced by such a small random sample could change

dramatically from one experiment to another, and would be unlikely to capture any valid statistical properties of the actual distribution of performance.

Interpretation of this symmetrical distribution of representation across the treatment levels in a random-selection diversity experiment is further complicated by the 'variance reduction effect' (Huston 1997), which is another consequence of the same probabilities for randomly selected sets that produce the 'selection probability' or 'sampling' effect (Aarssen 1997; Huston 1997; Tilman *et al.* 1997b). Just as the probability of drawing (without replacement) any particular species out of a fixed set of species increases as the number of species drawn increases, the probability of the same combination of species appearing in different draws increases as the number of species drawn increases. The probability of all draws having the same species composition reaches 100% when the number of species drawn (without replacement) equals the total number of species in the pool from which the draws are made.

The variance reduction effect complicates the interpretation of representation and replication in these experiments, because both of these properties reach a maximum in treatments with high species richness. Consider the cases of two-species and 16-species treatments in a random-selection experiment with a total species pool of 18 (Fig. 5.3). Both have the same number of unique species compositions, 153, so if both treatment levels have the same number of randomly assembled plots, both will have the same proportional representation of all possible combinations. However, as a consequence of the variance reduction effect, the 35 plots with 16 species are much more similar to one another than the 35 plots with two species. This occurs because the each of the 16-species plots contains most of the 18 possible species. Each plot must share a minimum of 12 species with each other plot, and any individual species is expected to occur in 31 out of the 35 replicates. In contrast, any particular two-species plot does not necessarily share any species with another plot, and each individual species is expect to appear in only four out of the 35 plots.

This pattern of increasing similarity (i.e. reduction in variance of species composition and related properties) with increasing levels of species richness can be simply illustrated using Jaccard's Index of similarity, a standard method for evaluating the similarity of species composition of pairs of ecological communities (i.e. species lists). Jaccard's index is equal to $a_{11}/(a_{11} + a_{10} + a_{01})$, where a_{11} is the number of species shared between list 1 and list 2, a_{10} is the number of species that occur only in list 1, and a_{01} is the number that occurs only in list 2. The index varies between 0 and 1.0.

Figure 5.3 shows the monotonic increase in pairwise similarity among randomly drawn sets of species as the sets include larger numbers of species. Similarity is 0 among unique single-species sets, and 1.0 among all 18-species sets. The critical distinction is between the two-species and 16-species sets. Mean pairwise similarity among two-species plots is 7.6%, while mean similarity among 16-species plots is 80%. Although there is the same number of compositionally distinct combinations (153) at both treatment levels, all 16-species plots are very similar to one another, while all two-species plots are very different from one another.

Consequently, the 16-species plots are almost complete replicates of the same species composition, and should have many of the same properties as true replicates in plant growth experiments, possibly including a normal distribution of values, depending on the factors determining the experimental error (i.e. unexplained variance in a regression or within-subgroup mean squares in ANOVA). In contrast, all of the two-species plots are quite distinct from one another, and are unlikely to show the same response to error factors that would be observed in the 16-species plots.

Each unique species composition should produce a characteristic (or 'true') response under constant or homogeneous conditions. This true response could be estimated as the mean of replicate responses under heterogeneous conditions. Assuming that there is some normally distributed error applied to all plots in the experiment, this error should shift the value of each plot away from its true value, but only a small percentage of the plots should be shifted far away from their true value. Because two-species plots are so different from one another, they would be expected to have true values that cover a broad range, e.g. from high biomass to low

biomass. So, a normally distributed error should have little effect on the overall distribution, since most values that are displaced far from their true values are still likely to be within the wide range of values produced by the different species combinations.

In contrast, because the 16-species plots are very similar in species composition (i.e. nearly true replicates), they would be expected to have similar true values of the observed response. Application of a normally distributed experimental error to these plots is very likely to produce values outside the relatively narrow range of the true values. For example, with a normally distributed error with a mean of 1.0, one might have 10% of values in each tail of the distribution, e.g. below 0.7 or above 1.3. If the error is applied as a multiplier of the true value, as might be the case for error associated with variations in soil properties across an experimental field, this would result in an average of 3.5 high outliers and 3.5 low outliers out of 35 plots.

This produces what is perhaps the most insidious hidden treatment in diversity/productivity research: the increasing probability of interpreting experimental error as a treatment effect at higher levels of species richness. The probability of high (or low) outliers that are outside the range of true values is much higher for high richness treatments than for low richness treatments. The 'experimental error' that produces these outliers may come from some undetermined and apparently random source of variability, or from an easily understood source, such as slight variation in soil properties over the experimental field. Combined with the higher mean value caused by the sampling effect at high species richness, this variability is more likely to produce very high outliers at high levels of species richness. This may lead to the incorrect conclusion that the treatment has produced an increase in response (e.g. biomass or productivity) at high levels of diversity. In reality, the presumed treatment effect is an error effect.

This interaction of 'variance reduction' with 'representation' and 'replication' produces reversed monotonic gradients in the causes of variation within treatment levels (i.e. experimental error) across the treatment gradient. This reversal in the causes of within-treatment variance confounds the interpretation and statistical analysis of random-selection diversity experiments. At low treatment levels of species richness, most of the variance in response is the consequence of differences in species composition among the quasi-replicates, and a relatively small proportion of the variance is due to other sources of variability, such as soil heterogeneity. In contrast, at high levels of the diversity treatment, very little of the variance in response is due to differences in species composition, because the quasi-replicates have converged with true replicates because of their high similarity in species composition. Thus, the primary source of response variance at high levels of species richness is not species composition, but rather whatever sources of error are present in the experiment, such as soil heterogeneity.

This phenomenon is clearly illustrated by the BIODEPTH experiments, where all plots of the highest diversity treatment were identical in species composition at the three sites designed to test for over-yielding (Portugal, Sheffield, and Sweden). At these three sites, each of the high diversity plots contained all of the species in the total species pool (i.e. all of the species that were grown as monocultures). Consequently, the variability in biomass production among the plots of the high diversity treatment could not have been due to differences in species composition, since all were identical in composition. This variability must have been caused by some other factor, of which underlying soil heterogeneity is the most likely candidate. The role of soil heterogeneity as the primary cause of variability in biomass production is strongly supported by the dramatically lower variability at Sheffield, the only site at which the soil was trucked in and homogenized (compare Portugal, Sheffield, and Sweden in Fig. 5.2). Note that the variability of biomass at the lower levels of species richness (1,2,4) is relatively similar at the three sites. This is expected because the primary source of variability in randomly selected small numbers of species is random differences in species composition among the plots. It is primarily at the highest level of species richness in these experiments, where all plots are identical, that the outliers from the error caused by soil heterogeneity across the field can be detected.

In summary, not only are the high diversity treatments most suitable for distinguishing the effects of soil heterogeneity (or other spatial factors) from the effects of species composition, they are also most susceptible to the confounding of error effects with treatment effects. The laws of probability produce a continuous gradient of: (1) increasing similarity among the quasi-replicate sets of species (e.g. different combinations of N species from a pool of 18 species) drawn at any given diversity level, and (2) increasing number of replicates of each particular subset of species (i.e. how many times exactly the same set of two or three species are drawn from 18) at any given diversity level, as the treatment level approaches the total number of species. The fact that high diversity plots are more similar to one another than are low diversity plots (the variance reduction effect) means that there are effectively more replicates of highly productive species combinations, and a greater chance of some of the replicates falling on portions of the field with anomalous soils. Therefore, there is a higher probability of misinterpreting the effect of soil properties, or other sources of experimental error, as treatment effects at high levels of species richness.

5.5 Simulating hidden treatment effects in a diversity–productivity experiment

One of the most contentious issues in the diversity–productivity controversy is whether the widely observed increase in mean productivity (or biomass etc.) with an increasing number of randomly selected species is (1) the consequence of interactions among many species, i.e. a true diversity effect caused by processes such as facilitation or complementary interactions among many species; or (2) the result of the chance presence of one or a few species, i.e. a statistical artefact of random sampling, called the selection probability (Huston 1997) or sampling (Tilman *et al*. 1997b) effect.

The sampling effect is the 'the increasing probability of selecting species with a specific property (e.g. large maximum height, stress tolerance, nitrogen-fixation ability, high seed germination rate) in samples of increasing number that are randomly selected from any group of species' (Huston

1997). This purely statistical phenomenon results from the random selection of any one, two, or more objects (e.g. species) from a group of any number of objects, according to the laws of probability. It can produce a pattern of increasing average productivity of mixtures with an increasing number of species as the result of (1) the increased probability of the inclusion of a single large or fast-growing species (Aarssen 1997; Huston 1997; Tilman *et al.* 1997b); or (2) the inclusion of specific groups of two or more species that interact positively either through facilitation (e.g. the fertilization effect of a nitrogen-fixing legume on a grass, Huston 1997; Hooper and Vitousek 1997; Grime 1997; Hodgson *et al.* 1998; Wardle 1999; Huston *et al.* 2000; Loreau 2000a) or through complementarity among groups of two or more species (i.e. niche effects) to produce more biomass than a smaller number of species could produce (Huston 1997; Tilman *et al.* 1997b; Hooper 1998; Loreau 2000a; Loreau *et al.* 2001; Fridley 2001; Dukes 2001b).

The two primary alternative hypotheses to explain the observed increase in mean performance with increasing species richness in random-selection diversity experiments are (1) the sampling effect for a small number of species (e.g. a highly productive individual species, or a highly productive species pair, such as a legume and a grass); and (2) multi-species complementarity (niche effects) among a large number of species. The definitive pattern for distinguishing between these two alternatives is not the increase in mean production, which is expected to be the same both for interactions among many species (niche effects) and for interactions among only a few species (sampling effects), but rather the shape of the upper boundary of the response distribution.

'With niche complementarity, the upper bound increases with diversity because no monoculture is as productive as some combinations of two species, and *no combination of N species is as productive as some combinations of N + 1 species*' (Tilman *et al.* 2001, italics added). This definition requires a continuous increase in the maximum value, as well as of the mean of the measured property (e.g. biomass production). In contrast, the 'signature' of the sampling effect is a constant upper bound for maximum biomass (or any other measured property) that remains unchanged while the mean response increases with the number of randomly selected species.

The most conservative criterion for distinguishing between these alternative hypotheses is a metric from agricultural research known as 'transgressive over-yielding'. Transgressive over-yielding occurs when the productivity of some mixture of species exceeds the maximum productivity of the monocultures of all of the component species or of mixtures with fewer species (Trenbath 1974; Mead and Riley 1981; Vandermeer 1989; Garnier *et al.* 1997; Fridley 2001). For example, transgressive over-yielding above the two-species level occurs when some plots with more than two species exceed the maximum production of all plots with two species.

In the case of multi-species niche complementarity, there is over-yielding above each sequential level of species richness, beginning with monocultures, and continuing to the maximum number of species at which niche complementarity has a significant effect on production. For few-species sampling effects, there is no over-yielding above the number of species for which the sampling effect occurs. If the increase in productivity is determined by a sampling effect for one or more pairs of species (e.g. one or more specific legume–grass combinations), no plots with more than two species will exceed the highest production level achieved by two-species combinations. The most productive two-species combination(s) is assumed to be present in the most productive plots at all higher levels of species diversity.

In most of the published experiments of this type (Naeem *et al.* 1995; Tilman *et al.* 1996; Hector *et al.* 1999), as well as the first several years of the long-term experiment reported by Tilman *et al.* (2001), none of the multi-species treatments exceeded the maximum productivity of monocultures, which demonstrated that there was no over-yielding above the single-species level. This result was quite surprising, since several of these experiments included nitrogen-fixing legumes, which are known to increase the productivity of plants grown with them as a result of the nitrogen fertilization effect. There were undoubtedly positive interactions between legumes and other species, but none that resulted in transgressive over-yielding.

However, in the last few years of the longest-running diversity–productivity experiment, there was clear transgressive over-yielding above the maximum production of monocultures (Tilman *et al.* 2001), and published analyses demonstrate the expected strong effect of legumes (particularly *Lupinus perennis*) growing with C₄ grasses. The critical question for interpreting these results is whether or not there was over-yielding above the maximum productivity at the two-species level. If there was no over-yielding above the level of two species, the observed increase in mean productivity with increasing species richness is most likely the result of the sampling effect for the

combination of a legume and a grass, a common and well-understood response, rather than the niche effects of complementary interactions among many (e.g. 9 to 13) species, as concluded by Tilman *et al.* (2001).

There is no doubt that the most recent experimental results show several plots with above-ground and total biomass that exceed the maximum amount at the two-species treatment level (Fig. 5.4). However, there are only a few plots (~10% of the 94 plots with four or more species) that exceed the two-species overyielding level. Does this represent the increasing upper bound expected for multi-species niche effects, or just scattered outliers above

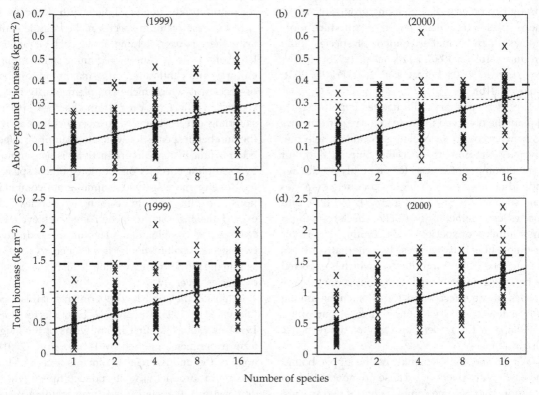

Figure 5.4 Biomass production of all replicates at five levels of species richness in a random-selection diversity experiment (Tilman *et al.* 2001). Data are shown for two years, and two measures of biomass: (a,b) above-ground biomass (approximating annual biomass production); (c,d) total biomass, including both above- and below-ground (the accumulation of roots which had grown over the 6–7 years of the experiment). Solid line indicates log-linear increase in average biomass with increasing number of species. Dotted line indicates the minimum biomass value for transgressive over-yielding above the level of one species (set at the mean of the two highest biomass values, because there are two replicates of each monoculture, and the mean is assumed to be the best estimate of the response of the most productive species). Upper dashed lines indicate the minimum biomass value for transgressive over-yielding above the level of two species (set at the level of the highest biomass value, because each of the two-species plots represents a unique species combination). The high biomass values in the 16-species treatments are most likely the result of experimental error caused by spatial variation in soil moisture or fertility across the experimental field (see text).

the maximum production level for the two-species mixtures as expected for the sampling effect?

We address this issue by simulating this experiment using the assumption that the *only interactions* among species are those between a legume and a C_4 grass. Of the 18 different species used in the experiment reported by Tilman *et al.* (2001), the only combinations that consistently produced over-yielding were those that included one or more of the four C_4 grasses plus one or more of the four legumes. To evaluate the consequences of the simplest possible explanation for the observed results, we restricted the mechanisms available to produce the treatment response to only the sampling effect for the presence of one legume plus one C_4 grass. Our question is whether this sampling effect alone can reproduce the observed responses, or whether some other mechanisms, such as multi-species niche effects involving more than two species, are necessary.

We assume that, in the absence of any positive or negative interactions with other species, each species has a characteristic maximum rate of productivity (or maximum accumulation of total biomass) that is equivalent to the maximum production of that species in monoculture (Table 5.1). This rate sets the maximum productivity of any mixture of non-interacting species to that of the species with the highest rate of productivity, which is the basic assumption for a single-species sampling effect. While we do not expect that the species with the highest monoculture productivity will dominate all mixtures or achieve the same biomass in mixtures as it did in monoculture, we assume

that a mixture will not be more productive than its most productive species (i.e. no over-yielding above the single-species maximum) *unless* facilitation or complementarity is occurring.

The only interaction among species in our model is the facilitation through nitrogen fertilization that occurs when a legume grows with a C_4 grass, so that the characteristic productivity of the grass is multiplied by some factor greater than 1.0 when a legume is present. Each of the four legumes species present in the experiment is assumed to have a different 'nitrogen fertilization multiplier', with the most abundant legume, *L. perennis*, having the highest multiplier (Table 5.1). If more than one legume and more than one C_4 grass species is present in a plot, we assume that the productivity of the plot is the rate of the most productive grass species multiplied by the highest fertilization factor among the legumes that are present. Thus, the productivity of all mixtures is constrained to result from a strictly two-species sampling effect.

Given these simple assumptions, we can ask what sorts of response patterns are likely to be observed. Each particular combination of species has a deterministic level of productivity, based on monoculture biomass that is augmented by the presence of legumes for C_4 grasses only (four species out of the total of 18). This mechanism produces the expected over-yielding above the single-species level and no over-yielding above the two-species level (Figs 5.5(a)–(c)). When there is a sufficiently high level of quasi-replication, the interaction of representation with variance reduction produces a maximum response that is identical at all treatment

Table 5.1 Parameter values used in simulation of diversity–productivity experiment. The total of 18 species includes four legumes and four C_4 grasses plus ten other species

Species type	Legume	C_4 grass	Other species
Biomass (kg m^{-2})	0.7 0.5 0.45 0.4	1.14, 0.9, 0.8, 0.4	0.45, 0.41, 0.37, 0.33 0.3, 0.28, 0.24 0.2, 0.16, 0.12
Legume multiplier	1.48 1.3 1.2 1.1		
Soil heterogeneity multiplier	Mean = 1.0, SD = 0.2		
Actual species number per treatment	16 species treatment: Mean = 11.5, SD = 1.4 8 species treatment: Mean = 7.0, SD = 1.4		

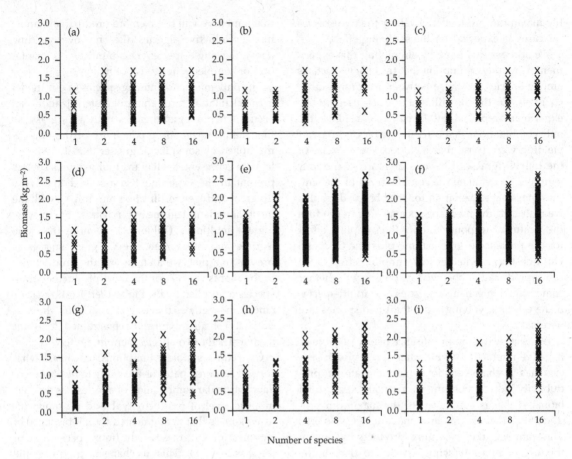

Figure 5.5 Results of simulation of a diversity–productivity experiment using the assumption that the most complex interaction is that between two species, specifically a C_4 grass and a legume. (a) 1000 quasi-replicates at each treatment level, note classic sampling effect of increasing mean with no values higher than the maximum production at the two-species level; (b) and (c) same model as (a), but with 36, 35, 29, 30, and 35 quasi-replicates at treatment levels 1, 2, 4, 8, and 16, respectively; (d) same model as (a) with 300 quasi-replicates at each treatment level; (e) and (f) same model and number of quasi-replicates as (d) with multiplication by an error distribution with a mean of 1.0 and SD of 0.2 (simulating soil variation across experimental field), randomly applied to each quasi-replicate at each treatment level; ((g)–(i)) same model as (e) and (f), but the same number of quasi-replicates as in (b) and (c).

levels greater than one species (Fig. 5.5(a)). However, when quasi-replication is reduced to the low levels used in most experiments, the effects of random selection of species produce a highly variable treatment response (Figs 5.5(b,c)). The primary effect is that the most productive species compositions tend not to be represented in the small random samples (low levels of species richness). This generally leads to an increase in the number of high values (none of which exceed the true maximum productivity of two-species combinations) at high levels of species richness.

Addition of an error distribution, representing heterogeneity in soil properties across the experimental field, to the simulation produces a pattern that resembles true over-yielding, but is in reality the interaction of experimental error with the variance reduction effect. Figures 5.5(e) and (f) show the effect of random variation in soil conditions (field effects) added to the base case of deterministic two-species interaction, for 300 quasi-replicates (compare with Fig. 5.6(d), with 300 quasi-replicates but no error effect). Note that the number of high outliers increases at higher levels of species richness because (1) the field effects are superimposed on a larger number of replicates because the quasi-replicates approach true replicates at the highest treatment levels as a consequence of the variance reduction effect; and (2) the replicates all have high rates of productivity because all contain at least one legume and at least one C_4 grass as a consequence of the sampling effect at the highest treatment levels. Note also that the actual pattern of outliers in any experiment will be determined by the actual distribution of soil properties across the experimental field, and will not necessarily match the normal distribution that we have used for generality. Because the pattern of soil properties is influenced by topography, soil depth, underlying geology, and other factors that have non-random spatial patterns (e.g. Robertson *et al.* 1988), the actual distribution is unlikely to be symmetrical.

The interaction of low levels of quasi-replication with field effects from soil heterogeneity produces great variety in the possible outcomes of the experiment, depending on which species combinations happen to be selected at low and intermediate levels of richness, and their random spatial locations in relation to soil properties (Figs 5.5 (g)–(i)). Many of these outcomes resemble the response observed in recent years in the Cedar Creek experiment, which shows only a few outliers that exceed the maximum productivity of two-species mixtures, most of which occur at the highest level of species richness (Fig. 5.4).

Thus, simple simulations based only on two-species interactions, combined with soil heterogeneity, reproduce the basic pattern of observed responses. This pattern occurs because the constant and low level of quasi-replication (e.g. ~35 plots per treatment level) results in very low representation of the most productive species combinations at low and intermediate levels of species richness, where similarity among the possible combinations is quite low. This correlation between treatment level and the causes of response variation is an inherent feature of random-selection diversity experiments. An increased number of plots (quasi-replication) at all levels of species richness, with highest numbers at the lower levels, would help solve the problem of inadequate representation. However, compensating for the increasing similarity among different species combinations that is the inevitable consequence of the variance reduction effect would require sufficiently high levels of true replication at lower levels of species richness to sample soil heterogeneity and other sources of experimental error to the same extent that it is sampled by the high richness treatment, where all quasi-replicates are very similar in species composition. Such a design is likely to be logistically intractable. Alternative approaches include careful quantification of soil conditions in each plot for use as co-variates in analysis, and/or use of a large number of replicates of a few selected species compositions to quantify the spatial pattern of relevant soil properties across the experimental field.

5.6 Conclusions

The inherent statistical problems associated with random-selection diversity experiments make them highly susceptible to misinterpretation of cause-and-effect relationships. Hidden treatments that result from the combinatorial probabilities of random selection create internal patterns in the properties of the quasi-replicates that are correlated with treatment level, and thus can potentially be confused with treatment effects. The primary hidden treatments are: (1) the sampling effect which increases the probability of productive species combinations being present at high levels of species richness, and thus confuses cause-and-effect by misinterpreting the effect of a few species (the productive combinations of two or three species) as the effect of many species (the high diversity treatment level); (2) the quasi-replication effect, which results in under-representation of species combinations at low and intermediate treatment levels, and thus confuses cause-and-effect by underestimating the maximum productivity of low treatment levels because the most productive species combinations are more likely to be missing at low treatment levels than at high treatment levels; and (3) the variance reduction effect, which produces a continuous increase in similarity among quasi-replicates with increasing treatment level, causing the quasi-replicates to converge with true replicates and allowing the effects of experimental error, such as soil heterogeneity, to appear most clearly at high treatment levels, thus potentially confusing cause-and-effect by interpreting experimental error as a treatment effect. None of these hidden treatments can be separated from presumed 'treatment effects' using multiple regression and ANOVA methods in the way that have been used in the past to analyse these experiments (cf. Schmid *et al.*, Chapter 6).

Given the inherent statistical problems with random-selection experiments that make them extremely difficult to analyse and interpret consistently, it does not seem advisable to proceed with more experiments of this design. Detailed retrospective analyses of the patterns of species abundance in these experiments are likely to lead to interesting insights (see Schmid *et al.*, Chapter 6). However, all conclusions and recommendations based on these experiments to date should be carefully re-evaluated.

Our understanding of the mechanistic basis of any relationships between species diversity and ecosystem function is likely to advance more rapidly using alternative experimental designs and analytic approaches. Experiments can be designed to address specific mechanistic hypotheses using such approaches as single-species deletions (Fridley, in press), strategically nested designs (Dukes 2001b), and designs that explicitly manipulate the spatial and temporal variation in environmental conditions (Austin *et al.* 1988; Keddy *et al.* 1999; Dukes 2001b; Fridley, in press).

Maintaining appropriate levels of net primary productivity is essential for maintaining all life on Earth, and thus for preserving biodiversity. A better understanding of the global patterns and controls on net primary production in relation to the patterns of biodiversity can provide the basis for a scientific approach to conservation and sustainable resource management.

MAH was supported by the US Environmental Protection Agency's STAR Grant Program, and the US Department of Energy's Carbon Dioxide Research Program, Environmental Sciences Division, Office of Biological and Environmental Research. Oak Ridge National Laboratory is managed by UT-Battelle, LLC, for the US Department of Energy under contract DE-AC05-00OR22725.

CHAPTER 6

The design and analysis of biodiversity experiments

B. Schmid, A. Hector, M. A. Huston, P. Inchausti, I. Nijs, P. W. Leadley, and D. Tilman

6.1 Introduction

Biodiversity changes have been recognized as one of the major components of global change (Chapin *et al.* 2000). The investigation of the potential consequences that changes in biodiversity can have on ecosystem processes has aroused considerable interest (Naeem *et al.* 1994a; Tilman *et al.* 1996; Leadley and Körner 1996; Hooper and Vitousek 1997; Hector *et al.* 1999; Wardle *et al.* 1999; Van der Putten *et al.* 2000) as well as debate on the ecological interpretations and the scale of the inference that can be drawn from recent experiments (Grime 1997; Huston 1997; Wardle 1999; Huston *et al.* 2000). Despite these challenges, we believe that the experimental approach can help us to understand the links between biodiversity, community- and ecosystem-level processes, and abiotic factors better (Bazzaz 1996; Lawton 2000; Loreau *et al.* 2001).

This chapter is based on a joint analysis workshop held by the authors at the Paris Conference in December 2000. We discuss the analysis and interpretation of experiments designed to assess the impact of species richness manipulations on ecosystem functioning, focusing on experiments involving synthetic plant communities drawn from a local or regional pool of species of temperate grassland ecosystems. Biodiversity experiments involving aquatic or terrestrial microcosms are reviewed by Petchey *et al.* (Chapter 11). Biodiversity experiments in which the number of plant species was deliberately manipulated over a larger range and other variables were held constant,

started in the 1990s in England (Naeem *et al.* 1995, 1996), Switzerland (Leadley and Körner 1996) and USA (Tilman *et al.* 1996). In a few cases, diversity was manipulated indirectly by environmental factors such as fertilization (Tilman and Downing 1994) or by deliberate species additions or removals from natural communities (Tilman 1997a; Wardle *et al.* 1999; Symstad and Tilman 2001). We use all these first-generation biodiversity experiments to outline the basic lessons that can be drawn and the essential challenges that remain concerning the design and analysis of biodiversity experiments.

6.2 Characteristics of the first-generation biodiversity experiments

6.2.1 The forgotten roots of biodiversity experiments in plant ecology

Although contemporary experiments in biodiversity and ecosystem functioning research often make little reference to earlier work in vegetation and agricultural sciences, questions about the species composition and functioning of plant communities have interested plant ecologists for a long time. It is, therefore, appropriate to very briefly review some of this earlier work. Clements (1916), e.g. saw the community as a well-organized assemblage of particular species. In contrast, Gleason (1926) considered communities as more random assemblages of species occurring together at the same place and time. Species distributions within and among sites and along gradients were also studied to explain composition and performance in mixtures (Mueller-Dombois and Ellenberg 1974; Whittaker 1975; Austin 1982; Bazzaz 1987). Interestingly, although the first antecedents of 'true' biodiversity experiments date back to the early nineteenth century (Hector and Hooper 2002) and were mentioned by Darwin (McNaughton 1993), later experimental approaches concentrated on explaining the behaviour of mixtures of two species from monocultures of the component species (Harper 1977). In fewer cases, single species or groups of species were added to (Goldberg and Werner 1983) or removed from existing vegetation (Silander and Antonovics 1982). The intercropping literature from tropical agriculture provides another source to assess effects of low plant diversity on plant production (Vandermeer 1989).

6.2.2 Different aspects of biodiversity

Most biodiversity experiments so far have manipulated the number of species and have kept other aspects of biodiversity (e.g. evenness, higher and lower taxonomic levels such as genetic or generic diversity, and spatial and temporal heterogeneity) constant or left them uncontrolled. Seed mixtures have been usually adjusted to obtain constant overall density and maximum evenness among species. These are typical features for substitutive competition experiments (Harper 1977). An additional aspect of biodiversity is the degree to which species in a community differ in traits such as growth, physiology or life-history strategy. This 'difference' aspect has often been recognized, implicitly, via the distinction of functional groups. In theory, the difference aspect can be viewed as the degree of niche separation or overlap among species. However, designing communities with different levels of niche separation requires prior information about the niches and this is often difficult to obtain (Silvertown et al. 1999). Even more difficult to measure and manipulate is the 'biotic-interaction' aspect of biodiversity (Thompson 1996). Obviously, two communities with the same number of and average difference between species may still differ in the number and strength of interactions among species. Interactions among species are often hard

to determine from the behaviour of monocultures (Bell 1990) and require considerable experimental research. Another somewhat overlooked experimental factor is spatial heterogeneity of plant positions or clumping, which in biodiversity studies is either left uncontrolled (random sowing) or set to zero (planting in a regular grid). Heterogeneity may interact with diversity because, similar to overall density, it determines the intensity of plant–plant interaction (Schmid and Harper 1985; Stoll and Prati 2000).

6.2.3 The purpose of biodiversity experiments and their corresponding designs

The basic question being addressed in biodiversity experiments is whether changes in diversity lead to changes in ecosystem functioning and what shape the potential relationship might have. This question may be asked at the level of a particular community, e.g. a typical grassland ecosystem at a particular locality, containing perhaps some tens of species, or at the level of an ensemble of such local communities over a larger region. In the first case generalizations cannot be made beyond the particular community.

The specific questions asked in biodiversity experiments depend on the scenario of species loss considered most relevant for testing a hypothetical relationship derived from theoretical considerations. Some examples of such questions include:

1. Does species redundancy within functional groups affect ecosystem functioning? In this case, the number of functional groups is kept constant while the number of species is reduced (e.g. Naeem et al. 1994a; Leadley and Körner 1996; Van der Putten et al. 2000).
2. Does random species loss matter for ecosystem functioning? In this case, the number of species, and often coincidentally the number of functional groups, are reduced (Tilman et al. 1996; Hector et al. 1999).
3. Does a particular species or species group affect ecosystem functioning? In this case, the simulated decline in biodiversity is obtained by removing single species or functional groups (Symstad et al. 1998) or combinations thereof (Hooper and Vitousek 1997).

4. Does ecosystem functioning respond when some species suffer relative declines in abundance while others become more dominant? This question can be addressed by varying species evenness (Wilsey and Potvin 2000).

It may seem desirable to develop experimental designs in which all but the last question can be addressed at the same time and indeed this is theoretically possible for experiments in which all possible species combinations drawn from a particular community can be produced (see e.g. Spaekova and Leps 2001). As soon as this particular community contains more than six or seven species, or if the base of inference is extended to a set of communities, the size of the experiment needed becomes intractable. It is here that the type of experimental design and method of statistical analysis becomes crucial to use best the information contained in the data and interpret it (Allison 1999).

6.2.4 The type of experimental manipulation used to create the diversity gradients

The major difference between observational studies and experiments concerning the relationship between diversity and ecosystem variables is the deliberate manipulation of diversity in the experiments. Most first-generation experiments simulated species loss by creating synthetic communities by random sampling from a species pool. In some of the experiments species deletion was incremental, leading to nested sets of increasingly simpler communities, and only one sequence of incremental loss was simulated, but this particular sequence was replicated several times (Naeem et al. 1994a; Leadley and Körner 1996). In other experiments species deletions were not incremental, i.e. a species could disappear in a medium-diversity and reappear in a low-diversity synthetic community, and in addition all communities had different species compositions (Naeem et al. 1995; Tilman et al. 1996). Finally, there have been experiments in which both the species pools (using multiple sites) and the individual species compositions within diversity levels were replicated (McGrady-Steed et al. 1997; Hector et al. 1999; Petchey et al. 1999; Van der Putten et al. 2000b). Such experiments

demonstrated diversity effects, but they were not specifically designed to elucidate underlying mechanisms. Nevertheless, there is often information in the design and data that can be used to address questions about mechanisms by careful statistical analysis.

6.2.5 The conditions under which experiments were carried out

The biodiversity experiments that we review in this chapter used grassland communities growing on substrates ranging from sandy (Tilman et al. 1996) or limestone (Leadley and Körner 1996) nutrient-poor soils to more mesic, nutrient-rich soils (the majority of the sites in Hector et al. 1999; Van der Putten et al. 2000b). Since soil fertility may determine the intensity of plant–plant interactions, these differences in fertility might have modified diversity effects (see e.g. Schmid 2002). The spatial scales of biodiversity experiments have varied considerably. Plot sizes ranged from about one to more than $100\,m^2$. Differences in biodiversity effects may thus be confounded with scale effects (see e.g. Groppe et al. 2001). The experiments were set up by planting young plants (Leadley and Körner 1996) or by sowing seed mixtures, lasted several years and usually were maintained by weeding and cutting (no cutting in Tilman et al. 1996). The technique used to install the plots (random sowing, or planting in a regular matrix) could modulate the diversity effect by producing a different degree of heterogeneity in plant position.

6.2.6 The response variables in biodiversity experiments

The focus in most first-generation biodiversity experiments was on demonstrating effects at the ecosystem level, in particular, plant cover and accumulated above-ground biomass (an indicator of net primary productivity). Other system variables measured were light interception, water retention, soil nutrient levels, decomposition variables, weed invasion and diversity of unmanipulated trophic groups such as insects or soil biota (Schmid et al. 2002). Population-level variables such as cover and biomass of component plant species are time consuming to measure but can help assessing changes between the original and realized species numbers and abundance distributions. Where they have been measured, it is possible to examine mechanisms that may have led to biodiversity effects on ecosystem processes (Hector *et al.*, Chapter 4).

It is important to note here that ecophysiological traits of component species such as growth rate, nutrient-uptake capacity and light-use efficiency, have almost never been measured or manipulated, although they are the basis of the main mechanisms proposed so far (Tilman *et al.* 1997a; Loreau 1998a; Nijs and Roy 2000; Nijs and Impens 2000). As we will discuss later, this makes it difficult to settle some disagreements over interpretation of results.

6.3 Analysis issues: getting the most out of data from biodiversity experiments

6.3.1 Statistical models

Biodiversity experiments would be easy to analyse if they only addressed one simple question, which, in statistical terms, consists of a hypothesis that can be tested with a single degree of freedom. But most ecological experiments ask more complicated or multiple questions, leading to complex statistical models with several parameters and many degrees of freedom, which can be partitioned in various ways. The appropriate, or at least most popular, parametric statistical analyses for such situations are general linear models (Searle 1971; McCullagh and Nelder 1989).

It is useful to distinguish two different, yet mathematically fully equivalent, methods which are special formulations of general linear models. The first is regression (simple or multiple) analysis which has the pedagogical advantage that it builds up from simple, single hypotheses to complex or multiple hypotheses. This method is preferred whenever the focus is on continuous functional relationships between variables. Discrete variables or treatment factors can also be represented as sets of indicator variables and be incorporated in regression analyses (Neter *et al.* 1996). The second approach is analysis of variance (ANOVA), which can better capture a design of different experimental factors

especially when treatment levels are discontinuous. Factors can be crossed or nested in a variety of ways enabling researchers to test many different kinds of hypotheses. Continuous variables are included as covariates in ANOVA. They can either have the same status as experimental factors or can be used simply to adjust the sums of squares (SS) in the analysis. While the latter option is standard in most statistical analysis packages, the former is preferred by statisticians (Bingham and Fienberg 1982; Hendrix *et al.* 1982).

In a typical general linear model, several terms reflect in their sum the systematic component of variation and only a single term reflects the error component, the so-called residual or unexplained variation (McCullagh and Nelder 1989). Here, a restrictive requirement is that this unique error variation is the same for all units in an experiment. However, in biodiversity experiments, there are typically several terms associated with error variation. These terms relate to multiple sites, blocks, species compositions and plots as illustrated in Fig. 6.1. It is therefore necessary to distinguish two partial models, a *treatment model* for the systematic variation and an *error model* (*block model*) for the error variation (Payne *et al.* 1993). The advantage of the ANOVA over the regression method in biodiversity experiments is that it can provide a comprehensive analysis in which the treatment and error models are fitted together. This comprehensive model indicates how different sources of variation in an experimental design relate to each other and how terms from the treatment model have to be tested against terms from the error model in *F*-tests (Snedecor and Cochran 1980; see e.g. Spehn *et al.* 2000), in order to avoid the pitfalls of pseudo-replication (Hurlbert 1984).

In biodiversity experiments, the error model includes all the spatial information available from the geographical layout of plots and blocks, whereas the treatment model includes the diversity manipulations. The overall mean of a dependent variable is a part of both the error and treatment models. If each diversity treatment is represented by different species compositions, then these compositions should be included in the error model, because differences between diversity levels (treatment model) should be compared with differences among

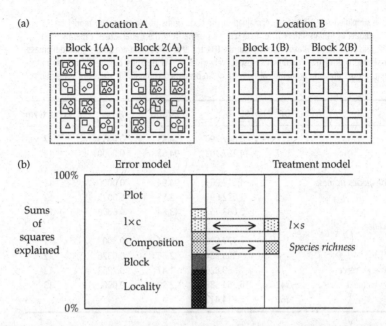

Figure 6.1 (a) Layout of a biodiversity experiment with four species (different symbols) occurring at different levels of biodiversity (one, two, three or all four species), two localities with two blocks each and 12 plots per block. Within each block every plot—except for the four-species mixtures—contains a different composition of species, but exactly the same species compositions are repeated between blocks. (b) The design shown in (a) allows distinguishing the effects of the treatment species richness from variation due to different compositions of species on ecosystem functioning, at the same time the geographical variation from the locality to the plot level can be estimated. The separation of the treatment model from the error model is illustrated by the two columns in (b). See text for further explanation of error and treatment models in ANOVA.

different combinations within diversity levels (error model). Only if 'within differences' are on an average clearly smaller than 'between differences', can it be concluded that diversity treatments differed over and above the differences that would be expected due to compositional differences in the absence of diversity effects (Givnish 1994; Huston 1997). This clearly shows that the effects of species richness and species composition can be separated in biodiversity experiments with random species selection (Tilman 1997b).

Some words about assumptions of general linear model approaches must be added here. For each term in the error model, the units for which the variation is assessed (e.g. blocks or plots) must be drawn from a single normal distribution, i.e. the variance must be the same for all units included in the term. If these assumptions are violated, it may be necessary to transform the dependent variable, to define further error terms whose variation can be separately estimated, to use special error

distributions (McCullagh and Nelder 1989) or to use non-parametric statistics. Non-parametric approaches should be used with caution, however, because they are less powerful than parametric approaches if the assumptions of the latter are met or nearly met. In addition, also non-parametric models have assumptions, e.g. that units are independently drawn from random distributions with the same shape.

To illustrate the comprehensive ANOVA approach, we use a simplified but original example with combined data from the Cedar Creek experiment in USA (Tilman *et al.* 2001) and the BIODEPTH experiment in Europe (Hector *et al.* 1999), on which we worked in a joint analysis workshop during the Paris Conference. The analysis is shown in Table 6.1 and a diagrammatic interpretation is given in Fig. 6.1.

The error model for this example is

$$y_{ikl} = m + l_{i..} + c_{.k.} + p_{ikl}$$

Table 6.1 Example of a simplified ANOVA model, explained in the text. Terms of the treatment model are in italics; terms of the error model in roman letters. The analysed data come from the Cedar Creek experiment in USA (Tilman *et al.* 2001), and the BIODEPTH experiment in Europe (Hector *et al.* 1999) and are above-ground biomass values in the fourth or second year of the experiments, respectively (note that due to the higher soil fertility in the European experiment, these different chronological time frames represented comparable stages of vegetation development)

Source	DF	SS	F-ratio	P	Error term
1 Continent	1	8,847,290	4.33	0.0759	2
2 Locality within 1	7	14,287,153	94.52	0.0000	13
Species richness:					
3 Log-linear effect of species richness	1	4,512,263	94.94	0.0000	12
4 Deviation from 3	9	1,092,696	2.55	0.0075	12
5 Legume presence	1	2,083,755	43.84	0.0000	12
Continent × Species richness:					
6 Continent × 3	1	727,520	15.31	0.0001	12
7 Continent × 4	4	505,964	2.66	0.0326	12
8 Continent × Legume presence	1	69,826	1.47	0.2263	12
12 Species composition within 1	342	16,254,588	2.20	0.0000	13
13 Residual	260	5,614,417			

where y_{ikl} is the value of the dependent variable in a particular plot, m is the overall mean, $l_{i..}$ is the effect of locality i (see Hector *et al.*, Chapter 4), $c_{.k.}$ is the effect of the particular species composition k, and p_{ikl} is the residual effect of plot l with composition k at locality i. Locality × composition interactions have not been included in the formula and will therefore automatically be pooled with the final error term.

The corresponding treatment model is

$$y_{noq} = m + k_{n..} + s_{.\sigma.} + g_{..q} + (k \times s)_{n\sigma.}$$
$$+ (k \times g)_{n.q}$$

where $k_{n..}$ is the effect of continent n (North America versus Europe), $s_{.\sigma.}$ is the effect of species richness level σ (in this example, decomposed into a log-linear contrast and deviation from log-linearity), $g_{..q}$ is the effect of legume presence or absence q, $(k \times s)_{n\sigma.}$ is the interaction of continent and species richness level (also decomposed into contrast and deviation) and $(k \times g)_{n.q}$ is the interaction of continent with legume presence or absence.

Combining the two models yields a single ANOVA table (Table 6.1) in which the sequence of terms from treatment (italic letters) and error

(roman letters) models should be arranged such that any nesting, e.g. of plots within localities, or of compositions within species richness, is respected, and interactions follow the main effects or simpler interactions. Each term from the treatment model is tested against the next lower term from the error model using the variance ratios between mean squares (*F*-values), e.g. continent/locality within continent. Terms from the error model may also be tested, e.g. to obtain information about significant spatial effects, by forming *F*-ratios between nested terms such as locality/plot within locality. We recommend that whenever possible researchers account for spatial variation at various scales (locality, block, position of plots within block etc.) in biodiversity experiments.

All terms containing more than one degree of freedom may in principle be partitioned into independent contrasts. In the example, we have formed a log-linear contrast for the species richness term; and legume presence or absence can be considered as a contrast of the composition term (Table 6.1). In the latter case, we had to move the contrast term from the error model to the treatment model. The remainder of the composition term, which is left in the error model, now estimates the variation among compositions within the two groups formed by the legume contrast. In fact, it would also be possible to consider the species richness term as a contrast of an all-inclusive composition term, because obviously species richness and composition are not mutually exclusive: once the species richness term is in the treatment model, the composition term in the error model automatically refers to differences between species compositions within species richness levels (and, in the example, within legume presence or absence). Contrasts can also be made for interactions, as shown in the example. In fact, the information contained in a complex design could be used fully for focused hypothesis tests if every line from the treatment model in the ANOVA table used a single degree of freedom (cf. Rosenthal and Rosnow 1985).

Is such an approach, which uses all the information contained in an experimental design, needed if the question is simply whether species richness has an effect? If there is only the null hypothesis of no such effect, then a single two-sided test for a positive or negative association would be sufficient. It would be enough to randomly select species compositions at two levels of species richness without recording the particular compositions. But none of the biodiversity experiments done so far was designed so restrictively. Taking three or more levels of species richness already allows one to further ask whether the diversity effect is linear. Typically, biodiversity experiments are designed to address several hypotheses and additional hypotheses are suggested by the data, justifying an exploratory rather than strictly confirmatory statistical viewpoint.

6.3.2 Alternative analysis sequences when no unique solution exists

The ideal comprehensive design with completely balanced factor combinations in the treatment model can rarely if ever be achieved in practice. Unbalanced designs do not present an unsolvable problem, but they affect the meaning of the hypotheses, which depends on the way the statistical analysis is carried out.

Unbalanced designs arise when different treatments or treatment combinations are represented by unequal numbers of replicates. Entire treatment combinations are often missing in biodiversity

experiments because there are theoretically impossible combinations. For example, it is impossible to have fewer species than functional groups. Designs also become unbalanced if continuous variables, i.e. covariates, are included in the treatment model. For example, if diversity is measured during the experiment, rather than taken as planned initially, it is usually treated as a continuous covariate in the analysis. Many ecologists seem to be unaware of the fact that introducing a covariate into an ANOVA results in an unbalanced design (Bingham and Fienberg 1982; Hendrix et al. 1982). To retain orthogonality after the inclusion of a covariate requires that the covariate is uncorrelated with all treatment factors in the model. Such 'secret balance' is very unlikely (Payne et al. 1993). The fact that a design becomes unbalanced because of a covariate does not mean that one should refrain from using covariates, rather one should be aware of the way they change the meaning of hypotheses.

The consequence of imbalance is that the model terms become correlated. Thus, the variation explained by different terms cannot be uniquely separated into additive components because they either overlap (Fig. 6.2(a)) or, in special cases, are too distant from one another (Fig. 6.2(b)). There are two contrasting approaches to the problem of analysing data from unbalanced designs. The first tries to maintain the orthogonality of hypotheses by treating the data as if they came from a balanced design and replacing the missing by estimated values. This leads to special ways of constructing sequences of SS that are no longer orthogonal (Yates 1934; Frane 1986; SAS 1990). The second approach is to allow hypotheses to become conditionally dependent on each other by using sequential SS which are always orthogonal. Below, we advocate the use of the second approach because the imbalance in biodiversity experiments is often so large, or even intrinsic to the underlying biology as in the case of species richness by functional-group richness combinations, that it would not make sense to uphold the idea that the data could have come from a balanced design.

When the terms of the treatment model are no longer orthogonal, the sequence in which they are entered into a multiple regression analysis or an ANOVA affects the amount of variation explained

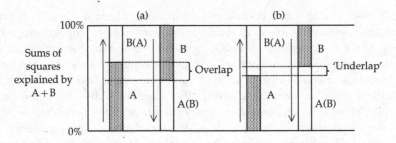

Figure 6.2 Graphical representation of the variance explained by two explanatory terms A and B which are correlated with each other to a certain degree. The arrows indicate the sequence in which the terms are incorporated into a multiple regression analysis. (a) Commonly, correlation leads to overlap, i.e. part of the variation cannot be attributed uniquely to one of the terms. The term fitted first in the analysis, i.e. ignoring the existence of a second one, takes the shared variation (grey bars), the term fitted second, i.e. after eliminating all variation explained by the first one (e.g. B(A) or B given A), only takes what is left (white bars). (b) In special cases, a term can explain more variation if it is fitted second rather than first, i.e. the grey bars do not overlap and leave a gap.

by each term. A term appearing before other terms 'takes' all the variation it can explain, ignoring the possibility that later terms might also explain some of this variation. Conversely, a term entering after other terms can only take whatever variation is left to explain, i.e. eliminating the possibility that previous terms might also take part in this explanation (McCullagh and Nelder 1989). The special case where the variation left to explain by a second term after eliminating a first is larger than the variation the second term could explain if the first is ignored (Fig. 2(b)) occurs if there are negative correlations between the different explanatory terms themselves and between these and the dependent variable (Shaw and Mitchell-Olds 1993).

Thus, when fitting general linear models, the order in which different terms are included may change the statistical significance attached to them (Neter *et al.* 1996). Since there is no unique answer to which sequence is best, it will often be useful to use different sequences to explore the data even in an experimental situation, because in statistical terms there are no longer precise predefined questions to be answered in a complex unbalanced design (cf. Frane 1986). The 'ignoring' and 'eliminating' terminology corresponds to the 'among' and 'within' terminology in nested models. For example, when in the joint analysis of the two biodiversity experiments at Cedar Creek (USA) and BIODEPTH (Europe) we entered legume presence or absence after entering log (species richness), it was as if species richness were artificially held

constant and we measured the (constant) difference between plots with and without legumes within a given species-richness level (Fig. 6.3). If we invert the sequence, we can look at the effect of log (species richness) within (holding constant) legume presence or absence. In the analysis shown in Fig. 6.3 (similar to the one in Table 6.1, but omitting deviations from the log (species richness) term, interactions and the term for species composition), log (species richness) and legume presence or absence jointly explained 12.6% of the total variation. However, log (species richness) only uniquely explained 1.8% if the legume presence or absence was eliminated, and legume presence or absence only uniquely explained 4.3% if log (species richness) was eliminated. Thus the majority of this variation, $12.6\% - 1.8\% - 4.3\% = 6.5\%$, was shared (see Hector 2002 for a further example).

We argue that it is dangerous to apply techniques that try to force a single solution to situations of overlapping variance (the above-mentioned first approach to analysing data from unbalanced designs), although this is recommended in some widely used statistical packages and referred to as type II, III or IV SS (SAS 1990, 2000). These methods may be appropriate for the situations for which they were originally developed, i.e. for cases of slight and accidental imbalance (Yates 1934), where they allow one to maintain orthogonality of hypotheses at the expense of non-orthogonal SS that no longer add up to meaningful totals (Frane 1986). Instead of using non-orthogonal SS, we recommend

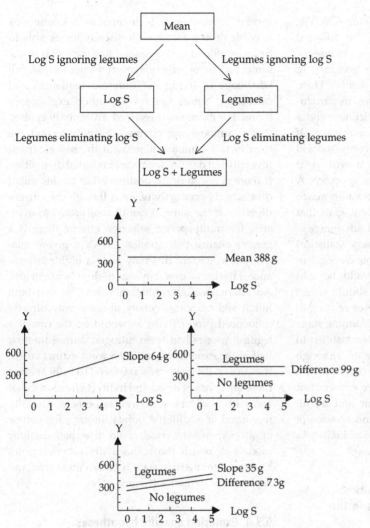

Figure 6.3 Illustration of model sequences leading via two different simple models to the same final model. If each of the two explanatory terms, the logarithm to the base 2 of species richness (log S) and legume presence, is analysed separately, ignoring the information contained in the other, its effect is larger than in the combined analysis. The analysed data come from the Cedar Creek experiment in USA (Tilman *et al.* 2001), and the BIODEPTH experiment in Europe (Hector *et al.* 1999) and are above-ground biomass values in the fourth or second year of the experiments, respectively (see Table 6.1).

the use of sequential, i.e. orthogonal SS (the above-mentioned second approach to analysing data from unbalanced designs), and to try out different sequences according to the different ecological questions raised by the researcher (Searle 1995). The advantage of this approach is that comparison of different sequential models allows one to identify the 'unique' (conservative type II) and the shared effects of each explanatory term on a dependent variable (see Fig. 6.2). The danger of the more rigid methods, especially the type III solution, is that they leave out all shared effects from the contributions of individual explanatory terms, leading to the awkward situation that the full model (and the shared effects) may be highly significant while at the same time none of the individual terms is. The tests for these individual terms are then very conservative and have almost no statistical power to reject false null hypotheses (Aitken 1995). The problem is particularly severe if covariates are included in an analysis.

Once a model has been chosen, the parameter estimates obtained at the end of any analysis sequence are the same (Fig. 6.3). An alternative analytic approach can, therefore, be to consider all the parameters of a model as levels of a single explanatory term and first calculate a simple one-way ANOVA. For the example in Fig. 6.3, all of the different treatment combinations of species richness× legume presence or absence could be assigned to an explanatory term 'diversity treatments'. Then, we could fit contrasts within this term, in particular, a log-linear contrast for species richness and a further contrast for legume presence or absence. Of course, the sequence of fitting the contrasts will again be relevant, and the contrasts will yield exactly the same SS as in the previous approach. A 'psychological' advantage of the contrast approach is that it raises less suspicion with ecologists that adhere to type III SS and a practical advantage is that it can be used with almost every statistical package (Neter *et al.* 1996; for an example see Stephan *et al.* 2000). A word of caution should be said regarding the fitted contrasts. These should reflect as much as possible a priori hypotheses relating to the questions originally asked at the planning stage of a biodiversity experiment. It is often justified to try a limited number of *ad hoc* contrasts, although this has the penalty that the significance tests are no longer confirmatory but become more exploratory. It would be dangerous to go too far and try all possible contrasts for the presence and absence of every particular species or species combination in the analysis of a biodiversity experiment.

6.3.3 Incorporating realized diversities, evenness and dominance measures in the explanatory statistical models

Most data sets have been analysed using the relationship between ecosystem functions and planned diversity levels (e.g. the number of species sown at each diversity level). However, it may be more informative to analyse data using observed species numbers or diversity indexes. Such a 'realized diversity' variable, measured at the same time as the dependent variables, can then be included as an additional explanatory variable in the statistical analysis, together with the planned species richness which has been measured at the beginning of an experiment. In cases where this has been done, positive relationships were found between realized diversity variables and productivity (Hector *et al.*, Chapter 4; Pfisterer and Schmid 2002). Realized

diversity, whether it is treated as a continuous variable or as a factor with discrete levels, should be incorporated in the general linear models in the same way as other explanatory terms to take full advantage of trying out different sequences and calculating interactions with other explanatory terms. For example, if realized diversity is entered first in a model and planned diversity second, we then in fact compare plots with the same current diversity but derived from various initial diversities. If there is a significant positive effect of this initial diversity on productivity, even though the current diversity is the same, it can be interpreted as evidence for multi-species selection effects: there is a greater chance that species sets of a given, relatively low, current diversity have a high performance if the initial diversity was high rather than low. An alternative interpretation would be that both initial and current diversity in some way directly influenced productivity, as would be the case if a legume species had fixed nitrogen during the first half of the experiment but then went extinct before the current diversity was assessed. To fully analyse these questions of direct and indirect effects of initial and current diversity, as well as perhaps diversity measured at additional points through the course of an experiment, one could use path-analytic models in which the realized diversity variables have an intermediate state between measured and explanatory variables (Kenny 1979).

6.3.4 Exploring specific hypotheses about mechanisms

A useful approach to explore specific hypotheses about mechanisms leading to biodiversity effects is to analyse only subsets of data from a complex biodiversity experiment. This avoids the automatic inclusion of all data in the estimation of residual variation. Subsets could represent orthogonal combinations of explanatory terms (such as species-richness levels 4 and 8 × functional-group richness levels 2 and 3) or plots with a vegetation cover of at least 95% etc. Of course, the disadvantage of the latter type of subset formation is that omission of plots below a certain 'veil line' will usually make a design even more unbalanced. There are many further possibilities of selecting subsets of data for analysis. However, we recommend that all of these should be used mainly in an exploratory or heuristic manner. Throwing away treatment combinations after an experiment has been completed is not the same as conducting a restricted-design experiment, because one could try throwing away all possible single treatment combinations or sets of treatment combinations until one reaches a subset that gives pleasing results. Confirmatory statistical tests are, strictly speaking, only possible if all data are included, and no covariates are added to the original design variables, during the analysis.

One important type of specific question is whether the presence of particular species has a particular effect on ecosystem functioning. This can be addressed by calculating contrasts between plots with or without a particular species and the same can obviously be done for the presence or absence of particular sets of two or more species. However, experiments may need to be carefully designed to do this with good balance and power, and *post hoc* analyses of experiments not designed in this way must be interpreted with care. It is also possible to include several contrasts and even their interactions in a single analysis sequence. If there were two species with positive effects, the null expectation under a purely competitive scenario would be that the interaction must be negative. If it is not, i.e. if the two positive effects are purely additive or even synergistic, we could take this as a strong indication for complementarity or even facilitation mechanisms (Tilman *et al.*, Chapter 3). One point to remember, however, is that a significant contrast for the effect of the presence of a particular species does not necessarily imply that it is in fact the particular species that contributes most to the measured ecosystem variable. For instance, the presence of a legume could increase the biomass production of other species in a system as much as its own. Furthermore, these contrasts do not consider the different proportions with which a particular species occurred in the original mixture. As a consequence, if species richness is fitted for a subset of systems in which the particular species A occurs, species richness may actually simply represent the proportion of A in the mixtures: e.g. 100% A for species richness 1, 50% A for species richness 2,

25% A for species richness 4 etc. We, therefore, do not recommend fitting species richness terms within contrasts for particular species.

While single-species contrasts can be used to test for potential selection effects of single species (see also the method of Loreau and Hector 2001; Hector *et al.*, Chapter 4), multiple-species contrasts can be used to test for selection effects due to multiple species. If more than one species needs to be considered to explain the diversity effects, then complementarity occurs within the set of selected species. If all species need to be selected, then this generalized selection effect becomes a pure complementarity effect. How many species need to be selected for an optimal explanation? The authors, when analysing data from the Cedar Creek and BIODEPTH experiments, very crudely guessed that in these experimental systems a set of not more than seven complementary species would probably often be sufficient to explain most of the short-term diversity effects in any given variable.

6.3.5 Summary on analysis issues and outlook on further development of methods

There is a wealth of possibilities to obtain still more insights from the complex information contained in designs and measurements of first-generation biodiversity experiments. We explored some of these possibilities in our joint data analysis workshop during the Paris Conference (examples in Table 6.2). The experience of the joint analysis workshop was that many of the issues raised about the appropriate interpretation of biodiversity experiments are readily resolved if the data are available to all for analysis. Thus, the ability to resolve issues of interpretation primarily requires data to be accessible to the research community, preferably in a standardized form. Of course, mechanisms for the protection of the rights of those who collected the data need to be established.

There are several promising new approaches for the analysis of biodiversity experiments that we have not touched upon. We can only give a very brief outlook here:

1. Spatial analysis may be used to assess the influence of soil heterogeneity and neighbour

Table 6.2 Some explorations of BIODEPTH (Hector *et al.* 1999) and Cedar Creek data (Tilman *et al.* 2001) carried out by the authors in the joint analysis workshop at the Paris Conference in December 2000

Analysis	Result
Effects of realized diversity (species richness or Shannon–Wiener index)	Explains a large part but not all of the variation attributed to initial richness
Effects of the presence of particular species in communities	Often significant, despite their different frequencies
Variation between replicate species compositions	Remains constant across increasing levels of species richness
Omitting plots with low cover (<80%, <95%, <100%)	Relationship species richness—above-ground biomass production still positive
Omitting plots with the most abundant species contributing >70% biomass	Relationship species richness—above-ground biomass production more positive
Correlation between species ranks at different diversity levels	Positive but relatively weak, weaker the further the diversity levels are apart
Correlation between degree of dominance and above-ground biomass production	Negative for eight-species mixtures across sites

relationships (serial correlations, spread of weeds and pathogens) between plots in a field experiment (Diggle 1983; Kempton and Lookwood 1984).

2. The analysis of mechanistic diallels (McGilchrist 1965) may be used to assess the contributions of particular species and of their general and specific combining abilities to the performance of mixtures. There is a rich knowledge about the analysis of genetic diallels (Mather and Jinks 1982), including incomplete diallels, which should be generalizable to mechanistic incomplete 'multi-allels'. This would allow one to obtain estimates about species contributions and combining abilities even in the absence of all possible monocultures, two-species mixtures, three-species mixtures, and so forth.

3. Recently, Loreau and Hector (2001) have developed a new method that partitions selection and complementarity effects (Hector *et al.*, Chapter 4). Like the mechanistic diallèl approach, this approach should also be generalizable to multi-species selection effects.

4. Multivariate approaches such as principal component or factor analysis may be used to re-express correlated explanatory terms in uncorrelated synthetic explanatory terms, thereby avoiding the problems with shared variation discussed at length above.

5. Path analysis (Kenny 1979) may be applied to all situations in which biodiversity experiments are analysed with multiple regression, and where

specific hypotheses can be stated about the sequence of effects between explanatory terms, covariates and dependent variables.

6.4 Design issues: the next generation of biodiversity experiments

6.4.1 General recommendations

What can we learn from the analyses of the first-generation biodiversity experiments for future experiments? These experiments demonstrated that there are significant relationships between plant diversity and ecosystem functioning. Controversies began when alternative interpretations could be made for demonstrated effects and when incongruences between experimental and observational studies became apparent. Although they were not specifically designed to distinguish between mechanisms or to mimic patterns described in observational studies, the experiments have helped us to understand how best to design the next generation of experiments to address more specific questions. As we discussed at the beginning of this chapter, the first biodiversity experiments were limited to one particular scenario of species loss. Later designs replicated richness by repeating mixtures with the same number of species but with different randomly chosen sets of species. Richness effects can then be tested against compositional effects (Tilman 1997b). However, after controlling for species richness, compositional effects form some unknown part of the residual variation along with random errors. Therefore, it is now recommended (Cottingham *et al.* 2001) that, if possible, designs should replicate both richness and composition as in some of the more recent experiments (e.g. Hector *et al.* 1999). This has the additional advantage that one can test if the assumption of constant variance between plots replicating composition (homoscedasticity) holds across all diversity levels, as it did for the BIODEPTH data (Table 6.2).

There is currently a tendency to stress the importance of growing all species of a biodiversity experiment in monoculture. Although this approach has proved useful in allowing comparisons to test for selection versus complementarity effects, it is now clear that it is not the only approach. As can be seen from the generalization of the selection effect to subsets of more than one species, the logical conclusion would be that all possible 2-, 3-, ..., *n*-species subsets should also be included in a design starting from simple to complex communities. An alternative approach is to analyse variations in dominance or evenness within given levels of species richness (Connolly *et al.* 2001). Another alternative would be 'synthetic removal experiments': the design starts with an intact community and then omits certain species or sets of species to see if their absence causes effects or whether they are redundant. This would allow one to identify species that may be redundant under given conditions or with regard to given measures of ecosystem functioning.

One of the disadvantages of approaches that stress the importance of including all monocultures or as many low-diversity mixtures as possible, is that at high diversity levels random sampling of different species combinations is restricted, often to the extent that there is only one combination at highest diversity. Furthermore, (1) the overlap of different species compositions between plots of the same number of species, as well as, (2) the average similarity between species within compositions (problem of species packing), always increase with increasing species richness if there is a fixed species pool of the same size as the highest diversity level tested. Huston (1997) has referred to problem (1) as the variance reduction effect. Problem (2) is

explained more fully in Schmid *et al.* (2002) and refers to an implicit negative correlation between what we called the 'difference' aspect of diversity— e.g. measured as the complement of average niche overlap along environmental gradients (Bazzaz 1987) or the among-species variance component in characters related to resource use (Roughgarden 1974; Nijs and Roy 2000)—and the 'numbers' aspect of diversity.

To avoid these problems and to increase the inference base of biodiversity experiments it would first be necessary to use multiple communities at high diversity levels, e.g. by testing hypotheses for different species pools as in multi-site experiments (Hector *et al.*, Chapter 4). Second, we recommend that information about functional roles of individual species should be incorporated into analyses. Such inclusion would allow disentangling the effects of functional-trait diversity and taxonomic diversity (Schmid *et al.* 2002). For example, one could test whether the relationship between species richness and ecosystem functioning is stronger in communities assembled from species pools with higher functional trait diversity (e.g late versus early successional communities, Bazzaz 1987). However, it may be difficult to measure the relevant aspects of functional diversity for a natural species pool and a lot of *a priori* information and classification may have poor predictive value, especially when interactions between species modify functional traits. Another idea would be to compare biodiversity–ecosystem functioning relationships in experiments using species with either shared or different evolutionary history (e.g. native versus non-native species), because functional differences between species are expected to result from evolutionary processes.

Besides the difference aspect, there are many further aspects of diversity that have not yet been sufficiently considered in biodiversity experiments. These include evenness, spatial pattern of species within mixtures, successional sequence, interaction strength, and genetic and demographic (e.g. age structure) diversity within species. Furthermore, both the abiotic and biotic conditions under which biodiversity experiments are carried out should be considered more explicitly. It is well known in plant population biology, e.g. that density has strong positive effects on the skewness of size distributions and on mortality (Harper 1977), and these may interfere with diversity effects in a way that the latter become suppressed with increasing overall density. Finally, we stress that we only discussed approaches to biodiversity experiments that look at the relationship between different types of constructed diversity gradients and ecosystem functioning. They do not relate in any way to the scenarios by which diversity gradients are formed in the real world, e.g. if extinction is biased towards species with particular attributes or functional traits (Schmid et al. 2002). Obviously, coupling biodiversity experiments with extinction scenarios is an extremely important and relatively new field that deserves attention.

In summary, the general recommendation for the next-generation biodiversity experiments is to use a diversity of approaches. Future research should explore as many of the different aspects of biodiversity as possible and move beyond species richness as the primary manipulated factor. It is important to conduct this research at the appropriate level of resolution such that sufficient statistical power is available to test the specific hypothesis under investigation. There may be situations where as few as three different species combinations may be appropriate if there are many replications of them (as in the first experiments described in Naeem et al. 1994a and Leadley and Körner 1996), but often the appropriate number of different combinations will be above 20–30, yielding residual degrees of freedom in the range above 15–20.

6.4.2 Refining the types of questions asked in diversity experiments

The recent debate over the relationship between biodiversity and ecosystem functioning may in part be due to a lack of precise definition of goals and terminology. For example, justifications for looking at particular ecosystem properties and value judgements associated with the identification of 'poor' or 'good' ecosystem performance are often lacking in discussions of diversity theory and experiments. We give some examples of how this may be responsible for some misunderstandings and suggest possible improvements in the definition of goals and terminology for future experiments.

Above-ground biomass production (or surrogates such as plant cover) are the most commonly reported measures of ecosystem performance, and 'poor' ecosystem performance has often been defined as low above-ground biomass (Tilman et al. 1996; Hector et al. 1999). This choice of ecosystem properties can be justified in some cases because above-ground biomass production is often correlated with other desirable ecosystem properties such as food production, soil fertility, and carbon storage. However, optimization of plant above-ground production is not a universal goal. Grassland of high conservation value is often managed for high species richness by reducing soil fertility, accepting reduced biomass production (Bakker and Berendse 1999, but see Schmid 2002). It is essential to be explicit when defining 'optimal' ecosystem performance, and we suggest that researchers always need to distinguish between these two major approaches (optimizing function versus optimizing diversity) and clearly define the goals in future experiments.

Thus, there are different, distinct approaches to developing diversity experiments, which differ fundamentally in how they should be interpreted. The first approach is to pose specific, management-oriented questions such as: 'are realistic, low-diversity plant communities less productive than more diverse communities?' and then design experiments to address this question. This type of question would be particularly applicable to ecosystems where humans control species composition and define high biomass production as good ecosystem performance. For example, species-rich permanent pastures have been replaced in many parts of Europe by species-poor temporary mixtures of grasses and legumes, with optimal yield in the first or second year after seeding. It would be of considerable interest to both farmers and decision makers to know if this has been a good or bad strategy for improving herbage production. Most existing diversity experiments cannot address this type of question directly because the experimental designs do not allow for statistically powerful tests to be made between monocultures of species that are commonly sown for their high productivity and mixtures of species commonly used in permanent pastures (Huston *et al.* 2000). Most existing experiments in agricultural systems do not go very far in addressing this type of question either, because they have traditionally focused on very species-poor communities, i.e. a comparison between monocultures and two-species mixtures (Trenbath 1974; Harper 1977).

The second approach is to pose much more generalized, 'idealized' questions such as: 'does random species loss alter above-ground biomass production and to what extent?' This type of question would be applicable to ecosystems where realistic species-loss scenarios cannot easily be defined. This is not only of considerable theoretical interest but also of practical interest for researchers modelling plant productivity and its links with other processes such as global change. Many existing experiments are well designed statistically to respond to this type of question, but the results may be difficult to apply directly in ecosystem management because of complicating factors such as soil disturbances that occur when establishing and maintaining the planted communities (for a successful example see Schläpfer and Erickson 2001).

An additional difficulty in setting the objectives for diversity experiments is that there are many ways in which optimal ecosystem performance could be defined. Soil nutrients can serve as an example. High nitrate concentrations in soil solution have been viewed as being indicative of poor ecosystem performance, because high concentrations are associated with nitrate contamination of ground water (Tilman *et al.* 1996; Niklaus *et al.* 2001). However, nitrogen-fixing legumes often raise soil nitrate concentrations (Scherer-Lorenzen 1999) and thus play an essential role in the restoration or maintenance of soil nitrogen fertility. The resolution of conflicting objectives is not a simple task (see Hooper and Vitousek 1998 for an attempt to take into account multiple criteria in evaluating ecosystem performance). To a certain extent the problem can be resolved either by defining a small number of clearly justified objectives—e.g. using diversity to limit pesticide use and pest infestations in rice (Zhu *et al.* 2000)—or by avoiding the use of value-laden assessments of ecosystem performance.

Initial experiments on the functional significance of biodiversity have helped articulate the major issues, but, perhaps more importantly, have also served to illuminate the complexities involved in addressing such issues. Considerable progress has been, and continues to be made in developing statistical methods that may be used for testing hypotheses concerning the relationship between biodiversity and ecosystem functioning. The future lies in refining these hypotheses and statistical tools as well as conducting new kinds of diversity manipulations to identify the mechanisms involved in diversity-functioning associations.

We thank M. Loreau and S. Naeem for many helpful comments on this chapter, as well as for the organization of the stimulating meeting in Paris, where we held our joint analysis workshop.

PART III

New perspectives on ecosystem stability

CHAPTER 7

A new look at the relationship between diversity and stability

M. Loreau, A. Downing, M. Emmerson, A. Gonzalez,
J. Hughes, P. Inchausti, J. Joshi, J. Norberg, and O. Sala

7.1 Introduction

The relationship between biodiversity and ecosystem functioning has emerged as one of the most exciting and controversial research areas in ecology over the last decade. Faced with the prospect of a massive, irreversible loss of biodiversity, ecologists have begun investigating the potential consequences of these dramatic changes in biodiversity for the functioning of natural and managed ecosystems. These investigations have been motivated by both the scientific challenge and the need to understand better the role biodiversity plays in providing sustainable ecological goods and services for human societies. An increasing amount of evidence suggests that changes in biodiversity can have adverse effects on the average rates of ecosystem processes such as primary production and nutrient retention in temperate grassland ecosystems (see chapters by Hector et al., Chapter 4; Tilman et al., Chapter 3; Wardle and van der Putten, Chapter 14). Most of this evidence, however, comes from relatively short-term experimental studies (see, however, Petchey et al., Chapter 11) under controlled experimental conditions, which are little informative about sustainable functioning.

The temporal variability of natural environments over a broad spectrum of time scales from days to centuries (Halley 1996), as well as increasing anthropogenic pressures (Sala et al. 2000), inevitably generate temporal changes in both population sizes and ecosystem processes. It is therefore of considerable interest to understand how biodiversity loss will affect long-term temporal patterns in

ecosystem functioning. Will ecosystem functional properties and services become more variable and less predictable when species diversity is reduced? Are species-rich ecosystems more capable of buffering environmental variability and maintaining ecosystem processes within acceptable bounds than species-poor ecosystems?

To answer these questions appropriately and avoid 'reinventing the wheel', it is important to realize that they address in a new form a long-standing debate in ecology concerning the relationship between the complexity (loosely defined as a combination of species diversity and the number and strength of species interactions) and stability of ecological systems. The study of this relationship has had a long and controversial history (May 1974; Pimm 1984, 1991; McCann 2000). To delineate the differences between contemporary issues and the historical debate, we first briefly revisit the central components of this debate, and propose a new, integrated conceptual framework derived both from lessons from this debate and insights newly arising from current research on biodiversity and ecosystem functioning. We then examine, within this framework, how recent theoretical and experimental work provide new insights into the complexity–stability debate. Our treatment therefore complements some recent reviews of the topic (Loreau 2000a; McCann 2000; Schwartz et al. 2000; Cottingham et al. 2001) which have not used the framework we present. Finally, we discuss the need to develop new theoretical and methodological approaches and to strengthen the link between theory and experiments in this area. We conclude with some implications for policy and management.

7.2 Historical and conceptual background

The early view that permeated ecology until the 1960s was that diversity (or complexity) begets stability. This view was formalized and theorized by people such as Odum (1953), MacArthur (1955) and Elton (1958) in the 1950s. Odum (1953) and Elton (1958) observed that simple communities are more easily upset than rich ones, i.e. they are more subject to destructive population oscillations and invasions. MacArthur (1955) proposed, using a heuristic model, that the more pathways there are for energy to reach a consumer, the less severe is the failure of any one pathway. These conclusions were based on either intuitive arguments or loose observations, but lacked a strong theoretical and experimental foundation. Probably because they represented the conventional wisdom ('don't put all your eggs in one basket') and the prevailing philosophical view of the 'balance of nature', they became almost universally accepted.

This 'conventional wisdom' was seriously challenged in the early 1970s by theorists such as Levins (1970), Gardner and Ashby (1970) and May (1972, 1974), who borrowed the formalism of deterministic autonomous dynamical systems from Newtonian physics and showed that, in these model systems, the more complex the system, the less likely it is to be stable. Stability here was defined qualitatively by the fact that the system returns to its equilibrium or steady state after a perturbation. The intuitive explanation for this destabilizing influence of complexity is that the more diversified and the more connected a system, the more numerous and the longer the pathways along which a perturbation can propagate within the system, leading to either its collapse or its explosion. This conclusion was further supported by analyses of one quantitative measure of stability, resilience (Table 7.1), in model food webs (Pimm and Lawton 1977; Pimm 1982). This theoretical work had a number of limitations. In particular, it

was based on randomly constructed model communities. More realistic food webs incorporating thermodynamic constraints and observed patterns of interaction strengths do not necessarily have the same properties (DeAngelis 1975; de Ruiter *et al.* 1990). Also, there have been few direct experimental tests of the theory, and many of the natural patterns that agree with theoretical predictions can be explained by more parsimonious hypotheses such as the trophic cascade model (Cohen and Newman 1985). Despite these limitations, the view that diversity and complexity beget instability, not stability, quickly became the new paradigm in the 1970s and 1980s because of the mathematical rigour of the theory.

There are other limitations in this theory which are critical for the questions that we address here. First, stability is really a meta-concept that covers a range of different properties or components. Summarizing the debate, Pimm (1984) recognized a number of these properties and concluded that the relationship between diversity and each of them need not be the same. In Table 7.1, we attempt a classification—albeit imperfect, as any classification—of the different components of stability, which includes more recent notions. Second, each of these stability properties can be applied to a number of variables of interest at different hierarchical levels, such as individual species abundance, community species composition, or ecosystem-level processes or properties (Table 7.1). Again, the relationship between diversity and any stability property may be different for different variables (Pimm 1984). This creates a large matrix of potential combinations of stability properties and variables of interest, of which the new theory concerned only a small part. Specifically, May's (1974) and Pimm's (1982) theory concerned the qualitative stability and resilience of communities as ensembles of populations, not ecosystem-level aggregate properties.

Third, the formalism of autonomous, deterministic dynamical systems, which describes a fixed set of variables with time-independent parameters, inherently excludes a number of phenomena that characterize biological and ecological systems. In particular, it does not allow for the fact that these systems are subject to continuous environmental changes at various temporal scales and have the

Table 7.1 Concepts and definitions

Stability property	Definition
Components of stability	
Qualitative stability	Property of a system that returns to its original state after a perturbation. Generally used for an equilibrium state, though it could also be applied to systems that return to non-equilibrium trajectories.
Resilience	A measure of the speed at which a system returns to its original state after a perturbation* (Webster *et al.* 1974). Generally used for an equilibrium state, though it could also be applied to systems that return to non-equilibrium trajectories.
Resistance	A measure of the ability of a system to maintain its original state in the face of an external disruptive force (Harrison 1979). Generally used for an equilibrium state.
Robustness	A measure of the amount of perturbation that a system can tolerate before switching to another state. Closely related to the concept of ecological resilience *sensu* Holling* (1973). Can be applied to both equilibrium and non-equilibrium states.
Amplification envelope	Describes how an initial perturbation from an equilibrium state is amplified within a system (Neubert and Caswell 1997).
Variability	A measure of the magnitude of temporal changes in a system property. A phenomenological measure which does not make any assumption about the existence of an equilibrium or other asymptotic trajectories.
Persistence	A measure of the ability of a system to maintain itself through time. Generally used for non-equilibrium or unstable systems before extinction occurs.
Variables of interest	
Individual species abundances	
Species composition	
Ecosystem processes or properties	
Sources of stability/instability	
Internal: species interactions, demographic stochasticity	
External: environmental changes, biological invasions, extirpations	

* Some confusion surrounds the term resilience in the ecological literature. Though the term was first introduced into ecology by Holling (1973), it has most often been used in the sense defined by Webster *et al.* (1974). We follow here the common usage without any judgement on the relative merits of the two definitions.

ability to react or adapt to these changes through asynchronous population fluctuations, species replacement, phenotypic plasticity and evolutionary changes. By ignoring these features, most of the theory on the complexity and stability of ecological systems has focused on deterministic equilibria and ignored much of the potential for functional compensation, both within and between species, which is the basis for the stabilization of aggregate ecosystem properties. Functional compensation between species or types occurs when changes in the level of functioning contributed by one type are associated with opposite changes in the level of functioning contributed by another, whether these changes be of a dynamical, phenotypic or genetic nature.

During the golden period of the new paradigm (the 1970s and 1980s), few dissenting voices were heard. Those proposing an alternative viewpoint were ecosystem ecologists emphasizing functional compensation between species as the mechanism that stabilizes ecosystem processes against a background of wider variability of individual populations (Patten 1975; McNaughton 1977, 1993). Though often ignored, these ideas are the basis of the new wave of theoretical, experimental, and observational work that developed in the late 1990s. The new interest in the functional consequences of biodiversity changes in the 1990s has moved the focus from populations, communities and food webs to ecosystems and the interplay between community-level dynamical processes and ecosystem-level functional processes (DeAngelis 1992; Jones and Lawton 1995; Loreau 2000a). This shift is particularly clear in the recent development of theory, which requires formalization of concepts and hypotheses. New approaches explicitly address the link between the variability of individual species and that of aggregate ecosystem properties, and explicitly incorporate population dynamical responses to environmental fluctuations, and even evolutionary adjustments (Ives 1995; Doak *et al.* 1998; Yachi and Loreau 1999; Lehman and Tilman 2000; Norberg *et al.* 2001).

These new approaches have generally emphasized the potential stabilizing influence of diversity on ecosystem properties, in agreement with the conventional wisdom of early ecologists. This refocusing of the diversity–stability debate does not contradict the previous findings of May (1974) and others, but it does considerably restrict their generality. Previous work focused on qualitative stability and resilience as the stability properties, on species-level population abundances as the variables of interest, and on deterministic autonomous systems at equilibrium, in which only the internal forces of species interactions came into play. In contrast, new work is focusing on variability as the main stability property, on ecosystem-level properties as the variables of interest, and on systems subject to environmental fluctuations, in which the species' responses to these external fluctuations interact with the internal forces of species interactions. The two perspectives are not necessarily contradictory (Tilman 1996; Ives *et al.* 2000). Another avenue of research which has received renewed interest concerns the invasibility or invasion resistance of communities or ecosystems (see Levine *et al.*, Chapter 10), which can be interpreted within our conceptual framework (Table 7.1) as the resistance (stability property) of species composition (variable of interest).

7.3 What we have learned from theory

The insurance hypothesis (Yachi and Loreau 1999) proposes that biodiversity buffers ecosystem processes against environmental changes because different species or phenotypes respond differently to these changes, leading to functional compensations

among species or phenotypes, and hence more predictable aggregate community or ecosystem properties (Patten 1975; McNaughton 1977). In this hypothesis, species that are functionally redundant for an ecosystem process at a given time show temporal complementarity (Loreau 2000a). There have been a number of variations on this theme during the last years (Doak *et al.* 1998; Naeem 1998; Tilman *et al.* 1998; Ives *et al.* 1999, 2000; Rastetter *et al.* 1999; Tilman 1999a; Walker *et al.* 1999; Lehman and Tilman 2000).

Although the assumptions, degree of generality and technical approaches differ considerably among models, a few generalities do emerge from this recent theoretical work. There is often a tension between the destabilizing influence of strong species interactions within the system and the stabilizing influence of asynchronous species responses to external forcing on ecosystem properties. As diversity increases, the number of interactions may increase, leading to the classical result of decreased resilience and increased variability of individual populations (May 1974; Tilman 1996). This destabilizing effect, however, may be reduced for aggregate ecosystem properties (May 1974; Tilman 1996; Hughes and Roughgarden 1998, 2000; Yachi and Loreau 1999; Ives *et al.* 1999, 2000; Lehman and Tilman 2000), and counteracted by decreased mean interaction strength or presence of weak interactions (McCann *et al.* 1998; Ives *et al.* 2000), which are the rule rather than the exception in many natural communities (Paine 1992; Raffaelli and Hall 1994). In contrast, variability of ecosystem processes driven by external environmental factors generally decreases as diversity increases because of the buffering effect of asynchronous species responses (Yachi and Loreau 1999; Ives *et al.* 1999). The net result is generally a smaller variability of aggregate ecosystem properties at a higher diversity (Ives *et al.* 1999; Lehman and Tilman 2000), in agreement with the insurance hypothesis. Hughes *et al.* (Chapter 8) discuss further how variability driven internally by species interactions and variability driven externally by environmental fluctuations interact to determine ecosystem-level stability. Although most of this new theory has been developed for competitive communities, the same conclusions seem to hold for multi-trophic systems (Ives *et al.* 2000).

Differences among species or phenotypes in their responses to environmental changes can not only lead to decreased variability, but also to increased average magnitude of ecosystem processes. When selection processes such as competition favour species or phenotypes within a functional group that are better adapted to current environmental conditions, a higher diversity of types permits a greater adaptability of the system, and hence an enhanced performance, at the functional group level (Yachi and Loreau 1999; Norberg et al. 2001). Although a high phenotypic trait diversity can lead to a lower instantaneous productivity because many sub-optimal types are present, a diverse system can have a higher long-term productivity than any single type because better adapted types tend to replace less adapted ones. It can even be shown quantitatively that the rate at which succession towards the current optimal type proceeds is proportional to phenotypic diversity (Norberg et al. 2001), which provides an ecological analogue to the fundamental theorem of natural selection, and a potential approach to defining and measuring the ability of ecosystems to adapt to the environment. Given this analogy with evolutionary selection, Loreau (2000a) and Loreau and Hector (2001) have coined the term 'ecological selection' to describe changes in dominance and species composition driven by differences in species traits. This analogy allows for employing theoretical approaches developed in evolutionary genetics for disentangling selection from complementarity effects (Loreau and Hector 2001; Hector et al., Chapter 4). The ecological processes that generate adaptability at the ecosystem level also emphasize the importance of regional species richness for ecosystem functioning since external inputs, such as immigration of individuals or propagules, are essential to maintain a wide range of phenotypic traits within an ecosystem, and it is this phenotypic diversity that provides adaptability.

In contrast to studies on variability, theoretical studies on resilience and resistance of ecosystem processes after a perturbation have been scarcer. The results of Hughes and Roughgarden (1998, 2000) and Ives et al. (1999, 2000) imply that the resilience of some ecosystem properties may be independent of species richness in systems with special symmetries amongst species interactions, but Loreau and Behera (1999) found that phenotypic trait diversity generally tends to decrease resilience at both the population and ecosystem levels. Loreau and Behera (1999) also showed that phenotypic diversity can have a variety of effects on the resistance of ecosystem properties. They suggested, however, that positive ecological selection, by which species with favourable traits become dominant, should generally yield a positive effect of diversity on the resistance of ecosystem processes at the primary producer level in the case of 'negative' perturbations (i.e. perturbations, such as drought, that have an intrinsically negative effect on the production of most species), while the opposite should be true for 'positive perturbations' (i.e. perturbations, such as nitrogen addition, that have an intrinsically positive effect on the production of most species).

The effect of species diversity on invasion resistance is another area that has received increased attention recently, although there have been very few theoretical studies on this issue. It is commonly hypothesized that species-rich communities are more resistant to invasion than species-poor communities because they use resources more completely (Elton 1958; MacArthur 1970; Levine and D'Antonio 1999; Tilman 1999). This pattern may be expected when reduced species richness is indeed accompanied by reduced saturation of niche space—a hypothesis for which there is some experimental evidence (see below). Otherwise, theory is mixed in its conclusions about species richness as a predictor of invasion resistance. The nature of the relationship between species richness and invasion resistance is expected to depend critically on the coexistence mechanisms that cause variation in species richness (Moore et al. 2001).

7.4 What we have learned from experiments

A number of recent experimental studies have investigated the relationship between species diversity and various stability properties. Experimental manipulations of diversity within a single trophic level have mostly concerned plants in grassland ecosystems (Table 7.2). The studies reviewed in

Table 7.2 Effects of experimental manipulations of species diversity within a single trophic level on ecosystem stability properties

Reference	Diversity gradient[a]	Species comp.[b]	Ecosystem type	External drivers[c]	Disturbance direction	Time scale	Plot size (m²)	Diversity levels	Stability property	Specifications to stability property	Type of diversity effect[d]
Berish and Ewel 1988	Succ., F	nr	Plantation, forest succ.	n; – R	Neg.	1 y	256	1,40,50,60	Resistance	r. of fine-root bm	None
Joshi et al. 2000[1]	Exp., F	rr	Grassland	bi	Neg.	4 m (3 y)	0.25	1–32	Resistance	r. to loss of above-ground bm	↑ with funct. group no.
Leps et al. 1982	Succ., F	nr	Grassland	n; – R	Neg.	4 y	no inf	4–20	Resistance	Comparison with pre- and post drought y	↑
Mellinger and McNaughton 1975	Succ., F	nr	Old field	e; +N	Pos.	1 y	1500	~35/~50	Resistance	r. in bm to N-pulse	↑
Mulder et al. 2001	Exp., F	rr	Bryophyte community	e; – R; +L	Neg.	5 d (15 m)	0.24	1–32	Resistance/Resilience	Decrease of bm after drought compared to control	↑
Pfisterer et al. (submitted)[1]	Exp., F	rr	Grassland	e; +Hi	Neg.	2 w (5 y)	0.09	1–32	Resistance	r. to loss of above-ground bm	↑
Tilman and Downing 1994	Nutr., F	nr	Grassland	n; – R	Neg.	2 y	16	1–26	Resistance	Decrease of bm in drought rel. to normal y	↑
Tilman 1996	Nutr., F	nr	Grassland	n; – R	Neg.	2 y	16	1–26	Resistance	Decrease of bm in drought rel. to normal y	↓
Brown and Ewel 1987	Succ., F	nr	Plantation, forest succ.	n	Neg.	2 y	256	1,40,50,60	Variability	v. of herbivory	↓
Dodd et al. 1994	Nutr., F	nr	Grassland	n	—	42 y	1000–2000	8–45	Variability	v. in bm	↓ (tendency)
Emmerson et al. 2001	Exp., M	nr	Marine benthic invertebrates	—	—	15 d	41	1–3	Variability	v. in nutrient flux	↓
Tilman 1996	Nutr., F	nr	Grassland	n; – R	Neg.	8 y	16	1–26	Variability	v. in bm in non-drought ys	↓
Leps et al. 1982	Succ., F	nr	Grassland	n; – R	Neg.	2 y	no inf.	4–20	Resilience	Prop. return during 2 y following drought	↓
Tilman and Downing 1994	Nutr., F	nr	Grassland	n; – R	Neg.	2 y	16	1–26	Resilience	Deviation 4 y after drought from pre-d. bm.	↑ (optimum)

[a] Exp.: experimentally newly created diversity gradient; Nutr.: gradients created by different nutrient levels; Succ.: gradients created by different successional stages; F: field study; M: microcosm/mesocosm study.

[b] r: random mixture; rr: random mixture with restrictions; n: nested design; nr: other non-random mixture.

[c] n: natural perturbation; e: experimental perturbation; bi: biological invasion; +Hi: addition of an insect herbivore; +N: increased N-supply; – R: drought. Time scale refers to either the duration of the perturbation (in the case of experimental perturbations—duration of study in brackets) or the duration of the study (in the case of natural perturbations).

[d] Presence/absence and direction of the observed ecosystem process. ↑: positive relationship between diversity and stability property; ↓: relationship negative; none: no significant relationship; id.: identity (species identity or species-mixture identity most important).

[1] Studies were conducted in the same experimental system.

Table 7.2 were selected as described in Schläpfer and Schmid (1999) and Schmid *et al.* (2002); they are restricted to those studies that observed effects of either experimentally or naturally imposed disturbances on the stability properties of communities differing in diversity within a single trophic level. These studies have provided some evidence that the temporal variability of various ecosystem properties decreases with increasing diversity, in agreement with the insurance hypothesis (Brown and Ewel 1987; Dodd *et al.* 1994; Tilman 1996; Emmerson *et al.* 2001; see also Schmid *et al.* 2002). When external perturbations were imposed on the system, plant species diversity had a positive effect on the resistance of above-ground biomass in all the studies listed (Leps *et al.* 1982; Tilman and Downing 1994; Tilman 1996; Mulder *et al.* 2001). The one study that measured the resistance of fine-root biomass, however, did not find a diversity effect (Berish and Ewel 1988). The evidence provided by most of these experiments, however (with the exception of Mulder *et al.* 2001), is inconclusive because of the presence of potential confounding factors (Givnish 1994; Huston 1997). For example, in Tilman and Downing's (1994) study, variations in diversity resulted from a fertilization gradient with plots receiving the highest fertilization having the lowest diversity. Fertilization itself could have resulted in the larger response to drought that was observed in the low-diversity treatments, although reanalysis by Tilman (1996) suggests that the effect of diversity was significant even after controlling for fertilization.

All but one study that tested the effects of external perturbations used negative perturbations (*sensu* Loreau and Behera 1999), mainly drought. The impact of positive perturbations such as nitrogen addition was only studied in Mellinger and McNaughton (1975). Given the anthropogenically induced global change in atmospheric nitrogen deposition (Vitousek *et al.* 1997) the effect of species diversity on the resistance of ecosystem and community properties under positive perturbations would merit more attention. There have been few studies on the influence of species diversity on the resilience of ecosystem processes (Leps *et al.* 1982; Tilman and Downing 1994). The one study that found a positive effect of species diversity on

resilience (Tilman and Downing 1994) used an inadequate measure of resilience (it incorporated resistance by ignoring differences in the magnitude of the initial effect of the perturbation). The effect disappeared after accounting for confounding factors (Tilman 1996). Thus, overall, the experiments performed so far provide results that do not contradict theory (Loreau and Behera 1999).

Experiments that test the effect of species diversity at multiple trophic levels on ecosystem stability properties (Table 7.3) might reflect realistic extinction scenarios of complex, highly connected ecosystems which have to face direct but also secondary extinctions (Williams and Martinez 2000). The studies reviewed in Table 7.3 are restricted to experiments in which species diversity at multiple trophic levels were manipulated and ecosystem stability properties were measured (see also Schläpfer and Schmid 1999; Schmid *et al.* 2002). Two such experiments found decreasing variability of ecosystem properties with increasing diversity (McGrady-Steed *et al.* 1997; Naeem and Li 1997), in agreement with theory. The interpretation of these experiments, however, has been debated because of the presence of confounding factors: in one study (McGrady-Steed *et al.* 1997), ecosystem variability was confounded with variability among replicates; in the other (Naeem and Li 1997), variation in species diversity was confounded with variation in similarity among replicates (Wardle 1998).

Resistance of ecosystem processes after a press perturbation increased with diversity in one study only (Griffiths *et al.* 2000). Studies that measured resistance of community (Petchey *et al.* 1999) or ecosystem properties after pulse perturbations found either no (Downing, submitted; Petchey *et al.* 1999) or a negative (Hurd and Wolf 1974) relationship with increasing diversity. There is no appropriate theory, however, with which these results can be compared.

Lastly, a number of experiments have recently been performed on the effects of species diversity within a single trophic level on invasion resistance. Studies reviewed in Table 7.4 were selected based on a search on ISI web of science (1988–2001) in June 2001 using 'biodiversity' and 'invasion' as search terms (see also Hector *et al.* 2001a; Levine *et al.*, Chapter 10, for reviews). The majority of these

Table 7.3 Effects of experimental manipulations of species diversity at multiple trophic levels on ecosystem stability properties

Reference	Diversity gradient[a]	Species comp.[b]	Ecosystem type	Time scale	Plot size	Groups with diversity[c]	Diversity levels	Number troph. gr.[d]	Level of effect	Stability property	Specifications to observed variable	Type of effect[e]
Downing submitted	Exp., M	r	Aquatic	—	300 l	p/c1/c2	1–5/1–5/1–5	3	All	Resistance	r. in resp. rates after pH-pulse perturbation.	—
Griffiths et al. 2000	Rem., M	—	Pasture soil	1 y	—	c1/c2/dec/b/f	—	5	All	Resistance	r. in decomposition after heavy metal press-pert.	↑
Hurd and Wolf 1974	Succ. F	nr	Old field	6 m	1500 m²	p/c1	~35, ~50 (p)	2	c2	Resistance	r. to N pulse-pert.	↓
Petchey et al. 1999	Exp., M	n	Aquatic	7 w	100 ml	p/c1/c2/b	1–5/1,3/0–3/1,3	4	All	Resistance	Extinction risk due to temperature elevation (press-pert.)	—
Smedes and Hurd 1981	Succ. F	nr	Marine benthic	2 y	0.01 m²	c1/c2/dec	~30, ~35 (total)	3	Several	Resistance	r. to predation	↓
Wardle et al. 2000	Exp., M	n	Grassland	14 m	0.006 m³	p/c1/c2	1–4/1–2/0–1	3	All	Resistance	r. to biomass loss and decomposition after drought press-pert.	—
Downing submitted	Exp., M	r	Aquatic	1 y	300 l	p/c1/c2	1–5/1–5/1–5	3	All	Resilience	r. in resp. rates after pH pulse-pert.	↑
Griffiths et al. 2000	Rem., M	—	Pasture soil	—	—	c1/c2/dec/b/f	—	5	All	Resilience	r. in decomposition after heat pulse-pert.	↑
Smedes and Hurd 1981	Succ, F	nr	Marine benthic	2 y	0.01 m²	c1/c2/dec	~30, ~35 (total)	3	Several	Resilience	r. after predation	↓
McGrady-Steed et al. 1997	Exp., M	rr	Aquatic	42 d	100 ml	p/c1/c2	3–31	4	Several	Variability	v. in ecosystem respiration	↓
Naeem and Li 1997	Exp., M	r	Aquatic	57 d	50 ml	p/c1/c2	1–3/1–3/1–3	5	All	Variability	v. of biomass per trophic group	↓

[a] Exp.: experimentally newly created diversity gradient; Rem.: diversity gradient created by selective removing of species from existing ecosystems; Nutr.: gradients created by different nutrient levels; Succ.: gradients created by different successional stages; F: field study; M: microcosm/mesocosm study.

[b] r: random mixture; rr: random mixture with restrictions; n: nested design; nr: other non-random mixture.

[c] p: primary producer; c1: primary consumer; c2: secondary consumer; dec: decomposer; f: fungivore; b: bacterivore.

[d] Number of trophic groups varied.

[e] Presence/absence and direction of the observed ecosystem process. ↑: positive relationship between diversity and stability property; ↓: relationship negative; none: no significant relationship.

Table 7.4 Effects of experimental manipulations of species diversity within a single trophic level on invasion resistance

Reference	Type of exp.[a]	Species comp.[b]	Ecosystem type	External drivers[c]	Time scale	Plot size (m²)	Diversity levels	Species pool	Stability property	Specifications to stability property	Type of diversity effect[d]
Crawley et al. 1999	F	nr	Grassland	bi	7y (8y)	9.0	1–4, 80	4, 80	Resistance	Weed invasion resistance	None/identity
Dukes 2001	M	rr	Med. grassl.	bi	1y	0.03	1–16	16	Resistance	Weed invasion resistance	↑ with funct. group no.
Hector et al. 2001a	F	rr	Grassland	bi	4y	4.0	1–11	47	Resistance	Weed invasion resistance	↑
Joshi et al. 2000[1]	F	rr	Grassland	bi	1y (3y)	0.25	1–32	48	Resistance	Weed invasion resistance	↑ with funct. group no.
Knops et al. 1999[2]	F	r	Grassland	bi	2y (4y)	9.0	1–24	24	Resistance	Weed invasion resistance	↑
Lavorel et al. 1999[3]	F	nr	Med. grassl.	bi	1m (8m)	4.0	3,6,18	18	Resistance	Weed invasion resistance	None
Palmer and Maurer 1997	F	nr	Crops	bi	4m	5.0	1,5	5	Resistance	Weed invasion resistance	↓/none
van der Putten et al. 2000	F	rr/n	Grassland	bi	2y	100	4,15	37	Resistance	Weed invasion resistance	↑ (id.)
Troumbis et al. submitted	F	rr	Med. grassl.	bi	1y (4y)	1.0	1–18	23	Resistance	Weed invasion resistance	↑
Diemer and Schmid 2001[1]	F	rr	Grassland	bi	2y (4y)	4.0	1–32	48	Resistance	Weed invasion resistance (phytometer study)	↑
Levine 2000	F	r	Med. riparian primary succ.	bi	1y (2y)	0.04	1–9	9	Resistance	(sown) Weed invasion resistance	↑ (id.)
McGrady-Steed et al. 1997	M	rr	Aquatic	bi	2w (8w)	100 ml	3–16	27	Resistance	Invasion resistance (controlled addition of *Euplotes* sp. (protozoa))	↑ (id.)
Naeem et al. 2000[2]	F	r	Grassland	bi	1y (3y)	9.0	1–24	24	Resistance	Weed invasion resistance (phytometer study)	↑
Prieur-Richard et al. 2000[3]	F	nr	Med. grassl.	bi	7m (1y)	4.0	3,6,18	18	Resistance	Weed invasion resistance (phytometer study)	(↑) /funct. gr. identity.
Stachowicz et al. 1999	F	r	Marine benthic	bi	65 d (—)	0.01	1–4	4	Resistance	Invasion resistance (controlled addition of an exotic ascidian species)	↑
Symstad 2000	F	n	Grassland	bi	2y (4y)	32	1–3 funct. groups	—	Resistance	(sown) Weed invasion resistance	↑ with funct. group no.

[a] Exp.: experimentally newly created diversity gradient; rem.: diversity gradient created by selective removing of species from existing ecosystems; nutr.: gradients created by different nutrient levels; succ.: gradients created by different successional stages; P: phytotron; F: field study; M: microcosm/mesocosm study.

[b] r: random mixture; rr: random mixture with restrictions; n: nested design; nr: other non-random mixture.

[c] bi: biological invasion. Time scale refers to either the duration of the perturbation (in the case of experimental perturbations—duration of study in brackets) or the duration of the study (in the case of natural perturbations).

[d] Presence/absence and direction of the observed ecosystem process. ↑: positive relationship between diversity and stability property; ↓: relationship negative; none: no significant relationship; id.: identity (species identity or species-mixture identity most important).

[1,2,3] Studies were conducted in the same experimental system.

studies showed a positive relationship between plant species or functional-group richness and resistance against naturally invading weeds. In addition, all studies investigating the impact of diversity within a trophic level on the performance of experimentally added invaders showed increased invasion resistance with community diversity. Only one experiment found the opposite effect (Palmer and Maurer 1997); this experiment investigated weed invasion in crop monocultures and five-species mixtures and found that the more diverse crop mixtures harboured a more species-rich (mostly annual) weed community. Weed invasion resistance in terms of weed biomass, however, was not affected by diversity. Another work by Lavorel *et al.* (1999) found no diversity effect on invasion resistance in Mediterranean grassland communities. In contrast to the other studies in Table 7.4, these two studies were relatively short-term experiments in which weed species were established at the same time as target communities. No predictable effect of diversity on invasion resistance was found either in an experiment by Crawley *et al.* (1999), in which species composition was non-random, just as in Palmer and Maurer (1997) and Lavorel *et al.* (1999). From the currently available evidence, it seems that more diverse communities are harder to invade in fully established communities with random, restricted random, or nested (one study) designs, with true replication of diversity levels and experimental introduction of invading species.

The mechanisms behind the positive relationship between diversity and invasion resistance in these small-scale experiments probably involve both better resource utilization in more diverse communities (Hector *et al.* 2001a; Levine *et al.*, Chapter 10) and selection processes since more diverse communities have a higher chance to contain species that benefit from altered environmental conditions (Schmid *et al.* 2002). Strong effects of species identity or species composition were detected in four of the 13 studies that found a positive effect of diversity on invasion resistance. Disentangling species diversity and species identity effects, however, requires true replication of diversity treatments with different species assemblages (Allison 1999), a requirement which was not fulfilled in all experiments. For example, few studies replicated the highest diversity level with different species mixtures. Another limitation of these studies is that they have typically used 'invaders' from the extant regional flora which have coevolved with those species that constitute 'invaded' communities. Invasion by new exotic species is likely to follow different dynamics.

7.5 Strengthening the link between theory and experiments

Perhaps for the first time in the history of the diversity–stability debate, we now have two essential ingredients for scientific progress: first, a conceptual framework that is sufficiently broad and clear—albeit certainly improvable—to avoid confusion and sweeping generalizations, and, second, a convergence of observational, experimental and theoretical approaches towards common objectives and questions. It must be borne in mind, however, that the current work focused on biodiversity and ecosystem functioning is addressing only part of the original debate, several aspects of which remain untested.

Theory has historically been prominent in the diversity–stability debate. But profusion of theory is no guarantee of clarity and relevance. As experimental and observational evidence accumulates, the weaknesses of past, abstract theories have become more apparent. Theory needs to evolve to provide better guidance for experiments. Most of the classical equilibrium approaches based on autonomous dynamical systems may be inadequate to understand stability properties such as variability, resilience and resistance at the ecosystem level.

Here, we have argued that, to understand functional compensations in ecosystems, new approaches should be developed that take into account the dynamics of diversity and the potential for adaptive changes through asynchronous species fluctuations, species replacement, phenotypic plasticity and evolutionary change. In other words, ecosystems must be fully treated as complex adaptive systems, as proposed by such scientists as Holling (1986) and Levin (1999). Most of the current theory is also borrowed from community ecology, with an emphasis on total plant biomass or primary production as the ecosystem properties investigated. Total plant biomass and primary production are easily related to individual plant or population-level properties by simple aggregation, but this might be less straightforward for other ecosystem processes. The historical separation between community ecology, which is demography oriented, and ecosystem ecology, which focuses on whole-ecosystem functional processes, demands new approaches to lay a bridge between these different perspectives (Loreau 2000a). There have been very few attempts to explore the effects of biodiversity on the functioning of full ecosystems comprising higher trophic levels, decomposers and nutrient cycling (Loreau 1996, 2001), and none as yet has considered stability explicitly.

It is very encouraging that experiments have started to test new ideas on the relationships between the diversity and various aspects of stability of ecosystems. A number of these experiments, however, have been debated because of the potential presence of confounding factors, which now need to be addressed by new experimental designs. To date, no experimental studies have directly manipulated long-term environmental variability to test the potentially important role that environmental fluctuations may play as both the creator and driver of the conditions necessary for the existence of compensatory dynamics. Several empirical studies suggest that this may be so. Some microcosm studies have demonstrated that certain types and frequency of environmental fluctuation may set levels of species richness and affect community stability (Eddison and Ollason 1978; Ollason 1977; Rashit and Bazin 1987). More directly, the work reported by Frost *et al.* (1995) and Klug *et al.* (2000) provides some short-term evidence demonstrating the operation of compensatory dynamics in lake communities in response to pH perturbations, and Morgan-Ernest and Brown (2001) provide long-term evidence for the existence of similar compensatory dynamics in arid grassland communities.

Establishing the general importance of the insurance hypothesis would require the demonstration of the assembly or evolution of an ecosystem functioning in this manner under controlled environmental fluctuations. Experiments of this kind would

define the subset of environmental and ecological conditions conducive to the establishment of such a mode of ecosystem functioning. Perhaps the main difficulty here is the production under experimental conditions of realistic environmental fluctuations with a controlled frequency structure. Methodological advances in this direction have been made (Cohen *et al.* 1998) and key microcosm experiments are starting to be conducted. Clearly, there is still scope for many innovative ideas in the design of experiments in this area.

To strengthen the link between theory and experiments, theoretical and experimental studies should attempt to adopt similar measurements of stability. Many past theoretical developments and predictions are difficult to directly test experimentally because equivalent measures of stability often do not exist in experimental systems. For example, there is not a straightforward experimental equivalent to an eigenvalue. Experimental approaches, in turn, must consider relevant theoretical work when designing and interpreting results. Experimental response variables could be chosen to correspond more closely to theoretical stability estimates. In addition, care must be taken to not misapply theoretical results to experimental results, particularly when the definitions of stability differ.

One of the difficulties of measuring stability in natural ecosystems is that natural ecosystems show a variety of complex dynamics. Many ecosystems experience predictable variations, such as the seasonal changes in the pelagic community of temperate lakes or succession in forests, or react to disturbance in a fairly predictable manner; algal blooms following eutrophication or re-establishment of forests after local clear-cuts are examples. Under normal environmental fluctuations, ecosystems often develop along a trajectory that is an environmentally determined dynamical attractor; systems starting out with different initial conditions then converge over time. Such non-equilibrium systems are stable and return to their attractor following a perturbation. In this case, tests of stability properties following perturbation could use the deviation between a perturbed system and a control system as a measure (e.g. Wardle *et al.* 2000a) or, alternatively, the relative difference in disturbance effects along a gradient of diversity (e.g. van der Putten *et al.* 2000). A caveat, however, is that perturbations may be initially amplified before returning to the original state (Neubert and Caswell 1997), in which case a sufficiently long experimental time period is necessary to ensure that the system does converge.

Ecosystems that exhibit more complex dynamics or flip between alternative stable states (Scheffer *et al.* 2001) will be harder to analyse because there is no single reference system. A disturbance can switch a system into a different configuration such that the 'recovered' system is vastly different than the unperturbed control. Such systems require a focus on what Holling (1973) termed 'ecological resilience', or what we here call 'robustness' (Table 7.1). An appropriate experimental design would employ a range of disturbance magnitudes, which allows defining the domain within which disturbed systems will not loose critical functional groups or processes. Complete similarity with the undisturbed system may not be necessary as a criterion for recovery; one might be interested in maintaining the same ecological processes, such as topdown or bottom-up control. Thus, different types of systems may require different kinds of concepts, measures and experimental design.

A future challenge will be to recognize various types of complex dynamics in natural ecosystems, and to incorporate them in theoretical work attempting to study the relationship between stability and diversity. Such theory would hopefully produce realistic patterns of diversity–stability relationships, provide testable mechanisms, and help to sharpen and focus experiments designed to explore diversity–stability relationships. As theoretical and experimental foundations become more solid, there will also be an increasing need for long-term empirical data in the field. Long-term monitoring of both biodiversity and ecosystems processes is critical to apply our basic scientific understanding to real ecosystems, both natural and managed. Such data will need to be scaled to the turnover time of the ecosystem processes being considered if we are to understand their implications and relevance in the context of the natural functioning of ecosystems. This means that in some ecosystems, such as forests, very long time series will be necessary. The critical challenges with natural ecosystems will be to untangle the effects of environmental factors that drive natural variations in diversity and of diversity itself, and to develop new theory that integrates the mutual interactions among biodiversity changes, ecosystem functioning and abiotic factors into a single, unified picture (Loreau *et al.* 2001).

7.6 Implications for policy and management

As human impact on ecological systems increases, scientists are increasingly challenged to communicate new knowledge to policy- and decision-makers (Lubchenco 1998). Does our current state of knowledge of diversity–stability relationships provide specific information for policy and management decisions?

Society depends on the steady and predictable inputs of ecosystem services (Daily 1997). Current evidence suggests that higher diversity may provide greater reliability in the production of ecosystem services such as food and fibre production, pollination levels, and nutrient cycling. Diversity may also decrease the probability of successful invasions of non-native species, many of which have had substantial economic, conservation, and societal consequences (Mooney and Drake 1986; Drake and Mooney 1989). Extinctions of native species may lead to a further decrease in stability that causes a cascade of other extinctions, accelerating the rate of community change (Pimm *et al.* 1988; Borrvall *et al.* 2000). Finally, declines in ecosystem stability may reduce our ability to predict or detect future environmental changes in a background of higher ecosystem variability, including the influence of slow processes such as climate change (Cottingham *et al.* 2000). Thus, the impact of biodiversity on ecosystem stability appears to be a relevant feature to consider in policy and management decisions.

We have shown, however, that there are a number of components to stability, and that changes in diversity may alter ecosystem stability in a variety of ways. Stability at one level may require change at another level; for instance, we have discussed how increased average magnitude and decreased variability of ecosystem processes come about through changes at the species level. Despite progress in our understanding of diversity–stability relationships, current research is still largely unable to provide specific policy recommendations due to the lack of abundant, consistent, and relevant long-term data on ecosystem processes for most biomes (Schwartz *et al.* 2000; Cottingham *et al.* 2001; Hector *et al.* 2001b; Lawler *et al.* 2002). Diversity appears to play some role in maintaining stability of certain processes in a handful of ecosystems, but more research is needed before we can confidently justify biodiversity conservation on the basis of its ability to enhance or protect ecosystem stability. Future work should begin to focus on scenarios that are immediately relevant to human society. For example, human impact causes non-random changes in diversity or composition (Petchey *et al.* 1999). Exploring the consequences of these non-random diversity changes for stability will become important. Research could also be profitably focused on economically important ecosystems, such as agricultural ecosystems, and on ecosystem goods and services of importance to society, including reliable supplies of clean fresh water, and crop and fisheries production.

Given what we currently understand about the potential of diversity to buffer ecosystems against environmental fluctuations, future management efforts should look towards preserving the already 'built-in' capacity of ecosystems to adapt to environmental perturbations. This approach would require an emphasis on preserving regional species diversity and the necessary habitat connectivity required for the assembly of local communities in order to maintain the potential for high local diversity. Until we have a better understanding of how diversity relates to stability, management strategies aimed at preserving diversity will at the very least increase the potential for ecosystems to respond to future, changing environments.

CHAPTER 8

Do species interactions buffer environmental variation (in theory)?

J. B. Hughes, A. R. Ives, and J. Norberg

8.1 Introduction

The interconnectedness of species within ecosystems underlies the popular notion that biodiversity contributes to a balance of nature. Species interactions are also fundamental to ecologists' formulations of how biodiversity and the stability of ecosystem functioning are related. Many have reviewed how the intricacies of stability definitions affect these relationships (e.g. McCann 2000; Pimm 1984). Fewer have examined how the varied aspects of species interactions, such as strength and number, affect diversity–stability relationships in models (e.g. Benedetti-Cecchi 2000; May 1973) and experiments (e.g. Aoki and Mizushima 2001; Mulder *et al*. 2001).

In this chapter, we investigate the role of species interactions in the buffering (dampening variation) of aggregate community biomass against environmental variation. Thus, we consider only one aspect of ecosystem stability, the variability of aggregate community biomass as measured by the coefficient of variation of total community biomass, or the CV of the sum of the abundances of all species in a community (CV_{sum}) (see Loreau *et al*., Chapter 7 for other definitions of stability). Community stability is inversely related to CV_{sum}; the lower the CV_{sum}, the greater the stability. This type of stability is of particular interest because of its potential to reflect the variability of primary production, which in turn is related to a variety of ecosystem processes. A caveat of this measure, however, is that biomass is the end product of many processes and does not necessarily reflect the

stability of the processes themselves (Naeem *et al*., Chapter 1).

In the first part of the chapter, we discuss the connection between a number of recent models. We then give a general formula for the relationship between diversity and stability that incorporates these models and discuss how different aspects of interactions affect the diversity–stability relationship. We limit our scope to communities of coexisting species that compete for common resources. In other words, the models represent communities with only one trophic level and do not address the issue of species coexistence and extinction. Furthermore, we examine communities that are subject to repeated, stochastic fluctuations, as opposed to other perturbations such as press disturbances (Ives 1995) or species removals (Naeem 1998; Petchey 2000).

We then discuss how interaction strength, the number of interactions, asymmetries among interactions, and interaction strength variability influence the relationship between diversity and stability. We conclude that the role of competition in stabilizing communities against environmental variability may be minor. In light of this, we consider the importance of species–environment interactions relative to species–species interactions.

8.2 How species interactions are incorporated

In the models considered below, the communities are made up of species that are assumed to compete for shared resources, although these resources are not explicitly included in the models. Instead, competition enters in two manners: directly, through per capita effects represented by interaction coefficients, and indirectly, through assumptions about the relationship between diversity and total community biomass. In the next sections, we discuss statistical models that incorporate interactions indirectly and dynamic models that incorporate interactions directly. These approaches are related, however, even if only conceptually.

Direct competitive effects can be conceptualized by considering the overlap of species along a resource axis, as in the classic literature on resource apportionment (e.g. MacArthur 1960; May 1975; Pielou 1975). At one extreme, species use exactly the same resources; they completely overlap on the resource axis (the left-hand side of Fig. 8.1(a)). At the other extreme, no species interact, as they all use unique segments of a resource axis (Fig. 8.1(b)). In the intermediate case, species partially compete with one another (Fig. 8.1(c)). The value of pairwise competition coefficients reflect the degree to which the species' resource utilization curves overlap.

Interactions can also be included (or excluded) indirectly by assumptions about the relationship between diversity and total community biomass. Consider again the three cases in Fig. 8.1. If every new species added to a community uses the same resources as in case (a), total biomass should be independent of species richness (the right-hand side of Fig. 8.1). As a result, as diversity increases, each species has access to a smaller share of resources. In a community where species do not interact (case (b)), the length of the resource axis used increases with diversity, and total biomass increases linearly with species richness. In the case of partial competition (case (c)), each species added extends the resources axis (i.e. resource complementarity occurs), and biomass increases at a decreasing rate with richness. Thus, even without explicitly defining the amount of competitive overlap among species, models that assume diversity–total biomass relationships make implicit assumptions about species interactions.

8.3 Statistical models with all or nothing interactions

Recently, Doak et al. (1998) formalized the idea that diversity may buffer environmental variation through a statistical averaging effect, even without interactions between species. They described a community where every species has the same mean abundance and variance, and total community biomass (T, the sum of species' abundances) is constant regardless of how many species (m) are present. Thus, the mean abundance of every species is T/m. They also assumed that the variance of a species' density (σ_n^2) scales with the square of its mean abundance, so that $\sigma_n^2 = c(T/m)^2$, where c is a constant. If species fluctuations are assumed to be independent of the fluctuations of other species, the CV of total biomass (CV_{sum}) is simply

$$CV_{sum} = \frac{\sqrt{m\sigma_n^2}}{T} = c^{1/2}\, m^{-1/2} \qquad (8.1)$$

Thus, CV_{sum} decreases and stability increases with increasing species richness. More generally, if we allow the correlation between the fluctuations

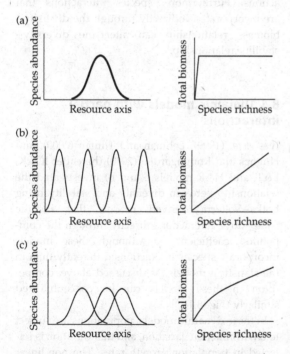

Figure 8.1 A schematic of three levels of resource competition and the parallel relationships between number of species and community biomass. (a) Complete competition, (b) No interactions and (c) Partial competition.

among all pairs of species to be ρ, then (Doak *et al.* 1998, eqn (5))

$$CV_{\text{sum}} = \frac{\sqrt{m\sigma_n^2 + \rho m(m-1)\sigma_n^2}}{T}$$
$$= c^{1/2} m^{-1/2} (1 + (m-1)\rho)^{1/2} \qquad (8.2)$$

(The constraint that $\rho \geq -1/(m-1)$ is simply the constraint on any covariance matrix, that it must be positive definite. For example, you cannot have the pairwise correlations between three species all equal to -1.)

Tilman *et al.* (1998) noted that the relationship between stability and species richness changes if one relaxes the assumption that the variance of a species' density scales with the square of its mean. Specifically, they let $\sigma_n^2 = c(T/m)^x$, where x is a constant that is not necessarily equal to 2. Under this assumption, eqn (8.2) becomes

$$CV_{\text{sum}} = c^{1/2} T^{(x-2)/2} m^{(1-x)/2} (1 + (m-1)\rho)^{1/2}$$
$$\qquad (8.3)$$

Thus, stability increases with increasing diversity when $x > 1$, but decreases with diversity when $x < 1$. Tilman *et al.* (1998) and Tilman (1999a) argue from field studies of prairie plant communities that x usually falls between 1 and 2.

Although competition is not included in these models directly, competition is included indirectly through the diversity–total biomass relationship. These models assume a fixed community biomass (T), regardless of the number of species added or subtracted to the community. Thus, the mean abundance of a species (k) depends on the number of other species present. This assumption is similar to the case depicted in Fig. 8.1(a), where all species compete for the same section along a resource axis.

What happens if these implied interactions are removed? If the species do not interact with one another, then k should be independent of species richness and total biomass should increase linearly as each species is added, i.e. $T = mk$ (the case in Fig. 8.1(b)). Equation (8.3) then becomes

$$CV_{\text{sum}} = c^{1/2} k^{(x-2)/2} m^{-1/2} (1 + (m-1)\rho)^{1/2}$$
$$\qquad (8.4)$$

Thus, without species interactions, stability always increases with diversity whenever $\rho \neq 1$,

regardless of how variance scales with mean abundance (see Appendix of Hughes and Roughgarden 2000).

Yachi and Loreau (1999) show that statistical averaging can also produce a positive diversity–stability relationship when species completely overlap in resource use. In their models, the abundance of each species is specified by a random variable, and the variances of the species are equal. Competitive interactions are reflected in the diversity–biomass relationship. For a community with extreme interspecific competition, community biomass is defined as the most productive species at any given time. As long as the correlation between species fluctuations is less than 1, the variance of total biomass decreases with increasing species richness. Because total biomass increases and then levels off with increasing diversity, the CV of total biomass also decreases as diversity increases.

In sum, statistical averaging can contribute to community stability with or without species interactions. Furthermore, species interactions that are incorporated indirectly through the diversity–biomass relationship can alter the diversity–stability relationship.

8.4 Dynamic models with partial interactions

Ives *et al.* (1999), Lehman and Tilman (2000), and Hughes and Roughgarden (2000) (hereafter IG&K, L&T, and H&R models a and b) investigated the relationship between diversity and stability using Lotka–Volterra derived models that include competitive interactions directly through the competition coefficient α. Although these models incorporate species interactions differently than in the statistical models, as discussed above, competition in these models can be conceptualized similarly (Fig. 8.1).

In these dynamic models, populations are subject to environmental variation, and this variation is reflected in population growth rates. The non-linear forms of these models are presented in Table 8.1. The linear approximation of all the models is similar, and the difference in density from equilibrium

Table 8.1 The difference equations of four dynamic models that track changes in the population density of species i at time t, $N_i(t)$, where ε is a random variable with mean zero and variance σ_ε^2. In all models, species have carrying capacities K and intrinsic rates of increase r, and α is the competition coefficient measuring the effect of one species on another. H&R model (a) is the diffuse competition case and model (b) is the limiting similarity case in Hughes and Roughgarden (2000)

Model	$N_i(t+1)=$	Diversity–stability relationship
IG&K	$N_i(t)\exp\left[r\left(1-\frac{N_i(t)+\alpha\sum_{j\neq i}^{m}N_j(t)}{K}\right)\right]\exp[\varepsilon_i(t)]$	Positive, independent of K
L&T	$N_i(t)+rN_i(t)\left[1-\frac{N_i(t)+\alpha\sum_{j\neq i}^{m}N_j(t)}{K+\varepsilon_i(t)}\right]$	Positive, dependent on K
H&R (a)	$N_i(t)+rN_i(t)\left[1-\frac{N_i(t)+\alpha\sum_{j\neq i}^{m}N_j(t)}{K}\right]+\varepsilon_i(t)$	Negative or positive, dependent on α and K
H&R (b)	$N_i(t)+rN_i(t)\left[1-\frac{N_i(t)+\alpha\left(N_{j-1}(t)+N_{j+1}(t)\right)}{K}\right]+\varepsilon_i(t)$	Positive, dependent on α and K

$(N^* - N_i(t+1))$ of species i at time $t+1$ is (Ives and Hughes, in press)

$$n_i(t+1) = \left(1-r\frac{N^*}{K}\right)n_i(t)$$

$$-r\alpha\frac{N^*}{K}\sum_{j\neq i}^{m}n_j(t)+z_i(t) \qquad (8.5)$$

for IG&K, T&L, and H&R(a), and

$$n_i(t+1) = \left(1-r\frac{N^*}{K}\right)n_i(t)-r\alpha\frac{N^*}{K}\left(n_{i-1}(t)\right.$$

$$\left.+n_{i+1}(t)\right)+z_i(t) \qquad (8.6)$$

for H&R(b), where K is the carrying capacity of a species, r is the intrinsic rate of increase, and $z_i(t)$ gives the environmentally driven variability in the population growth rate with standard deviation σ_z (the value of $z_i(t)$ is discussed below). When $0<r<2$ and $0<\alpha<1$, the equilibrium is positive and stable.

The CV_{sum} of all models can be summarized by the general equation

$$CV_{\text{sum}} = \sigma_p\left(\frac{1+(m-1)\rho}{m(1-(1-r)^2)}\right)^{1/2} \qquad (8.7)$$

where σ_p is the standard deviation of the environmental fluctuations measured as the change in per capita population growth rates from time t to $t+1$ (see derivation in Ives and Hughes 2002). The standard deviation of the per capita population

growth rates are related to the standard deviation of $z(t)$ by $\sigma_p = \sigma_z/N^*$. The difference between σ_p and σ_n from the statistical models in the preceding section highlights an important distinction between these two types of models. In the models in Table 8.1, environmental fluctuations are added to species' population growth rates. In contrast, the statistical models introduce environmental fluctuations as variation to species' densities. These terms are related, in the sense that increasing the environmentally driven fluctuations in population growth rates, σ_p, will increase the variation in population densities, σ_n. (Note, however, that an increase in σ_n does not necessarily indicate an increase in σ_p.) Nonetheless, for models explicitly including species interactions, the variation in population densities (σ_n) depends not only on how environmental fluctuations change growth rates (σ_p) but also on how these environmental fluctuations are perpetuated through species interactions (Ives et al. 1999).

The disparate diversity–stability relationships among models in Table 8.1 arise from differences in the relationship between σ_p and m, which result from three assumptions that differ among the models. First, the point-of-entry of an environmental perturbation ($\varepsilon(t)$) differs among the models. In IG&K the perturbations are multiplicative. In T&L the perturbations are added to the carrying capacity, and in the H&R models the perturbations are additive (Table 8.1). An environmentally driven fluctuation in a population's growth rate is given by z (the final term in eqs (8.5) and (8.6)) and is a function of an environmental perturbation ε. Specifically, $z_i(t) = gN^*\varepsilon_i(t)$, and it can be shown that the value of g depends on the placement of $\varepsilon_i(t)$ (Table 8.2; Ives and Hughes 2002).

Second, how the variance in population growth rates scales with the mean abundance of a species is important. In general terms, the variance of z can be written as a function of the mean abundance of the population as $\sigma_z^2 = \sigma_\varepsilon^2(gN^*)^x$, where as before, x scales the rate at which the variance in z increases with the mean density.

Third, the mean abundance of a species (its equilibrium value, N^*) may depend on the relationship between diversity and total biomass. As a general case, we let $N^* = T/m^y$, where y is a constant.

We can incorporate all three assumptions, which leads to the relationship

$$\sigma_p = \sigma_\varepsilon g^{x/2} \left(\frac{T}{m^y}\right)^{x-2/2}$$

Substituting σ_p into eqn (8.7),

$$CV_{sum} = \sigma_\varepsilon g^{x/2}$$
$$\times \left(\frac{T}{m^y}\right)^{(x-2)/2} \left(\frac{1 + (m-1)\rho}{m(1 - (1 - r)^2)}\right)^{1/2}$$

$$(8.8)$$

where g is given in Table 8.2. This equation is a general form that includes all of the above models as special cases, including the models without interactions. For instance, if $r = 1$ and T/m^y is given by a constant k, then eqn (8.8) reduces to eqn (8.4).

Table 8.2 The values of g, x, and y (eqn (8.8)) for the models in Table 8.1. The three models assume different values of g and x, and thereby result in a different relationship between the per capita environmentally driven variation (σ_p) and the mean abundance of a species (N^*). The value of y does not affect the IG&K and L&T models

Model	g	x	y	σ_p
IG&K	1	2	—	σ_ε
L&T	$\dfrac{r}{K}$	2	—	$\sigma_\varepsilon \dfrac{r}{K}$
H&R(a)	$\dfrac{1}{N^*}$	0	> 0	$\sigma_\varepsilon N^*$
H&R(b)	$\dfrac{1}{N^*}$	0	0	$\sigma_\varepsilon N^*$

How do interspecific interactions affect stability in this most general case? The interaction coefficient α is missing from eqn (8.8). Therefore, interspecific interactions only enter the equation indirectly, by their influence on mean abundances. Exactly how interactions influence stability depends on a combination of the diversity–biomass relationship, the point-of-entry of the perturbations, and the mean–variance scaling constant. Table 8.2 summarizes the three differences among the models presented in Table 8.1. Both IG&K and T&L assume that $x = 2$; therefore, the mean abundance of a species does not affect the diversity–stability relationship. In contrast, H&R assume $x = 0$, and the effect of species interactions on mean abundance explains the different diversity–stability relationships produced by their models (a) and (b) (Fig. 8.3).

8.5 Which aspects of species interactions influence stability?

The above models give some insight into how species interactions might buffer or propagate environmental variation through the dynamics of a community; in particular, they suggest that it is important to distinguish between different aspects of species interactions when determining whether interactions influence the diversity–stability relationship. There are at least four aspects of species interactions that can be distinguished: interaction strength, the number of interactions, asymmetries among interactions, and interaction strength variability.

8.5.1 Interaction strength

The strength of interspecific interactions (the magnitude of α) does not affect the direction of the diversity–stability relationship for competitive communities with symmetrical interaction strengths (Ives et al. 1999, 2000). In fact, as Doak et al. (1998) demonstrated, the occurrence of species interactions is not required for a positive diversity–stability relationship. Thus, in this sense, a community of strongly interacting species is not necessarily more stable than a community of weakly interacting species.

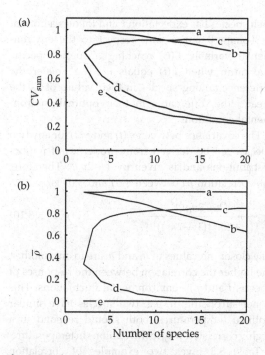

Figure 8.2 The change in (a) CV_{sum} and (b) $\bar{p}[m]$ with increasing numbers of species in competitive communities. Equation (8.10) was used to calculate each pairwise correlation between species, and (b) gives the average of these values. Values of $\bar{p}[m]$ from (b) were then used in eqn (8.11) to compute CV_{sum} and are plotted in (a), with $\sigma_p = 1$ and $r = 1$. See text for explanation of cases a–e.

8.5.2 Number of interactions

Interactions may influence stability indirectly through the relationship between diversity and the biomass of individual species. The shape of the diversity–biomass relationship will be determined in large part by the number of non-zero interactions or the connectance (Gardner and Ashby 1970; May 1973) of a community. For instance, in the H&R model (a), every species interacts with every other species, and therefore total biomass generally increases at a decelerating rate with diversity. At the other end of the spectrum, in the H&R model (b), each species interacts with two species, regardless of community diversity, and total biomass increases linearly with diversity.

The shape of the diversity–biomass relationship, however, only affects the diversity–stability relationship if the per capita effect of environmental

fluctuations on individuals of a species changes with the mean density of the species (i.e. when σ_p is a function of N^*). Recognizing the independence of the diversity–stability and the diversity–biomass relationships is important for disentangling the effects that diversity has on ecosystems.

8.5.3 Asymmetries among interactions

We have only considered models in which species interactions are symmetric ($\alpha_{ij} = \alpha_{ji}$) and equal for all species. Simulations of communities with moderate asymmetries among interactions suggest that the general conclusions still hold (e.g. Hughes and Roughgarden 1998; Ives *et al.* 1999). In the case of extreme asymmetries, however, the community will be dominated by one or a few species, and accordingly the community should have the dynamics corresponding to these few dominant species rather than the dynamics of an equally diverse community with a more even distribution of species abundances.

8.5.4 Interaction strength variability

Environmental fluctuations could affect the strength or presence of interactions and thereby affect the diversity–stability relationship. The models we have analysed examine stability in the face of environmental fluctuations that affect population growth rates. However, environmental fluctuations may also vary interspecific interaction strengths, particularly by altering the abundance and quality of different resources. For example, a resource may or may not be shared between species, depending on other resources available to each species. Thus, as resource availability fluctuates, the amount of resource overlap between species may vary. Benedetti-Cecchi (2000) constructed a conceptual model to examine the importance of variability in the strength of interactions between consumers and resources. Although this model is not directly comparable to competition models, it demonstrated that variance in interaction strength affects the variability of species abundances. This same idea deserves further attention in other models.

In sum, interspecific competitive interactions do not necessarily buffer community biomass against environmental variation (eqn (8.7)). Moreover, diversity may increase stability in the absence of competitive interactions (eqn (8.4)). Thus, we suggest that the role of competition in stabilizing some aggregate measure of a community (such as biomass) against environmental variability may be relatively minor, particularly in comparison to species–environment interactions discussed in the next section. This conclusion does not exclude the possibility that other types of interspecific interactions such as predation and mutualism play a larger role in community stability (Aoki and Mizushima 2001; Mulder *et al.* 2001).

8.6 Species–environment interactions

Given the indirect role played by competition in the diversity–stability relationship, the importance of how a species responds to environmental fluctuations, rather than to other species, emerges. The correlated responses of species growth rates (ρ) gives a measure of species–environment interactions for the community as a whole, and this term appears in the stability equations above (Eqs (8.4), (8.7), and (8.8)).

To demonstrate the importance of species–environment interactions for the diversity–stability relationship, we developed a stochastic model inspired by the deterministic model analysed by Norberg *et al.* (2001). Consider a community of species that compete for the same set of resources. The growth rate R of species i is $\exp[r + \gamma_i(t)]$, where r is a constant and

$$\gamma_i(t) = -(u_i - E(t))^2 + u_i^2 + 1 \tag{8.9}$$

$E(t)$ is a time-varying environmental variable such as rainfall or predator abundances that affects all species in the community, and u_i is the species' phenotype, the position along the environmental spectrum where the species reaches its maximum growth rate. We assume that $E(t)$ follows a normal distribution, and that there is no serial correlation. Without loss of generality, we can also assume $E(t)$ follows a $N(0, 1)$ normal distribution.

In eqn (8.9), $\gamma_i(t)$ has an expectation of zero and reaches a maximum value when $E(t)$ equals u_i. Thus, the log per capita population growth rate of

each species has expectation r and follows a normal 'bell-shaped' curve with respect to the environmental variable $E(t)$, reaching a species-specific maximum when $E(t)$ equals u_i. The greater the difference among species in their values of u_i, the greater the differences in their optimal environmental conditions.

The covariance between $\gamma_i(t)$ and $\gamma_j(t)$ for any two species can be derived from properties of normal distributions and is given by $1 + 2u_iu_j$. Therefore, the correlation ρ_{ij} between $\gamma_i(t)$ and $\gamma_j(t)$ is

$$\frac{(1 + 2u_iu_j)}{((1 + 2u_i^2)(1 + 2u_j^2))^{1/2}} \tag{8.10}$$

The closer the values of u_i and u_j are to one another, the higher the correlation between the responses of species i and j to environmental fluctuations. This is not surprising; if, e.g. two species have similar optimal temperatures, both should respond in a highly correlated way to fluctuations in temperature.

Table 8.3 gives two examples of correlation matrices for five species, with the values of u_i equally spaced and with u_3 for the middle species located at zero, the mean value of the environmental variable $E(t)$. In the first example (Table 8.3(a)), the species phenotypes are closely spaced, i.e. species are similar in how they respond to $E(t)$. As a result, the species growth rates are highly positively correlated (Table 8.3(b)). In the second case, the species

Table 8.3 Correlation matrices for the effects of environmental fluctuations $E(t)$ on the intrinsic rates of increase, $\gamma_i(t)$, for five species with values of u_i equaling (a) -0.2, -0.1, 0, 0.1, and 0.2 and (b) -2, -1, 0, 1, and 2.

Species	1	2	3	4	5
(a)					
1	1	0.99	0.96	0.91	0.85
2		1	0.99	0.96	0.91
3			1	0.99	0.96
4				1	0.99
5					1
(b)					
1	1	0.96	0.33	-0.58	-0.78
2		1	0.58	-0.33	-0.58
3			1	0.58	0.33
4				1	0.96
5					1

phenotypes are dissimilar with values of u_i spaced far apart. Their growth rates are less positively correlated than in the first case, and some species are negatively correlated. Thus, if two species have optima that are very different and are on opposite sides of the mean of the environmental variable, environmental fluctuations can drive negative correlations in population growth rates.

This simple example demonstrates that a single environmental factor can generate a range of positive and negative correlations among species responses to environmental fluctuations, even in the absence of interspecific competition. Moreover, the model suggests that environmentally driven correlations are probably not independent of species diversity, because the distribution of phenotypes in a community may depend on the number of species in that community. Thus, when correlations between species are not identical, eqn (8.7) becomes

$$CV_{sum} = \sigma_p \left(\frac{1 + (m-1)\bar{\rho}[m]}{m(1 - (1-r)^2)} \right)^{1/2} \quad (8.11)$$

where $\bar{\rho}[m]$ is the average of the pairwise correlations between species responses to environmental fluctuations. We have written $\bar{\rho}[m]$ as a function of m to emphasize that this average correlation could depend on the number of species in a community.

To demonstrate how assumptions about species–environment interactions may influence the diversity–stability relationship, consider five cases of the above model (Fig. 8.2). Each case differs in its assumption about how the added species differ in their responses to $E(t)$. In case (a) there are initially two species ($m = 2$) and the difference between species' optima ($u_1 - u_2$) equals 0.1. Species are then added such that the maximum difference between the optima of the species remains 0.1. In other words, the range of responses to $E(t)$ does not change with increasing numbers of species. In case (b), the difference between the optima of the two initial species ($u_1 - u_2$) also equals 0.1, but each additional species is added so that the spacing between all adjacent species ($u_i - u_{i+1}$) equals 0.1. In other words, each new species increases the total optima range by 0.1. Cases (c) and (d) are similar to (a) and (b), respectively, except that the initial difference between species is 1.0 rather than 0.1; therefore,

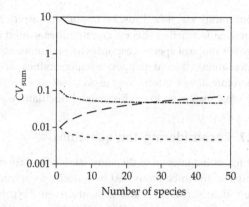

Figure 8.3 The diversity–stability relationship for the four models (IG&K, —; L&T, ······; H&R(a), – –; H&R(b), ----) using the eqn (8.8) and the parameter constraints given in Table 8.2. The remaining parameters used are $\rho_\varepsilon = 10$, $T = 1000$, $\rho = 0.2$, $r = 1$, and $K = 100$.

species are more dissimilar in their responses to the environment. Finally, case (e) gives the situation in which there is no correlation in species responses to environmental fluctuations ($\rho = 0$), i.e. when each species is influenced by different, independent environmental factors, rather than the single factor $E(t)$.

Figure 8.2(a) reveals that when increasing the number of species in a community does not increase the range of species' environmental optima, increasing the number of species either has little effect on CV_{sum} (case a) or actually increases CV_{sum} (case c). This pattern results because the average correlation between species' responses to environmental fluctuations ($\bar{\rho}[m]$) remains constant (case a) or increases with m (case c) when the range of species' optima remains fixed (Fig. 8.2(b)). Conversely, when the difference between adjacent species' optima remains fixed (cases b and d) so that the range of species' optima broadens with higher species diversity, increasing m decreases the average correlation between species' responses to environmental fluctuations. As a result, CV_{sum} decreases with increasing numbers of species in the community.

In sum, species diversity only increases stability (i.e. decreases CV_{sum}) if the additional species are different in how they respond to environmental fluctuations from those already present. The diversity of responses to environmental perturbations, rather than simply the number of species, influences community stability. Thus, an important empirical question is whether diverse communities exhibit a greater range of species responses to environmental fluctuations than depauperate communities and therefore have a lower average correlation between species' responses to environmental fluctuations.

8.7 Reconsidering the questions

Much of the interest in diversity–stability relationships seems to be motivated by a desire to predict how biodiversity loss and human-driven disturbances affect ecosystems. Our ability to apply theoretical results to natural communities rests on how well model assumptions reflect reality. Even if these models reflected reality perfectly, however, do they involve the type of stability, disturbance, and biodiversity in which we are interested? More specifically, do any of the models we have discussed actually address how changes in biodiversity and the rate of human-driven disturbance will affect the stability of ecosystem functioning in a manner that is ecologically or economically significant?

Earlier results demonstrated that similar studies could reach 'opposite' conclusions based on different definitions of stability (McCann 2000; Pimm 1984; Loreau et al., Chapter 7). For instance, the Lotka–Volterra models above predict that, in general, stability of community biomass increases with diversity, but that the stability of individual species decreases with diversity.

The question of which aspect of ecosystem stability is most relevant to understanding the consequences of biodiversity loss is inextricably linked to the question of whether current studies examine appropriate types of disturbances. It has long been established that the point of entry of random fluctuations influences the results of stochasticity on population growth models (e.g. Levins 1969; May 1973), and this conclusion is upheld for the diversity–stability models here. This result suggests that it matters if environmental fluctuations more directly affect a species' growth rate or its carrying capacity. In fact, the appropriate way to model environmental variability may be different for different types of disturbances. For instance, in some species, fluctuations in the availability of nesting sites might be best modelled as variation in carrying capacity, whereas fluctuations in food resources might be best modelled as variation in growth rates.

Similarly, the models in this chapter explore community responses to continuously occurring but serially uncorrelated fluctuations. With climate change and widespread pollution, however, perhaps it would be more appropriate to study the effects of long-term unidirectional changes such as temperature increases (Petchey et al. 1999) or large abrupt disturbances such as nutrient pulses (Cottingham and Schindler 2000; Vitousek et al. 1981). In these cases, one might be more interested in the magnitude of change of some ecosystem property rather than variability of the property over time. In sum, much remains to be uncovered about the nature of environmental variation and how species' experience this variation.

It is also not clear that using species richness as a measure of diversity is appropriate for understanding how biodiversity loss affects community stability. For instance, biomass variability is a product of the number of species weighted by their density, the biomass of an individual of that species, and their degree of independence in responses to environmental fluctuations. The models here address communities in which biomass is evenly distributed among species. Losing a species in a community of 100 species that is dominated by one species is not the same as losing a species in a community of 100 species that are evenly abundant. Thus, perhaps an evenness measure would be a more appropriate measure of species richness. The situation of uneven communities may be particularly important for understanding how invasive species affect community stability. Although the addition of a new species may increase species richness, it also often alters the relative composition of the community (Hobbs and Mooney 1998).

Measuring diversity as species richness also ignores the reality that species loss is non-random in natural communities. Some species will be more susceptible to extinction and may have traits that differentially affect stability. For instance, species with slow growth rates may be more susceptible to extinction; therefore, decreased diversity will change the distribution of growth rates. The models above demonstrated that variability in species' responses to environmental perturbations influence community stability. How the community responds to species extinction will depend in part on how the remaining species respond to environmental fluctuations compared to the species lost (e.g.• whether the species lost was in the centre or on the extreme of the range of species' phenotypes).

Finally, feedbacks between diversity and ecosystem function are notably absent from most diversity–stability theory. Most models change diversity, holding 'all else equal.' Yet environmental disturbances will directly affect the ecosystem properties that maintain community diversity. For example, nitrogen deposition will not only affect relative abundances of species, but also nutrient availability and the number of species that can coexist. To investigate these types of links, mechanistically based models are needed. Loreau (1998a) modelled how species richness was affected by the mobility of nutrients in a community of spatially separated plants. The mobility rate of the nutrient determined the degree of interaction between adjacent plants and thus also the number of species maintained in the community. Moreover,

environmental disturbances themselves may maintain species diversity and in turn, affect ecosystem functioning. For instance, Norberg et al. (2001) investigated the relationship between phenotypic diversity and community biomass in a community where all species compete equally for the same resource. In this study, the environmental fluctuations themselves allow phenotypic diversity to persist, and this diversity affects the average and variance of community biomass.

In conclusion, only through better understanding of the nature of disturbances that natural ecosystems face and the mechanisms that control the feedbacks between diversity and ecosystem function will we be able to predict the response of a natural community to disturbance. At this point, we cannot yet be complacent in the mathematical result that species diversity and stability are positively related.

We thank P. Inchausti, M. Loreau, S. Naeem, O. Petchey, and two anonymous reviewers for insightful comments and discussion.

CHAPTER 9

Biodiversity and stability in soil ecosystems: patterns, processes and the effects of disturbance

P. C. de Ruiter, B. Griffiths, and J. C. Moore

9.1 Soil biodiversity and soil processes

Soil harbours a large part of the world's biodiversity and governs processes that are regarded as globally important components in the cycling of materials, energy and nutrients (Wolters 1997; Griffiths *et al.* 2000). The taxonomy of soil organisms is however relatively poorly known and possibly the majority of the species is still to be identified (Mikola *et al.*, Chapter 15). Precise taxonomical descriptions of the soil biodiversity are hampered by technical difficulties in extracting and identifying the soil organisms and by the large variation in species richness among soil ecosystems (Lawton *et al.* 1996; Wall and Virginia 1999; Mikola *et al.*, Chapter 15). By far the most dominant groups of soil organisms, in terms of numbers and biomass, are the microbial organisms, i.e. bacteria and fungi (Andrén *et al.* 1990; Bloem *et al.* 1994). Besides the microbial organisms, soil ecosystems generally contain a large variety of faunal organisms, like protozoa (amoebae, flagellates, ciliates), nematodes (bacterivores, fungivores, omnivores, herbivores and predators), micro-arthropods such as mites (bacterivores, fungivores, predators) and collembola (fungivores and predators), enchytraeids and earthworms (Fig. 9.1).

Soil organisms are assumed to be directly responsible for soil ecosystem processes, especially the decomposition of soil organic matter and the cycling of nutrients (Wardle and Giller 1997). These processes are regarded as major components in the global cycling of materials, energy and nutrients. For example, arable soils may harbour 250 kg C up to 2500 kg C in living biomass in the topsoil layer (25 cm) per hectare (Table 9.1). This is equivalent to c. 500–5000 kg dry matter and 5000–50,000 kg fresh material. Given estimates of a biomass turn-over rate of once per year and an energy conversion efficiency of 50% (Hunt *et al.* 1987; de Ruiter *et al.* 1993b), the soil biomass process 10,000 up to 1,00,000 kg of fresh organic material each year. This processing includes the decomposition of dead organic matter by the microbes as well as the consumption and production rates in the soil community food web (Hunt *et al.* 1987; Moore *et al.* 1988; de Ruiter *et al.* 1994). With its large diversity and complexity the soil community has an impact on soil processes, and the way in which these processes may vary in time and space. As these processes also determine nutrient availability for plants to take up, the below-ground decomposer food web interactions also influence above-ground primary productivity and carbon sequestration (Wall and Moore 1999).

Several field and laboratory studies have been carried out to establish the role of the various groups of soil organisms in soil processes (Hendrix *et al.* 1986; Hunt *et al.* 1987; Brussaard *et al.* 1988, 1990; Moore *et al.* 1988; Andrén *et al.* 1990; Verhoef and Brussaard 1990; de Ruiter *et al.* 1993b) showing that microbes, because of their abundance, are the most important contributors to the soil processes. The faunal groups of organisms, however, are also considered to contribute considerably to the soil processes, despite their relatively low densities. For

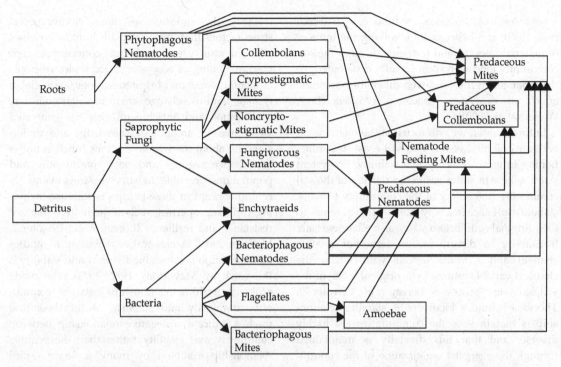

Figure 9.1 Diagram of the below-ground food web from the conventional farming system of the Lovinkhoeve experimental farm (de Ruiter *et al.* 1993a). Species are aggregated into functional groups, i.e. based on food choice and life-history parameters (Moore *et al.* 1988). Detritus refers to all dead organic material. Material flows to the detrital pool through the death rates and the excretion of waste products are not represented in the diagrams, but are taken into account in the material flow calculations and stability analyses.

example, in many soils microbes constituted more than 90% of the total soil biomass, while the contribution of the soil fauna to mineralization processes was estimated to be higher than 30%. Apart from their direct contributions, higher trophic level organisms may also indirectly contribute to soil processes by stimulating microbial activity, e.g. through reducing the microbial population sizes and hence enhancing the relative and absolute microbial growth rates (Coleman *et al.* 1978, 1983; Woods *et al.* 1982; Verhoef and Brussaard 1990), or through increasing the availability of limiting resources like oxygen or nutrients (Woods *et al.* 1982; Coleman *et al.* 1983; Mikola and Setälä 1998a; Scheu and Setälä 2001). Stimulation of microbial growth through increasing nutrient availability by the soil fauna links the above-ground and below-ground ecosystem compartments. The higher trophic levels in the below-ground decomposer food web enhance the availability of nutrients for microbial activity

and the decomposition of soil organic matter, as well as for plants, enhancing above-ground primary productivity and primary producer based food chain processes (Wall and Moore 1999; Hooper *et al.* 2000).

The range of biological soil processes depends on the activity of a variety of soil organisms. Some groups are exclusively responsible for particular processes in the N-cycle like nitrification and denitrification (Swift *et al.* 1998). Other processes such as the decomposition of soil organic matter depend on a large variety of soil organisms, as the ability to decompose the various organic compounds is related to the genetic properties and variation in the soil microbial community (Griffiths *et al.* 2000). A recent review on the possible relationship between soil biodiversity and soil ecosystem functioning has, however, not provided clear and unambiguous patterns (Mikola *et al.*, Chapter 15). In fact, experimental evidence indicates that soil ecosystems have a high level of functional 'redundancy' (Andrén *et al.* 1995), and hence traits involving the transfer of materials, energy and nutrients through the soil community food web seem broadly distributed, so that changes in the food web diversity have little effect on ecosystem processes (Wolters 1997; Wolters *et al.* 2000).

In this chapter, we will focus on the relationship between soil biodiversity and soil ecosystem functioning in terms of ecosystem stability. Ecosystem stability here includes *qualitative stability* of the soil community (the ability of the community to withstand disturbance in a way that all species return to their original equilibrium state), as well as *resistance* (sensitivity to disturbance in terms of relative change) and *resilience* (speed with which the changed variable returns to its original level) of the soil ecosystem processes (Loreau *et al.*, Chapter 7). The idea behind a biodiversity–stability relationship is that in soils the communities are highly diverse, and that this diversity is maintained through the energetic organization of the communities, e.g. in terms of pools and flows of energy, compartmentation in energy channels or patterned interaction strengths among the populations (Moore and Hunt 1988; de Ruiter *et al.* 1995; Moore and de Ruiter 1997). Disturbance is thought to affect this organization leading to reduced community stability and diversity and in consequence reduced stability of soil ecosystem processes. First, we will summarize observed patterns in soil community structure, and how these patterns relate to community stability. Then we highlight food web analyses as a way to connect soil community structure to community stability and present experimental evidence of disturbance effects on soil biodiversity and the stability of soil processes. Finally, we discuss these findings in the context of the relationship between biodiversity and stability in ecosystems (Loreau *et al.*, Chapter 7).

9.2 Patterns in community structure, community stability, and the effects of disturbance

The increasing awareness of environmental problems has directed much ecological research towards the question of how environmental stress-factors that are caused by human activities alter the structure of ecological communities and the functioning of ecosystems. Examples of environmental issues that inspired ecologists are global change, land-use change, environmental pollution, acidification and enrichment. This has generated an impressive amount of knowledge and understanding about the way ecosystems function under disturbance regimes and how communities and populations are able to survive stress-events. A central concept in these studies is stability, including various properties, such as qualitative stability, resistance and resilience (Loreau *et al.*, Chapter 7; McCann 2000). A classical contribution to studies on the relation between biodiversity and stability is the work by May (May 1972, 1973) who made ecologists rethink relationships between community complexity and stability, as his theoretical models indicate a negative relationship between complexity and stability rather than the positive relationship assumed by many ecologists since then (MacArthur 1955; Elton 1927). The models by May encouraged many theoretical and empirical studies on complexity, patterns in community structure and stability (Pimm and Lawton 1977; Cohen 1978; Paine 1980, 1992; Pimm *et al.* 1981; Yodzis 1981; McCann *et al.* 1988; Moore and Hunt 1988; Polis 1991; DeAngelis 1992; de Ruiter *et al.* 1995; Neutel *et al.* 2002). All these studies have provided indications that in real ecosystems, the structure of communities is organised in patterns that are important to community stability. These patterns may refer to properties of food web structure such as the rarity of omnivory and cycles (Pimm *et al.* 1981), the lengths of food chains (Pimm and Lawton 1977) and compartmentation (Pimm and Lawton 1977; Moore *et al.* 1993), as well as to patterns in the interaction strengths among the populations such as a few strong links embedded in a majority of weak links (Paine 1992) and the specific organization of interaction strengths over trophic levels and trophic loops (McCann *et al.* 1988; de Ruiter *et al.* 1995; Neutel *et al.* 2002). This has led to the argument that patterns in natural communities are critical to the sustainable preservation of biological diversity (Pimm 1981; Polis 1998).

9.3 Food web analysis as a way to connect community structure to processes and community stability

While in the above-ground ecosystem compartment biodiversity and ecosystem processes are importantly influenced by intraspecific and interspecific competition and habitat exploitation (Naeem *et al.* 1994a; Tilman *et al.* 1996), in soils the relationship between biodiversity and soil processes is thought to be primarily controlled by dynamics and interactions in the soil community food web (Moore and de Ruiter 1997; de Ruiter *et al.* 1998; Wall and Moore 1999). Food web (i.e trophic) interactions affect the distribution and abundance of organisms in fundamental ways, since the dynamics of populations are largely a function of energy gains and losses derived from predation by and on other populations. Food webs therefore provide a way to analyse the dynamics of populations in the context of the community structure as a whole. Since trophic interactions represent transfer rates of energy and matter, overall processes, such as the cycling of energy and nutrients, and food web structure are deeply interrelated (Hunt *et al.* 1987; Moore *et al.* 1988, 1993; DeAngelis 1992; Hairston and Hairston 1993; de Ruiter *et al.* 1994, 1995). Food webs therefore also provide a way to connect the dynamics of populations to the dynamics in pathways within the cycling of matter, energy and nutrients (DeAngelis 1992; de Ruiter *et al.* 1994). Hence, by means of the population dynamical descriptions of the trophic interactions in food webs, we can analyse the stability and disturbance effects on community structure as on ecosystem processes.

The aim to connect community structure to ecosystem processes, and the incomplete taxonomic descriptions of the soil biodiversity, has led to the functional group approach in soil food web research (Moore *et al.* 1988). Within broad taxonomical units (bacteria, fungi, protozoa, nematodes, micro-arthropods), functional groups are defined on the basis of similar food choice, predators, and physiological properties such as energy conversion efficiencies, and specific growth and death rates. The functional groups approach therefore implicitly assumes that species within groups respond

similarly to changes in community structure and environmental conditions. The functional group approach aims to describe community and processes in the same 'currencies', i.e. in terms of material pools (e.g. population sizes in $kg\,C\,ha^{-1}$) and material flows (e.g. feeding rates in $kg\,C\,ha^{-1}\,yr^{-1}$). Each functional group then represents a specific component in the cycling of materials, energy and nutrients. Effects of disturbances on particular groups can then be translated into effects in terms of altered contributions to ecosystem processes. This approach implies that diversity is defined 'functionally', in terms of the diversity in functional groups and the diversity in pathways in the cycling of materials, energy or nutrients.

9.4 Patterns and stability in soil food webs

During the last decades several large multidisciplinary research programs have been carried out directed to the development of sustainable agricultural management practices (Hendrix *et al.* 1986; Hunt *et al.* 1987; Brussaard *et al.* 1988, 1990; Moore *et al.* 1988; Andrén *et al.* 1990; de Ruiter *et al.* 1993b; Zwart *et al.* 1994). The outcome of these field programmes provided estimates of the population sizes for all functional groups, making it possible to construct qualitative (Fig. 9.1) and quantitative (Table 9.1) descriptions of the soil food web structures and relate these structures to soil organic matter decomposition and the mineralization of nutrients.

Based on observed population sizes (Table 9.1) and on literature values regarding death rates and energy conversion efficiencies, the feeding rates among the functional groups in the soil food webs were estimated through food web modelling (O'Neill 1969; Hunt *et al.* 1987; de Ruiter *et al.* 1993b), (Box 9.1(*a*)). The outcome of these calculations indicated that the feeding rates in the food webs show similar patterns as the population sizes (Table 9.1) in the form of trophic pyramids, i.e. decreasing population sizes and feeding rates with increasing trophic level (Fig. 9.2(a)). The modelled feeding rates have been verified at the level of overall carbon mineralization, indicating that the calculated rates were close to the observations

Table 9.1 Biomass estimates (kg C ha^{-1}) for the functional groups in the different food webs. Values refer to the 0–25 cm depth layer, except for the Horseshoe Bend webs (0–15 cm). CPER: a short grass prairie from the Central Plains Experimental Range, Colorado, USA (Hunt *et al.* 1987; Moore *et al.* 1988). LH: Lovinkhoeve Experimental Farm (Marknesse, The Netherlands); CF: conventional farming; IF: integrated farming; integrated farming differs from conventional farming with respect to the more frequent use of organic manure instead of inorganic fertilizer, reduced use of pesticides and reduced soil tillage (Brussaard *et al.* 1988, 1990; de Ruiter *et al.* 1993; Zwart *et al.* 1994). HSB: Horseshoe Bend Research Site, Georgia, USA; CT: conventional tillage; NT: no tillage (Hendrix *et al.* 1986, 1987). KS: Kjettslinge Experimental Field, Uppsala, Sweden; B0: barley without nitrogen fertilizer; B120: barley with fertilizer (Andrén *et al.* 1990)

	CPER	LH-IF	LH-CF	HSB-NT	HSB-CT	KS-B0	KS-B120
Microbes							
Bacteria	304	245	228	440	690	740	900
Fungi	63	3.27	2.12	160	150	1500	2300
VAM	7						
Protozoa							
Amoebae	3.78	18.9	11.5	40^2	50^2	110^2	34^2
Flagellates	0.16	0.63	0.53				
Nematodes							
Herbivores	2.90	0.35	0.19	0.40	0.50	0.18	0.29
Bacteriovores	5.80	0.36	0.30	0.46	1.40	0.45	0.50
Fungivores	0.41	0.13	0.08	0.12	0.08	0.20	0.12
Predators[1]	1.08	0.06	0.06			0.44	0.44
Arthropods							
Herbivorous Herbage Arthropods						0.10	0.14
Predatory Herbage Arthropods						0.15	0.19
Herbivorous Macro-arthropods						0.19	0.19
Microbivorous Macro-arthropods						0.25	0.25
Predatory Macro-arthropods						0.49	0.49
Predatory Mites	0.16	0.08	0.06	0.20^3	0.04^3	0.18^4	0.28^4
Nematophagous Mites	0.16	0.006	0.004				
Cryptostigmatic Mites	1.68	0.003	0.007	0.80	0.22		
Non-Cryptostigmatic Mites	1.36	0.04	0.02	0.90	0.39		
Bacteriovorous Mites	0.0003	0.001					
Fungivorous Collembola	0.46	0.38	0.47	0.30	0.09	0.17^5	0.17^5
Predatory Collembola		0.008	0.03				
Annelids							
Enchytraeids		0.21	0.43	0.10	0.30	4.20	3.40
Earthworms		63.5	—	100	20	13	13

[1] Including predators and omnivores.
[2] including amoebae and flagellates.
[3] including all predatory arthropods.
[4] including all predatory micro-arthropods.
[5] including all microbial feeding micro-arthropods.

(de Ruiter *et al.* 1993b). Hence, the structure of the soil food webs includes patterns that are common to most ecosystems: energy pools (population sizes) and the energy flows (feeding rates) form a pyramidal structure along trophic level (Odum 1963).

The strengths of the interactions, i.e. the effects of the populations upon each other, are regarded of central importance to food web stability (May 1972; Yodzis 1981; McCann *et al.* 1988; Paine 1992; de Ruiter *et al.* 1995). Interaction strengths can be mathematically defined as the entries of the

Box 9.1 Methods of calculation

(a) Modelling energy flow rates: feeding rates
Feeding rates are derived from population sizes and data on death rates and energy conversion efficiencies. The basic assumption underlying the calculation of feeding rates is that the annual (equilibrium) feeding rates should balance the annual death rate through natural death and predation (Hunt *et al.* 1987; O'Neill 1969):

$$F_j = \frac{d_j B_j + M_j}{a_j p_j} \tag{9.1}$$

where F_j is the feeding rate (kg C ha^{-1} yr^{-1}), d_j is the specific death rate (yr^{-1}), B_j is the average annual (equilibrium) population size (kg C ha^{-1}), M_j is the death rate due to predation (kg C ha^{-1} yr^{-1}), a_j is the assimilation efficiency, and p_j is the production efficiency.

For polyphagous predators, the feeding rate per prey type (F_{ij}) is based on the relative abundances of the prey types and on prey preference:

$$F_{ij} = \frac{w_{ij} B_i}{\sum_{k=1}^{n} w_{kj} B_k} F_j \tag{9.2}$$

where F_{ij} is the feeding rate by predator j on prey i, w_{ij} is the preference of predator j for prey i over its other prey types. The calculations of feeding rates start with the top predators, which suffer only from natural death, and proceeded working backwards to the lowest trophic levels.

(b) Modelling interaction strengths and stability
Interaction strengths are the entries of the community matrices (May 1972, 1973), referring to the per capita—in this case per biomass—effects upon one another in equilibrium. The interactions strengths are derived from the population sizes and energy flow rates by assuming Lotka–Volterra equations for the dynamics of the functional groups:

$$\dot{X}_i = X_i \left[b_i + \sum_{j=1}^{n} c_{ij} X_j \right] \tag{9.3}$$

where X_i and X_j represent the population sizes of group i and j, respectively, b_i is specific rate of increase or

decrease of group i, and c_{ij} is the coefficient of interaction between group i and group j. The matrix elements (α_{ij}) are defined as the partial derivatives near equilibrium: $\alpha_{ij} = (\partial \dot{X}_i / \partial X_j)^*$. Values for the interaction strengths are derived from the equilibrium descriptions by equating the death rate of group i due to predation by group j in equilibrium, $c_{ij} X_i^* X_j^*$, to the average annual feeding rate, F_{ij} (eqn 9.1) and the production rate of group j due to feeding on group i, $c_{ji} X_j^* X_i^*$, to $a_j p_j F_{ij}$ (de Ruiter *et al.* 1995). With equilibrium population sizes, X_i^*, X_j^*, assumed to be equal to the observed annual average population sizes, B_i, B_j, the effect of predator j on prey i is

$$\alpha_{ij} = c_{ij} X_i^* = -\frac{F_{ij}}{B_j} \tag{9.4}$$

and the effect of prey i on predator j is

$$\alpha_{ji} = c_{ji} X_j^* = \frac{a_j p_j F_{ij}}{B_i} \tag{9.5}$$

(c) Evaluating community stability
The stability of the matrices is established by evaluating the eigenvalues of the community matrices; when all real parts are negative the matrix is stable, and the food web is considered to be locally stable (May 1973).

In the stability analysis of the seven soil food webs (Fig. 9.3), the matrix element values in the 'real' matrices are sampled randomly from the uniform distribution with intervals [0, $2\alpha_{ij}$], in which α_{ij} is the value as derived from the observations (eqs (9.4) and (9.5)). Elements referring to the feedbacks to detritus are derived in the same way as the trophic interactions. The diagonal matrix elements referring to intragroup interference are set at three levels of magnitudes (s_i) proportional to the specific death rates (d_i) with $s_i = 1.0$, 0.1 and 0.01 for all groups equally (de Ruiter *et al.* 1995). In the disturbed matrices, the non-zero pairs of values of the matrix elements are randomly permuted (Yodzis 1981). This method leaves food web structure (placing of the non-zero terms) and the logical paring of the elements unchanged. The comparison is based on 1000 runs.

community matrices (May 1973), to be derived from population sizes and feeding rates (Box 9.1(*b*)). Estimates of the interaction strengths obtained this way for the soil food webs also revealed patterns along trophic position (Fig. 9.2(b)), but different from the population sizes and feeding rates, as the

patterning is characterized by relatively strong top-down effects at the lower trophic levels and relatively strong bottom-up effects at the higher trophic levels (Fig. 9.2(b)).

The patterns of interaction strengths in the soil food webs were found to be important to the

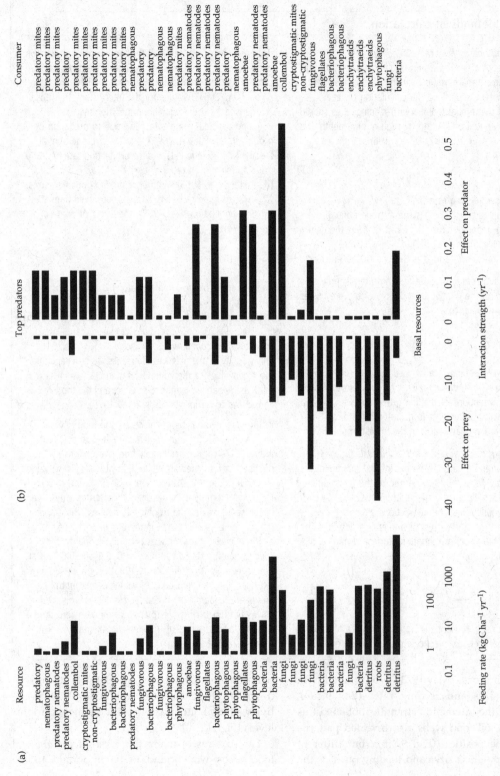

Figure 9.2 Pattern in feeding rates and interaction strengths along trophic level in the food web of the Lovinkhoeve experimental farm (conventional farming), serving as a representative example for seven food webs (Table 9.1), (de Ruiter et al. 1995; de Ruiter et al. 1998). (a) Feeding rates (kg C ha^{-1} yr^{-1}), (b) Interaction strengths (yr^{-1}). A sensitivity analysis was carried out to evaluate the effects of variation in the input parameters (as given in Table 9.1) showing effects on the estimates of the interaction strengths, but in terms of the stabilizing effect of the patterning these effects were relatively small (see legend Fig. 9.3).

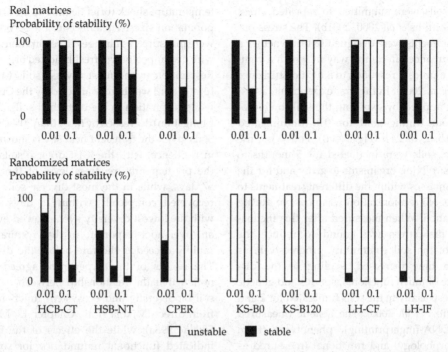

Figure 9.3 The effect of disturbing the patterning of interaction strength on the stability of seven food webs (de Ruiter *et al.* 1995). The black fraction in the bars denotes the percentage of stable matrices based on 1000 runs (Box 9.1(*c*)). In the real matrices, the element values are sampled randomly from the uniform distributions [0, $2\alpha_{ij}$], in which α_{ij} is the value as derived from the observations (Box 9.1(*b*)). The diagonal matrix elements referring to intragroup interference are set at three levels of magnitudes (s_i) proportional to the specific death rates (d_i) with $s_i = 1.0$, 0.1 and 0.01 for all groups equally. In the disturbed matrices, the non-zero pairs of values of the matrix elements are randomly permuted (Yodzis 1981). Abbreviations denoting the different soil ecosystems are given in the legend of Table 9.1. A sensitivity analysis (Neutel *et al.* 2002) showed that variation in the parameter values within intervals between half and twice the observed value (Table 9.1) led to variation in interaction strengths and in the probability of matrix stability, but this variation was relatively small compared to the difference in stability probability between the real and the random matrices, indicating the robustness of the stability analyses (Neutel *et al.* 2002).

community stability. A comparison was made between the stability of community matrix representations of the soil food webs using the empirically based values of interaction strengths ('real' matrices) and that of matrices in which these values were randomized (Yodzis 1981), (Box 9.1(*c*)). Stability of the community matrices was established by evaluating the signs of eigenvalues of the matrices; when all real parts were negative the matrix is stable, and the food web was considered to be locally stable (Box 9.1(*c*)). This approach follows the definition of local stability analysis (May 1973). The comparison showed that matrices including the realistic patterns of interaction strengths were far more likely to be stable than their randomized counterparts (Fig. 9.3). This result indicates that energetic organization and community stability

are inextricably interrelated, as the stabilizing patterns of the interaction strengths were the direct results of patterns in population sizes and feeding rates (Box 9.1(*a*)). Disturbing the patterning of interaction strengths (Fig. 9.3) or the energetic properties of the functional groups such as population size distributions (de Ruiter *et al.* 1995) caused a loss in community stability.

9.5 Experimental evidence of environmental effects on soil biodiversity and process stability

To obtain empirical information of the relationship between disturbance, biodiversity and stability in soils ecosystems, a European consortium (MICRODIVERS) carried out laboratory experiments in which soils were submitted to repeated stress regimes (Griffiths *et al.* 2000, 2001b). The stress-on-stress experiments were designed to test whether a (first) disturbance affects the way in which a community or a process responds to a next disturbance (Griffiths *et al.* 2000). In the first experiment, a first stress was applied by exposing the soil to chloroform vapour (fumigation) for 0 h (unfumigated control), 0.5, 2, or 24 h (Fig. 9.4(a)). Following the fumigation, soils were incubated for 5 months to allow the surviving organisms to grow and for the overall biomasses within the different treatments to equilibrate. No recolonization was possible during this incubation. When measured after this incubation, the first stress was found to reduce the diversity of the soil community progressively as fumigation time increased, leading to the disappearance of much functional, species and genetic variation, especially in the soils fumigated for 2 and 24 h (Griffiths *et al.* 2000). The genetic (measured through DNA-fingerprinting), phenotypic (from colony morphology) and functional (measured as ability to utilize sole carbon sources) diversity of the bacterial community decreased, as did the biodiversity (number of trophic groups, phyla within trophic groups, and taxa within phyla) of the microfauna (protozoa and nematodes; micro-arthropods not measured). Overall there was a 60% reduction in biodiversity. The fumigation also affected soil ecosystem processes, but to a more limited degree: many species disappeared but no ecosystem process was eliminated. Some process rates increased, e.g. microbial growth with a 30% increase in thymidine incorporation and a 10% increase in the decomposition rate of added plant residues, while other process rates decreased, e.g. nitrification by 90%, denitrification by 70% and methane oxidation by 95%. The results of the first stress therefore indicate a level of functional redundancy: although many groups of organisms disappeared, ecosystem processes were still able to continue, especially the process of soil organic matter decomposition in which a high diversity of decomposers is involved (Griffiths *et al.* 2000).

The second stress was applied as either a persistent disturbance by adding a heavy metal (copper in the form of $CuSO_4$), which reduced growth rates, or a transient disturbance, by applying a temperature shock (brief heating to 40 °C) reducing population sizes. The effects of the second stresses were measured as changes in respiration associated with decomposition of freshly added organic matter. Respiration in the most diverse soils (0 and 0.5 h fumigation) was hardly affected by the Cu addition, while respiration in the disturbed soils (2 and 24 h fumigation) decreased by more than 70% (Fig. 9.4(b)). Soils given the transient heat stress showed a trend in resilience, with the least diverse soils regained the pre-temperature-stress level of function after 57 days, while in the most diverse soils processes recovered completely within 15 days. The soils with the lowest diversity even showed evidence of an 'overshoot-response', in that respiration rates rapidly exceeded the rate before the disturbance. The results of this experiment agreed with the notion that the relationship between disturbance, soil biodiversity and ecosystem functioning is far from direct (Mikola *et al.*, Chapter 15; Mikola and Setälä 1998a): while the effects of the first stress indicated functional redundancy for soil organic matter decomposition and respiration, the effects of the second stress showed that the stability of these processes was reduced. Hence, there was loss of redundancy after the first stress; such losses of redundancy may be highly relevant for evaluating recovery in the course of multiple stress events.

These experimental results cannot be interpreted as direct effects of reduced biodiversity on process stability, however, as it might have been the disturbance that reduced both biodiversity and process stability. A second experiment was carried out to separate disturbance effects from biodiversity effects (Griffiths *et al.* 2001b). Basically, the experimental set-up was the same as the first experiment (Fig. 9.4(a)), but the first stress was now applied by inoculating sterile soils with serially diluted soil suspensions prepared from the parent soil (Griffiths *et al.* 2001b). Dilution factors were 10^0 (control) 10^2, 10^4, and 10^6 and there was a similar period for populations to recover from the first stress as in the first experiment. Hence, a comparison could be made among soils that were all disturbed, but differed with respect to biodiversity depending on the dilution factor. A possible 'hidden treatment' in such a dilution approach might be that dilution had differential effects on body size, because larger

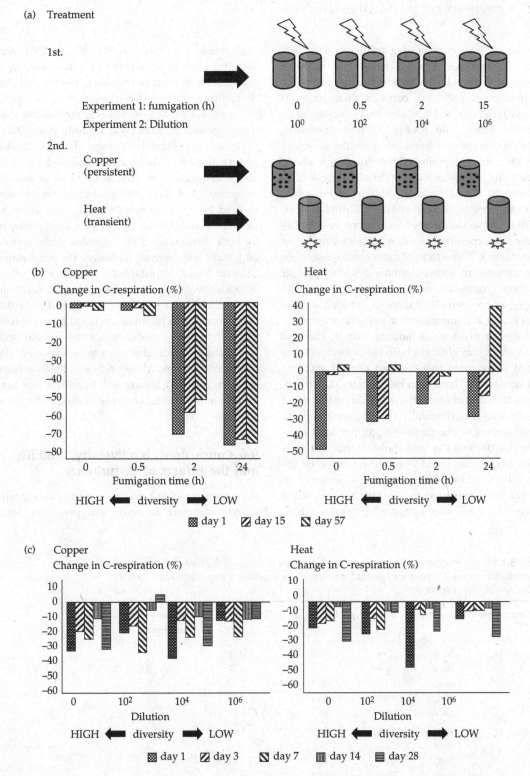

Figure 9.4 Design (a) and results (b, c) of two stress-on-stress experiments (Griffiths *et al.* 2000, 2001b). In the first experiment, the first stressor is soil fumigation of various duration, and in the second experiment it is inoculating sterile soils with serially diluted soil suspensions from the parent soil. In both experiments, the second stressors are a persistent (Cu addition) or a transient (heat-shock) disturbance. (b) and (c) the effects are measured in terms of relative change in carbon respiration measured 1, 15 and 57 days (experiment 1) or after 1, 3, 7, 14 and 28 days (experiment 2).

species are less common than smaller species, and that the variance between treatment replicates would be expected to be higher at higher dilutions. In practice, this did not occur as there were no trends observed of body size or variability with dilution factor. The results of this experiment showed that the first stress led to similar effects as in the first experiment, with progressively decreasing biodiversity (bacterial, fungal and protozoan) with increasing dilution factor, while process rates were less affected (Griffiths *et al.* 2001b). The second stress factors were the same as in the first experiment: either a copper treatment or a heat shock. The effects of the second stressors in this experiment showed similar responses in all dilution treatments, with the strengths of the responses comparable to those in the least diverse soils (2 and 24 h fumigation) in the first experiment. These results therefore indicate that in the first experiment it might have been the stress itself (the initial fumigation) that reduced process stability, not necessarily changes in biodiversity. The hypothesis that disturbance affects stability has further been tested experimentally by subjecting soils to combinations of the persistent (copper) and transient (heat) stresses as used above. In this case, soils from two management regimes, intensive or organic horticulture as described by (Griffiths *et al.* 2001a), were subjected to a copper or heat stress followed a week later by the other type of stress.

The results of this experiment showed that decomposition of organic matter was less stable in soils that had been previously stressed, but that destabilization depended on management regime: the organically managed soil was more stable than the intensively managed soil (Griffiths *et al.* 2001a).

These experimental findings all indicate that disturbance affects the structure and biodiversity of soil communities and the stability of ecosystem processes. The first stress-on-stress experiment showed effects on process stability that might be caused by disturbance, by reduced biodiversity or by both disturbance and biodiversity; the second and third experiment indicated the importance of disturbance. To establish a biodiversity effect, separated from any disturbance effect, observations are required on a series of undisturbed soils that vary (strongly) in biodiversity. It has however been difficult to find suitable series, as natural soils that differ in biodiversity are to be expected also to differ in organic matter richness, heterogeneity, soil structure and texture and humidity, that may all interfere with the analyses of biodiversity and stability.

9.6 Conclusions: biodiversity, stability and the effects of disturbance

Food webs provide a way to explicitly relate community structure to ecosystem processes, since

Table 9.2 The decomposition of grass residues added to intensively or organically managed soils receiving combinations of persistent (copper) and transient (heat) stresses. Decomposition was measured as the liberation of CO_2 over 24 h following addition

Soil management	Pre-stress	Stress	$\mu g CO_2\text{-}C\,g^{-1}d^{-1}$	Standard error	% change from unstressed
Intensive	None	None	45.6	0.68	
	None	Cu	41.5	1.69	−9
	Heat	Cu	38.4	0.97	−16
	None	Heat	46.4	1.36	2
	Cu	Heat	36.1	1.08	−20
Organic	None	None	50.2	1.01	
	None	Cu	51.6	1.01	3
	Heat	Cu	49.4	1.03	−5
	None	Heat	53.8	1.26	7
	Cu	Heat	53.6	3.32	7

food web interactions represent transfer rates that participate in the cycling of materials, energy and nutrients. Such a direct relation is important for untangling and understanding the complex relationship between biodiversity and ecosystem functioning. This is especially the case when looking at properties such as community stability and ecosystem process stability (Loreau *et al.*, Chapter 7). The results show that in soils, communities are energetically organized in a way that is important to stability, since energetic properties, such as population size distributions, energy conversion efficiencies and growth and death rates (Table 9.1), determined the stabilising patterns of the interaction strengths (Fig. 9.2(b)). The results also indicate that biodiversity itself does not necessarily influence stability, but that it might also be the environmental stress and disturbance that causes this effect. The stabilizing energetic set-up of the soil communities results from properties that seem common to many ecosystems, such as the trophic biomass pyramids in combination with omnivorous interactions (Neutel *et al.* 2002). Therefore, the present results might also apply to other kinds of ecosystems.

Although disturbance seems to affect both community stability and process stability, we should be careful in linking these two *variables of interest* (Loreau *et al.*, Chapter 7). For the soil communities it was the qualitative stability that was assessed, while for process stability the analyses focused on resistance and resilience. Also, process stability was assessed in soils in which community structures were strongly altered and biodiversity was reduced. On the other hand, there should be some link between community stability and process stability since population dynamics and interactions in food webs represent material flow rates and hence components in ecosystem processes. Relating community stability to process stability, and identifying key-properties in natural ecosystems that are critical to both community and process stability, should be focus of future research. This may provide the necessary scientific basis for understanding biodiversity–ecosystem functioning relationships and the adverse effects and risks of human activities on natural ecosystems on which decisions can be made about how best to treat our natural environment (Polis 1998).

CHAPTER 10

Neighbourhood scale effects of species diversity on biological invasions and their relationship to community patterns

J. M. Levine, T. Kennedy, and S. Naeem

10.1 Introduction

Concern over the consequences of species loss has generated tremendous interest in how species diversity affects ecosystem processes (Naeem *et al.* 1994; Tilman and Downing 1994; Chapin *et al.* 1998; Hector *et al.* 1999). Resistance to biological invasions has been a process of particular interest because exotic species are well documented to have large ecological and economic impacts (Mooney and Drake 1986; Drake and Mooney 1989; Vitousek *et al.* 1997). Furthermore, that diverse communities better resist biological invasions is a classic notion in ecology, credited to Elton (1958), but also implicit in island biogeography (MacArthur and Wilson 1967) and species packing (MacArthur 1970, 1972) theories. Recently, however, this commonly held notion has become controversial (Enserink 1999).

In this chapter, we synthesize recent theoretical, observational and experimental evidence for the hypothesis that species diversity enhances the resistance of ecological communities to biological invasions. We demonstrate an apparent contradiction between the results of the theoretical and experimental studies as compared to the observational results, but then resolve this apparent contradiction with a synthetic framework. We then present case studies from our own work that support the hypotheses implicit within our framework. First, from work at Cedar Creek Natural History Area, Minnesota, we show how species diversity

enhances resistance to biological invasions at the scale of species interactions or the neighbourhood. Second, from work along the Eel River in Northern California, we show how natural patterns of diversity and invasion are driven by factors correlated with diversity across communities, overwhelming the neighbourhood scale processes. We conclude by outlining promising areas for future research including underlying mechanisms, the role of spatial scale, the effects of diversity in multitrophic systems, and effects on exotic species impact.

10.2 Literature review

The hypothesis that species diversity enhances community resistance to biological invasions is generally credited to Charles Elton. In his book, *The ecology of invasions by animals and plants*, Elton (1958) argued that 'the balance of relatively simple communities of plants and animals is more easily upset than that of richer ones; that is more subject to destructive oscillations in populations...and more vulnerable to invasions.' Although nearly all papers on the relationship between species diversity and biological invasions cite this work, Elton never clearly defined an underlying mechanism. Rather, Elton based his conclusion on six independent lines of evidence including the fact that islands and agricultural systems, both typically depauperate communities, seemed highly susceptible to invasion. He also cited microcosm and modelling work that demonstrated how unstable simple communities were more likely to be invaded. Although lacking in mechanism, Elton's hypothesis was clearly influenced by the notion that diverse communities posed considerable challenges for invasive species in part because of the mélange of biotic interactions newly arriving species would have to face.

The mechanism invoked in most studies of diversity effects on invasions actually follows from the species packing models of MacArthur and colleagues (MacArthur 1970, 1972; May and MacArthur 1972). In these models, the more species present along a resource axis, the more thoroughly resources are utilized, and thus it is more difficult for new species to invade. MacArthur and colleagues were concerned primarily with processes that governed community structure rather than invasive species in the sense of exotics or weeds, but the principles they developed applied equally to any species attempting to gain a foothold in an established community. Although this mechanism comes with a number of caveats, the more general notion that diverse communities more thoroughly use available resources is analogous to the complementarity mechanism (Chapin et al. 1998; Naeem et al. 2000b).

10.2.1 Theoretical support

Recent models (Robinson and Valentine 1979; Post and Pimm 1983; Shigesada et al. 1984; Drake 1988, 1990; Case 1990, 1991; Law and Morton 1996; Kokkoris et al. 1999) consistently support the notion that diverse communities better resist biological invasions. This result is independent of whether communities are composed of species competing via interference (Shigesada et al. 1984) or exploitation (Case 1990, 1991), or if these species are members of a single or multiple trophic levels (Post and Pimm 1983; Drake 1988, 1990). We briefly review these studies below, but it is important to note that the evidence is much more equivocal if one examines the entire species packing/limiting similarity literature.

Robinson and Valentine (1979) and Case (1990, 1991) examined the invasibility of stable Lotka–Voltera communities varying in diversity. Although their models differed in how the elements of the community matrix were assigned, both found that the proportion of communities that could be invaded declined with diversity. In the assembly models of Post and Pimm (1983), Drake (1988, 1990), Law and Morton (1996), and Kokkoris et al. (1999) species were randomly drawn from a pool and 'introduced' to initially depauperate communities. Invasion was considered successful if the invader increased when rare, and maintained positive equilibrial abundance. Although these models differed in their invasion criteria, they all indicated that the rate of successful invasions declined as the communities accumulated more species. In a related result, Law and Morton (1996) found that increasing the richness of the species pool, independent of local diversity, enhanced community resistance to invasions.

Although the same general result emerges from all these models, this must be qualified with the fact that these studies are all Lotka–Volterra models (Levine and D'Antonio 1999). Recently, Moore et al. (2001) examined the effects of diversity on invasions in model systems where coexistence occurs via a range of alternative mechanisms including niche differentiation, resource heterogeneity, and/or recruitment limitation. They found that the mechanism of species coexistence determined whether diversity enhanced invasion resistance. Thus, understanding the underlying processes that maintain diversity was essential to understanding the relationship between species diversity and invasions.

10.2.2 Experimental invasions in controlled communities

Published experiments that directly manipulate species diversity also support the hypothesis that diversity enhances invasion resistance (Table 10.1). In these experiments, most of which have been published within the last two years, communities are randomly assembled from a pool of species such that species composition does not vary across the different levels of diversity. After establishment, the communities are invaded experimentally or naturally by species in the surrounding habitats (Knops et al. 1999; Hector et al. 2001). Because of the high degree of manipulation involved with these studies, the systems are typically small and highly

Table 10.1 Invasions into experimentally assembled communities

Study	System investigated
Diversity enhances invasion resistance in experimentally assembled communities	
McGrady-Steed *et al.* 1997	Protist microcosm
Knops *et al.* 1999[a]	Minnesota prairie plants grown in field plots
Stachowicz *et al.* 1999	Marine invertebrate fouling communities
Symstad 2000[a]	Minnesota prairie plants in greenhouse pots
Levine 2000a, 2001	Riparian plants growing on habitat islands in California
Van der Putten *et al.* 2000	European grassland succession
Prieur-Richard *et al.* 2000	Annual plants grown in field plots in Southern France
Naeem *et al.* 2000[a]	Minnesota prairie plants in greenhouse pots and in the field
Dukes 2001	California grassland plants in outdoor microcosms
Lyons and Schwartz 2001	California annual plants in field plots
Hector *et al.* 2001	Grassland plots in Southern Britain
Diversity reduces invasion resistance in experimentally assembled communities	
Robinson and Dickerson 1984 (as reanalysed in Levine and D'Antonio 1999)	Protist microcosm
Palmer and Maurer 1998	Agricultural fields with one or five species

[a] All conducted with plants from the same system.

controlled. For example, the systems include microbial microcosms (Robinson and Dickerson 1984; McGrady-Steed *et al.* 1997), marine invertebrates on fouling plates (Stachowicz *et al.* 1999), and plants in greenhouses (Symstad 2000; Naeem *et al.* 2000b), outdoor microcosms (Dukes 2001), and homogenized field plots (Knops *et al.* 1999; Prieur-Richard *et al.* 2000).

Eleven of the thirteen experimental studies that have directly manipulated species diversity have found that increasing diversity enhances invasion resistance. One study (Crawley 1999), with a different design than those in Table 10.1, found no effect of diversity. Overall, these results suggest that species loss, if it reduces diversity at these smaller spatial scales, will increase exotic species invasions. Indeed, Lyons and Schwartz (2001) found that reducing diversity through the removal of rare species decreased the invasion resistance of a California annual plant system. One important caveat with the experimental studies is that several invade communities with species native to the habitat (e.g. Knops *et al.* 1999). This is similar to many of the models where communities are invaded by species drawn from the same pool as the residents (e.g. Drake 1990). The effects of diversity on invaders qualitatively different than the residents

of the communities they invade remain poorly understood (Levine and D'Antonio 1999).

10.2.3 Natural patterns of diversity and invasion

Despite the strong support from experimental studies for the hypothesis that diversity enhances invasion resistance, these studies have been criticized for their small spatial scale and relatively homogeneous environmental conditions (Enserink 1999; Stohlgren *et al.* 1999). Invasions into natural communities, it is argued, are influenced by a whole host of factors other than local species diversity. These criticisms have been particularly pronounced because 'uncontrolled' studies that examine natural variation in diversity and invasions within or across entire communities tend to find that the most diverse communities are also the most invaded (Table 10.2). A similar result has been found comparing plant communities world-wide (Lonsdale 1999). Furthermore, some of the studies reporting negative correlations between diversity and invasion resistance (Woods 1993; Pysek and Pysek 1995; Morgan 1998) are likely to reflect the negative impacts of exotic species invasion on

Table 10.2 Studies examining natural invasion patterns or experimental invasions in communities varying naturally in diversity

Study	System investigated
Positive correlation between diversity and invasion	
Pickard 1984	Plant communities on the Lord Howe Islands
Fox and Fox 1986 (as reanalysed in Levine and D'Antonio 1999)	Heathlands and shrublands on Australian reserves
Kruger *et al.* 1989	Mediterranean plant systems in California
MacDonald *et al.* 1989	Plants in South African nature reserves
Peart and Foin 1989[a]	California coastal grassland
Damascos and Gallopin 1992	Shrub invasion of high elevation Argentina
Smallwood 1994	Birds and Mammals in reserves in California
Knops *et al.* 1995	Coastal plant communities in central California
Robinson *et al.* 1995[a]	California grasslands
Ullmann *et al.* 1995	Roadside vegetation in New Zealand
Planty-Tabacchi *et al.* 1996	Riparian plant assemblages in Washington and France
Wiser *et al.* 1998	Eastern US forests
Higgins *et al.* 1999	South African plant communities
Stohlgren *et al.* 1999	Several plant communities in central US
Smith and Knapp 1999	Plants in Kansas prairies
Stadler *et al.* 2000	Alien plants in north-western Kenya
Levine 2000	Riparian plants on habitat islands in California
Negative correlation between diversity and invasion	
Woods 1993	Honeysuckle invasion of Eastern US forests
Pysek and Pysek 1995	Invasion of a Czech Republic plant community
Tilman 1997[a]	Minnesota prairie plants
Morgan 1998	Urban remnant of Australian grassland
Stohlgren *et al.* 1999	Several plant communities in central US
Shurin 2000[a]	Zooplankton in temporary ponds in Michigan
No correlation between diversity and invasion	
Timmins and Williams 1991	Plants in forest and scrub reserves of New Zealand
Holway 1998	Ant systems in central California

[a] Experimental addition of invader propagules.

native diversity rather than the reverse (Levine and D'Antonio 1999).

10.3 Synthetic framework

The published work on the relationship between diversity and invasion resistance has produced seemingly contradictory results. Models and small-scale experimental manipulations of diversity (Table 10.1) have tended to support the hypothesis of Elton (1958), that diverse communities better resist biological invasions. In contrast, studies examining invasion of natural diversity gradients tend to find the reverse, that the most diverse areas are the most invaded (Table 10.2). Naturally, this apparent contradiction has lead to controversy in the literature and given an impression that the relationship between species diversity and biological invasions may depend on the system or how the problem is approached (Levine and D'Antonio 1999; Enserink 1999). Similar contradictions between observational and experimental studies have characterized the diversity–ecosystem functioning debates more generally. Here, we present a conceptual framework (Fig. 10.1) that resolves these different results by distinguishing the local or causal effects of diversity from the

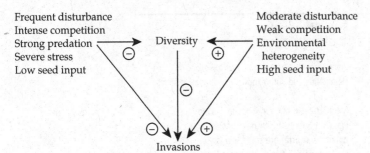

Figure 10.1 Synthetic framework for reconciling natural patterns of diversity and invasion with the results of theory and experiments.

effects of factors associated with diversity across communities.

Within our framework, the theoretical models and the experimental manipulations of diversity examine the intrinsic or causal effects of diversity on invasion resistance operating at neighbourhood scales—this is represented by the direct line between the two in Fig. 10.1. In contrast, the correlations between diversity and invasion resistance across entire communities result from these effects, but also the effects of factors covarying with diversity across natural systems. The covarying factors are simply factors known to influence native diversity such as disturbance, resource availability, and propagule supply. We, along with the authors of most of the observational studies (e.g. Pickard 1984; Wiser *et al.* 1998; Stohlgren *et al.* 1999), hypothesize that these correlated factors do not simply obscure the relationship, but rather predictably drive the positive correlations between diversity and invader abundance reported in the literature. This is because factors known to promote native species diversity, such as moderate disturbance or environmental heterogeneity, for example, are known to similarly favour the occurrence of invasions (Levine and D'Antonio 1999). Meanwhile, factors known to negatively influence native diversity, such as severe stress, intense competition, or low seed supply, are known to similarly inhibit exotic invasions. The consequences of all this is that if areas with low native species diversity are also associated with severe stress, intense competition, etc. they might therefore be difficult to invade. High diversity, in contrast, indicates favourable biotic and abiotic conditions (Stohlgren *et al.* 1999). Thus, diverse areas may be particularly susceptible to

invasion despite the intrinsic or causal effects of diversity operating at neighbourhood scales.

Our framework has implicit within it several key hypotheses. The first is that diversity enhances invasion resistance at the scale of species interactions or the neighbourhood. Secondly, however, these intrinsic diversity effects can be weak relative to factors covarying with diversity across entire communities. Thus, factors correlated with diversity are responsible for making the most diverse communities the most invaded. We next present work from our own studies that evaluates these hypotheses.

10.4 Neighbourhood effects of diversity on invasions at Cedar Creek, Minnesota

Working at the Cedar Creek Natural History Area, two of us (T. Kennedy and S. Naeem) evaluated the hypothesis that neighbourhood properties drive the inverse relationship between diversity and invasibility often reported in experimental studies (Table 10.1). Specifically, in experimental grasslands of varying diversity, we assessed the relationship between invader performance and several features of the surrounding plant neighbourhood including species richness, density, and crowding. In 1994, 147 3×3 m experimental grassland plots were established and seeded with 1, 2, 4, 6, 8, 12, or 24 species. The appropriate number of species was randomly assigned to each diversity treatment from a pool of 24. Thus, plots ranging in diversity from 1 to 12 species represented a random subset of the species present in the 24 species plots. In 1997, the year of our study, per cent cover estimates indicated that species richness was still highly correlated with the original diversity treatment ($r = 0.78$, $n = 147$, $P < 0.001$). For the three years prior to this study, invading plants were excluded by intensive weeding, but in 1997 we let the natural invasions from outside the plots proceed. For this study, invading plants were defined as any plant other than the 24 species sown into the experiment. Thus, even species found in plots in which they were not assigned, but part of the 24 species pool, were not considered invaders.

Mid (late June/July) and late in the growing season (August) of 1997 we drew high-resolution maps ($0.01\,cm^2$) of the location, size, and species identity of every plant in a $40 \times 125\,cm$ ($5000\,cm^2$) sub-plot within the larger $3 \times 3\,m$ experimental grassland. On these 294 maps, we recorded over 20,000 resident plants and over 5000 invading plants of 33 different species. These maps were digitized and analysed using image-analysis software. Analyses were simplified by using the centroid of each plant for its location and representing its occupied space by a circle.

For each invading plant, four features of the surrounding neighbourhood were measured: (1) the number of resident plants, (2) the species richness, (3) the nearest neighbour distance, and (4) an index of crowding that accounts for both the size and distance to all the plants within a neighbourhood. Crowding for each individual weed in a neighbourhood were measured as the angle subtended at the weed by another plant and the crowding index for the neighbourhood was calculated as the sum of all the angles for a given weed. An invader's neighbourhood consisted of all the resident plants within $7.5\,cm$ of the invader. To ensure that all analysed invaders had complete neighbourhoods, only those that were over $7.5\,cm$ from all edges of the mapped sub-plot were analysed. We selected this size for neighbourhood analyses because it struck a balance between smaller neighbourhoods that often contained no plants and larger neighbourhoods that would require excluding large numbers of invaders that were too close to the edge of the map to have a complete neighbourhood. This neighbourhood size is also comparable to that used in other published studies of neighbourhood analysis and plant competition (e.g. Pacala and Silander 1987).

First we present the relationship between patterns of invasion and parameters at the level of the plot. Then we present results suggesting that these plot level patterns are driven by differences in neighbourhood properties that affect invader establishment and success across diversity treatments.

Plot diversity had a highly significant and negative effect on the overall success of invading plants, as measured by total cover of invaders in each $5000\,cm^2$ subplot (slope $= -0.07$, $r^2 = 0.10$, $t = -4.07$, $P < 0.001$). 1996 data on invasions within the entire $3 \times 3\,m$ plots indicated a similarly negative relationship. Multiple regression using ecosystem parameters indicated that this relationship was probably due to the effects of diversity on local resource abundance. Specifically, plot level measurements of soil nitrogen and light penetration, which were both negatively associated with diversity, explained significantly more of the variance in total invader biomass for a plot than plot level diversity treatments (Knops *et al.* 1999).

Plot diversity had a positive and highly significant effect on the crowding and diversity of invader neighbourhoods (Fig. 10.2) (ANOVA results for effects = number of neighbours, species richness, nearest neighbour distance, and crowding, and treatment = species added, df = 2184 and $P \ll 0.001$ for all). That is, increases in the number of species planted led to more crowded and diverse neighbourhoods, and smaller nearest neighbour distances (Fig. 10.2). Furthermore, the performance of individual invaders was negatively affected by the degree of neighbourhood crowding and diversity (Fig. 10.3). That is, while neighbourhood characteristics might be poor predictors of invader size, they do set an upper bound on invader size and therefore affect invader success. Invaders in neighbourhoods that are crowded and diverse will never get very large compared to invaders in neighbourhoods that are uncrowded and species poor (Fig. 10.3—note the decrease, by up to two orders of magnitude, in the upper bound of invader size as neighbourhoods become more crowded and diverse).

In summary, invasions of the experimental plots ($9 \times 10^4\,m^2$) at Cedar Creek are negatively related to the diversity treatments, and ecosystem measurements indicate that this pattern is driven by

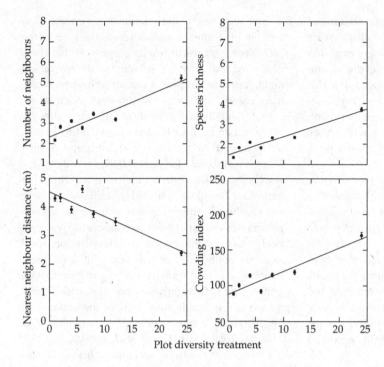

Figure 10.2 Properties of invader neighbourhoods by plot diversity treatment. Neighbourhoods become more crowded (i.e. more neighbours, shorter nearest neighbour distance, and higher crowding index) and more diverse (greater species richness) as plot level diversity increases. Data presented are means ± 1 SE and lines are a linear smoother (Systat v. 9.01) that approximates a least squares regression line.

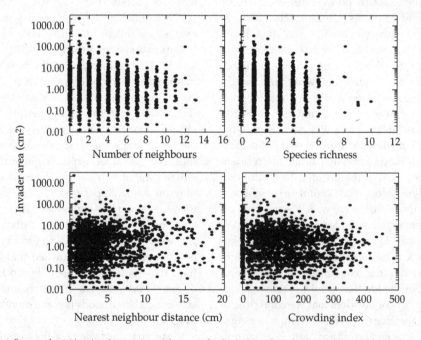

Figure 10.3 The influence of neighbourhood properties on the area of individual invading plants. Neighbourhood crowding and diversity limit the upper bound of invader area. This is evident in the two orders of magnitude decline in maximum invader area for invaders with uncrowded (large nearest neighbour distance, few neighbours, low crowding index) and species poor neighbourhoods relative to those with crowded and diverse neighbourhoods.

differences in resource availability. However, a plot level measurement of resource availability is actually the average of numerous small-scale, or neighbourhood level, measurements of these parameters. For example, in the Knops *et al.* (1999) study of the invasions into these plots, soil nitrogen values were calculated using four different 4 cm diameter soil cores that were composited. Thus, the lower soil nitrogen in diverse plots probably resulted from greater nitrogen utilization by the group of plants in the neighbourhood of the soil core. The greater the plot scale diversity, the greater the expected diversity and density of the neighbourhoods (Fig. 10.2).

We have shown that higher levels of plant diversity lead to crowded neighbourhoods and that crowded neighbourhoods significantly reduce the maximum size attainable by an invading plant. We did not explicitly measure how the availability of resources such as light varied by neighbourhood characteristics. However, it seems likely that neighbourhoods that are crowded and diverse are highly competitive environments, where resources such as light will be scarce. This suggests that while plot level patterns of invasion can be explained by measurement of the 'average' level of resources available in a plot, greater resource utilization in the crowded neighbourhoods common in diverse plots are actually what drives down resource availability and thereby limits the success of invading plants.

10.5 Relating neighbourhood process to patterns of diversity and invasion along the South Fork Eel river, California

The Eel river plant community provides an ideal natural system in which to test our second hypotheses that patterns of diversity and invasion are driven by factors correlated with diversity, maybe even despite the intrinsic negative effects of diversity operating at neighbourhood scales. The system is dominated by *Carex nudata*, a tussock forming sedge that provides the primary habitat for over 60 plant species in the community. Each tussock is a discrete micro-island colonized by up to 20 perennial herbaceous plants and bryophytes that depend on the tussock for stable substrate during winter

flows (Levine 2000b). The system is being invaded by Canada thistle, *Cirsium arvense*, common plantain, *Plantago major*, and creeping bent grass, *Agrostis stolonifera*, the propagules of which encounter numerous replicate tussocks containing varying numbers of native species. This system thus provides an ideal natural context in which to explore the relationship between diversity and biological invasions.

In a survey of 256 tussocks over an 8 km stretch of river, the incidence of all three exotic plants was greater on more diverse tussocks (Fig. 10.4). Specifically, the presence of *Agrostis, Plantago*, and *Cirsium* were significantly (logistic regression $P = 0.001, 0.008, 0.004$, respectively) and positively related to species richness (not including the invader). The tussocks were of similar size (400 cm²) and plant cover (90%). The positive correlations between species diversity and invader incidence are consistent with the observational studies in the literature (Table 10.2). Still the question remains: are these effects of diversity itself, or alternatively, the effects of ecological factors correlated with diversity?

To resolve these two possibilities, one of us (J. Levine) 'invaded' tussocks in which the number of resident species was manipulated *in situ*. This approach left other environmental factors free to vary (e.g. ambient nutrients and light, *Carex* stem number), but experimentally decoupled them from diversity. In late Spring 1998, I carefully removed all species from a randomly selected 65, 350 cm² tussocks at a single large riffle. Each tussock was randomly assigned to one of five species richness treatments: 1, 3, 5, 7, and 9 native species (not including *Carex* itself), corresponding to the natural range of richness found on tussocks of that size. I determined the composition of each replicate by first defining a pool of nine native species, and then assigned species to each tussock by a separate random draw from the pool. I added all species as small transplants of similar size, maintaining a constant cover of ~90%. In the spring of 1999, I added 200 seeds of each exotic plant to the surface of the experimental tussocks. For each tussock, I censused seedling number every three weeks during summer 1999 with a final census in summer 2000 (see Levine 2000a, 2001 for details of methodologies).

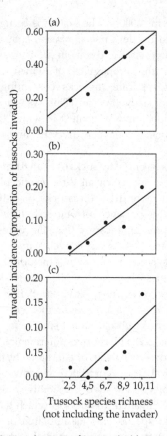

Figure 10.4 Natural patterns of invasion by (a) *Agrostis stolonifera*, (b) *Plantago major*, and (c) *Cirsium arvense* relative to the richness of species on the tussocks. Data for adjacent richness levels were pooled to better estimate the incidences, since some richness levels contained as little as five tussocks. Presented are the best fit lines from simple linear regression (*Agrostis* $R^2 = 0.82$, $P = 0.035$, *Plantago* $R^2 = 0.89$, $P = 0.017$, *Cirsium* $R^2 = 0.67$, $P = 0.092$), though I also conducted more statistically powerful logistic regressions showing significant ($P < 0.01$) effects of richness on invasion by each of the three invaders. Reprinted with permission from *Science*, **288**, 852–4. © 2000 American Association for the Advancement of Science.

As species richness increased, the proportion of the propagules that germinated and survived the two growing seasons significantly declined for two of the three invaders, and the biomass per individual significantly declined for all three invaders (Fig. 10.5). Thus, even with the uncontrolled environment of a natural system, diversity can significantly enhance resistance to biological invasions, supporting the relevance of studies conducted in more homogeneous settings (Table 10.1) to a complex natural community.

Resident species cover likely mediated the effects of diversity on invasion success. Absolute cover at the time of invasion was greater in more diverse treatments ($R^2 = 0.36$, $P = 0.001$), and like diversity, negatively impacted invasion success (Levine 2000a). Furthermore, even three weeks into the experiment, results were very similar to those obtained at the end of the second growing season (Fig. 10.5), suggesting that the effects of diversity on invasion arose at the germination/seedling stage. These early life history stages are particularly vulnerable to shading by plant cover (Goldberg and Miller 1990). The results in Fig. 10.4 were unlikely the effect of a single key resident species more often found in the high richness tussocks. For the seven nine-species tussocks, four different species were the most abundant resident, and never was their relative cover >30%.

In this community, patterns of diversity and invasion do not reflect the intrinsic effects of diversity operating at neighbourhood scales, pointing to factors covarying with diversity such as disturbance, propagule pressure, and species composition, as determinants of the community-wide patterns. That these factors overwhelm the local effects of diversity is not surprising since species richness explained only up to 17% of the variation in invasion success in the experimental tussocks (note that Fig. 10.5 shows standard errors).

Thus, work along the Eel river supports our second hypothesis, that patterns of diversity and invasion are driven by factors correlated with diversity. These results are found despite the effects of diversity on invasions operating at neighbourhood scales. Further exploring the correlated factors is critical, because they are ultimately responsible for the relatively high frequency of invasions into naturally diverse communities.

For this system, a more detailed analysis of the patterns of diversity and invasion suggests that they could result from spatial variation in propagule pressure. Consistent with the predominantly downstream movement of water and seeds in river systems, it was the downstream tussocks that were the most diverse ($R^2 = 0.35$, $P = 0.001$) and the most invaded by *Agrostis*, *Plantago*, and *Cirsium* (logistic regression, $P = 0.001, 0.047$, and 0.025, respectively). Furthermore, seed addition experiments showed

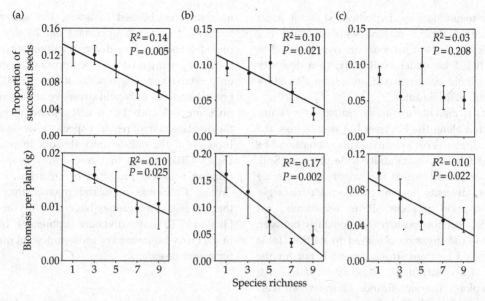

Figure 10.5 Results of a direct manipulation of species richness on the invasion success and biomass per plant of (a) *Agrostis stolonifera*, (b) *Plantago major*, and (c) *Cirsium arvense*. Presented are the proportion of seeds that germinated and survived to the end of the second growing season (2000) and the mean dry weight of three randomly selected individuals per species per tussock. Regressions were conducted on the raw data, not the means. Points show mean ± 1 SE.

that although the most upstream increases in diversity (from 0 to 3 km of the 8 km study stretch) may result from increasingly favourable physical conditions (Levine 2001), neither abiotic nor biotic conditions explained the increasing diversity from 3 to 8 km. Given that tussocks are seed limited (Levine 2001), and that waterborne seeds can only move downstream, differential seed supply is a likely explanation. This is further supported by results showing effective water dispersal of seed mimics downstream and prevailingly downstream winds (Levine 2001). In sum, for the Eel river system, variation in seed supply may be what overwhelms the intrinsic or causal effects of diversity (Levine 2000a).

10.6 Future directions

In both case studies, diversity enhanced invasion resistance at neighbourhood scales. The Eel river study showed, however, that this local effect might be weaker than other processes influencing invasions. Given these results and those of other studies (Tables 10.1 and 10.2), we see four promising areas for future research: (1) What are the mechanisms

by which diversity enhances invasion resistance? (2) How does the spatial scale of inquiry influence these effects? (3) How does diversity affect the invasion of multitrophic systems? And (4) How does diversity influence the susceptibility of communities to exotic species impacts?

Uncovering the diversity-based mechanisms of invasion resistance. Although many studies have now used experiments to directly test the effects of species diversity on invasion resistance, the mechanisms by which these effects arise remain poorly understood. To what extent are the results of these experiments (Table 10.1) the effect of a few or many species, or the effects of species versus functional diversity? We also need more information about the mechanistic pathways by which diversity influence invasion resistance. Do diversity effects on invasions occur via nutrient, light, or water availability? Already, we have correlational evidence showing that water or nitrogen availability, for example, is reduced in high-diversity plots and may be inversely correlated with invasion success (e.g. Knops *et al.* 1999; Dukes 2001). Still, this alone does not conclusively demonstrate that these are the mediating factors. A further step would be to directly manipulate the hypothesized factor in an experimental design crossed with diversity. For example, if effects of diversity are hypothesized to be mediated via water availability, then diversity effects should be eliminated or at least altered by sufficient water addition.

Effects of diversity at various spatial scales. Work conducted along the Eel river has shown how the local scale effects of diversity on invasion cannot be inferred from the community-wide patterns. Still, that does not mean that the neighbourhood-scale effects of diversity have no consequence for larger scale invasion patterns. It is reasonable, for example, that the positive relationships between diversity and invasion observed in nature (Table 10.2) would be even stronger were it not for the effects of diversity itself. Thus, we need research that explores the significance of the small-scale experimental results for the invasibility of entire communities. By exploring the relationship between local and community-wide diversity, this work will improve our understanding of how species loss influences invasion resistance at the scale of entire communities.

Expanding diversity–invasion research to multiple trophic levels. The majority of research presented here examines how diversity within a single trophic level (mostly plants) affects invasibility. However, Elton's (1958) original hypothesis and some of the theoretical work involved multitrophic systems. Do the results found for a single trophic level (Table 10.1) also apply to more complex systems? Within this general framework, we might ask how herbivore diversity influences plant invasions, or the converse, how plant diversity influences herbivore invasions. By extension, other trophic interactions, such as the role of pathogens and predators that regulate herbivory, and the microbial and animal decomposer communities may also influence diversity effects on invasion.

Interactions between diversity and post-invasion impacts. Nearly all work on the relationship between species diversity and biological invasions examines how diversity influences the establishment of exotic species. Many communities, however, have

numerous established invaders, but suffer little impacts. An equally important but largely unexplored question is how diversity influences the susceptibility of organisms to the impact of an invader once established. The models of Case (1990, 1991) predict that not only will diversity act to enhance invasion resistance, but it will also act to prevent the displacement of resident species after invasions. In some of the only empirical work in this area, Dukes (2001) found that community evapotranspiration was less impacted by invasion in more diverse California grassland microcosms. Given that diverse communities have so many invaders (Table 10.2), how diversity influences invader impact may be among the most important question for future research.

10.7 Conclusions

Our chapter points to two important conservation messages about species diversity and biological invasions. First, the two case studies presented here and many other published experiments (Table 10.1) indicate that species diversity enhances invasion resistance. Work at Cedar Creek suggests that this common result arises from processes operating at neighbourhood scales. The conservation implication is that species loss, if it affects neighbourhood scale diversity, may erode invasion resistance (Lyons and Schwartz 2001). Nonetheless, factors correlated with diversity may be more important in driving the community-wide patterns (Table 10.2). Work along the South Fork Eel river shows how positive correlations between diversity and invasion may occur despite the intrinsic effects of diversity operating at neighbourhood scales. Furthermore, because we have shown that the most diverse natural areas are consistently the most frequently invaded, the second conservation concern is that diverse natural communities are at the greatest risk from biological invasions. Among the most important messages of this chapter is that these two conservation concerns are not contradictory.

PART IV

Extending the scope to other systems

Contributions of aquatic model systems to our understanding of biodiversity and ecosystem functioning

O. L. Petchey, P. J. Morin, F. D. Hulot, M. Loreau,
J. McGrady-Steed, and S. Naeem

11.1 Introduction

Biodiversity has been shown to affect ecosystem functioning in an assortment of ecosystems (McGrady-Steed *et al.* 1997; Naeem and Li 1997; van der Heijden *et al.* 1998b; Hector *et al.* 1999; Petchey *et al.* 1999; Hulot *et al.* 2000; Naeem *et al.* 2000a; Norberg 2000; Emmerson *et al.* 2001; Ruesink and Srivastava 2001). These studies continue to guide the development of theories that relate diversity to the magnitude and stability of ecosystem functioning (Loreau 2000a). In a simple world, the existence of comparable findings from a wide range of systems would suggest a general relationship between biodiversity and ecosystem functioning. However, reviews show that effects of diversity vary widely among systems (Schläpfer and Schmid 1999; Schwartz *et al.* 2000; Loreau *et al.* 2001). For example, increasing plant diversity increases grassland primary productivity across Europe (Hector *et al.* 1999), while the diversity of soil microbes has affected functioning in some experiments (van der Heijden *et al.* 1998b) but not in others (Mikola and Setälä 1998a).

Key differences between experiments include the kinds of ecosystems considered, the levels of biological complexity (numbers and types of interactions among species), degree of experimenter control over environmental conditions, and the different kinds of manipulations (direct or indirect, random or non-random) of biodiversity used. Here we show that differences in (1) duration, (2) trophic complexity, and (3) links with theory separate experiments with aquatic model systems from biodiversity–ecosystem functioning experiments in many terrestrial model systems (e.g. growth chamber, pot, greenhouse, and plot experiments). To make general points we primarily use examples of our own research because of our familiarity with it.

11.2 What are aquatic micro and mesocosms?

Microcosms and mesocosms are spatially delimited artificially constructed model ecosystems. They have enjoyed a long history of use in ecology (Byers and Odum 1993). These aquatic systems, which range in scale from bottles and beakers to tanks and artificial ponds (Lawler 1998; Morin 1998), are less complex than natural systems, a situation that poses special advantages and interpretive challenges (Peterson and Hastings 2001). Such systems may contain fewer species and trophic levels than their most complex natural counterparts, but nonetheless preserve key features of interest. That simplicity, combined with opportunities for replication and experimental manipulation, provides an important rationale for their use. Distinctions between micro-cosms and mesocosms are necessarily relative and somewhat arbitrary (Lawler 1998). Although the smallest microcosms are easily housed in laborat-ories and mesocosms are often placed in the field, larger size and greater biological and spatial com-plexity distinguish mesocosms from microcosms. The greater complexity of mesocosms can include greater environmental variability as well as greater total diversity when, for example, plankton com-munities are samples of natural communities (Hulot *et al.* 2000). The varying complexity of micro and mesocosm experiments, with mesocosms often reflecting natural systems more closely than micro-cosms do, is analogous to the range of realism present in theoretical models.

For many ecological issues, aquatic microcosms are ideal experimental systems. Initial composition can be controlled, processes can be manipulated, and many replicates can be readily constructed. Microcosms complement field and observational studies of natural systems. Skeptics, however, view them as unsatisfactory simplifications of the full biological and physical complexities of nature (Carpenter 1996). We note a continuum between microcosm and field studies in which the former represent controlled examinations of theory and mechanism while the latter represent closer approximations to natural systems (Lawton 1995). Observational studies of natural systems remain crucial for formulating hypotheses, but observa-tions and correlations alone cannot unambiguously sort causes from effects (Manly 1992; Naeem 2001). That is where experiments play an essential role.

Aquatic microcosm studies differ from other studies in several respects. Here, we focus on three of their most important differences (duration, trophic complexity, links to theory). These attributes of aquatic microcosm studies offer a unique insight into the functional role of biodiversity that might not be obtained from terrestrial experiments using communities that are dominated by long-lived organisms. Obviously, ecologists need to use the full range of systems and approaches at their dis-posal to understand how ecosystems work (Lawton 2000). Progress in ecology is made by integrating results from theoretical, microcosm, field, and observational studies.

11.3 Duration

Although aquatic microcosm studies have a shorter absolute duration (weeks/months) than many studies conducted in the field (years), they typically last for tens to hundreds of generations of the organisms involved. This means that populations can grow (or decline) to levels determined by resources and interactions with other species (Fig. 11.1). The average number of generations that elapse during an experiment provides a dimen-sionless scaling factor that can be used to evaluate whether patterns represent transient or long-term dynamics (Peterson and Hastings 2001).

A survey of 35 recent biodiversity–ecosystem functioning experiments (similar to the survey in Ives *et al.* (1996) and found by searching the liter-ature manually) illustrates the contrast in duration, measured in number of generations of the mani-pulated organisms, between experiments in artificial terrestrial and aquatic ecosystems. The duration of each study was scored as less than five generations (21 studies), between five and twenty generations (nine studies), or more than twenty generations (five studies) of the manipulated organisms. These broad categories make differences between genera-tion times of different organisms within a single experiment less influential, although the range of generation times within a single study may, in itself, be important. Where the generation times of organisms within an experiment differed greatly we used the longest generation time. The majority of terrestrial experiments lasted less than five generations, whereas the aquatic experiments most commonly lasted more than 20 generations (Fig. 11.2(a)). Indeed, with generation times often as short as 3–12 h (e.g. for ciliates in aquatic microcosms), a six week long experimental study of aquatic microbes provides the population dynamic equivalent of a study of annual plants running for 84–336 years!

These very different durations make integrating results across experiments problematic. Results of longer-term experiments reflect population dynamics, including differential population growth (Hulot *et al.* 2000), density compensation (McGrady-Steed and Morin 2000), and complex interactions within the food web (Naeem and

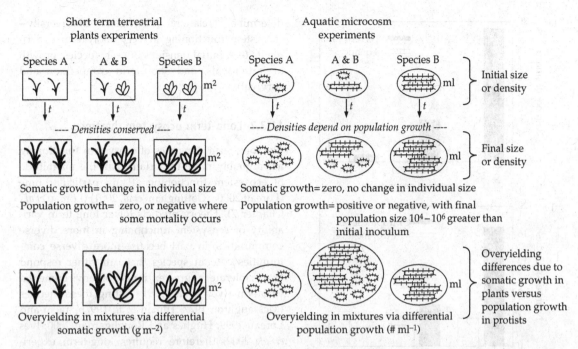

Figure 11.1 Differences between terrestrial plant experiments and aquatic microbe experiments involve the extent to which initial planting/ seeding densities or population growth determine results. Planting densities matter in terrestrial plant experiments because results are dominated by changes in the size of individuals. Population growth determines results and initial densities matter less in the longer-term aquatic microbe experiments. Recent experiments in terrestrial plant communities (e.g. Hector *et al.* 1999; Tilman *et al.* 2001) may not fit neatly into this dichotomy as they include several generations and some population growth.

Li 1997; Petchey *et al.* 1999; Naeem *et al.* 2000a). Results in shorter terrestrial experiments that run for one to several generations may be coloured by transient effects of initial species composition and organism densities. (Note that aquatic meso-cosm (i.e. large microcosm) experiments often contain longer-lived organisms without concurrent increases in absolute duration. Therefore, micro and mesocosms may also be separated along a continuum of generational duration).

11.3.1 Design issues in long-term experiments

Different durations of experiments pose challenges for optimal experimental design, in particular regarding the choice of initial conditions. Diversity manipulations use either a substitutive or additive design to determine initial density/abundance (Harper 1977; Naeem *et al.* 1994a; Hooper and Vitousek 1997; Symstad *et al.* 1998; Jolliffe 2000; Levine 2000a). Implicit in both designs is the assumption that the density of organisms set at the beginning of the experiment will influence the results, as happens in most experiments with vascular plants. Initial density is less relevant to microcosm experiments, where densities are low to start (e.g. tens of individual ciliates per 100 ml), but over the many generations of a microcosm experiment, can increase by several orders of magnitude, remain unchanged, or go extinct. Densities and population dynamics are not constrained by the initial densities of species added to the systems, except for the obvious situation where species initially absent remain so (unless contamination or deliberate invasion comes into play, e.g. McGrady-Steed *et al.* (1997)). Population densities therefore reflect the consequences of many generations of interactions within and between species, resource levels, and nutrient cycling.

These differences in sensitivity to initial densities between field experiments and aquatic microcosms reflect differences in the number of generations

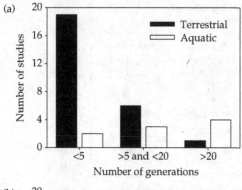

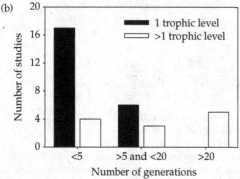

Figure 11.2 The distribution of biodiversity–ecosystem functioning experiments by their duration, (a) the type of ecosystem, and (b) the number of trophic levels that were manipulated. The survey was limited to direct and indirect experimental manipulations of diversity and excluded observational studies of relations between biodiversity and ecosystem functioning because it is difficult to define their duration. Using broad categories of experimental durations, less than five generations, between five and 20 generations, or more than 20 generations, made the results less sensitive to imperfect knowledge of the generation times of organisms.

these experiments typically cover rather than fundamental ecological differences (but see Hairston 1989). The duration of an experiment in generations is important because it measures the effective temporal scale of experiments, and changing temporal scales (as well as spatial scales) can change the dominant ecological processes and patterns (Levin 1992; Levin and Pacala 1997; Peterson and Hastings 2001). For example, relations between biodiversity–ecosystem functioning in microcosm experiments can be caused by changes in population size that result from intra- and inter-specific competition, predation, indirect interactions, and extinctions. A different set of processes and properties

determine relations between biodiversity–ecosystem functioning in very short-term experiments (e.g. initial densities of each species, growth of individuals, and short-term competition and survival).

11.3.2 Long-term ecosystem dynamics

The long-term dynamics of ecosystem functioning are arguably more important than the magnitude of ecosystem functioning at any particular time (Loreau 2000a; Cottingham *et al.* 2001) (Loreau *et al.*, Chapter 7). Theory predicts lower long-term variability of ecosystem functioning in more diverse communities, in part because more diverse communities contain species that can either respond to or tolerate a greater range of environmental variation (Ives *et al.* 1999). Testing these theories (McNaughton 1977; Tilman *et al.* 1997b; Yachi and Loreau 1999; Hughes and Roughgarden 2000; Ives *et al.* 2000) therefore requires long-term experiments, for which microcosms are ideal (e.g. McGrady-Steed *et al.* 1997; Naeem and Li 1997).

Studies of microcosms show that ecosystem-level predictability increases with increasing diversity. Here, predictability measures both temporal variation and between-replicate (spatial) variation in ecosystem processes (temporal variation is equivalent to the 'variability' defined in Loreau *et al.*, Chapter 7). Total autotroph biomass (Naeem and Li 1997) and carbon dioxide flux (McGrady-Steed *et al.* 1997) have been found to be more predictable in more diverse communities. These experiments used microbial communities containing algae, bacteria, protists, and (in the case of McGrady-Steed *et al.* 1997) micro-metazoans, and communities differed in species richness within and across functional groups.

Both approaches indicated that ecosystem functioning was more predictable in more diverse communities (Fig. 11.3(a)). Subsequent analysis showed that the temporal variability of biomass within functional groups usually decreased with diversity, while the temporal variability of individual population sizes was unrelated to diversity (McGrady-Steed and Morin 2000). These results mirror patterns seen in terrestrial systems, where temporal variability in total plant biomass

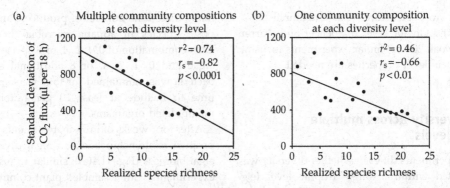

Figure 11.3 Effects of species richness on predictability of the magnitude of an ecosystem-level process. (a) Each data point represents a mean calculated across time, replicates, and different communities at each diversity level. Here, greater compositional similarity at high diversity levels compared to low diversity levels could contribute to the negative relation, in addition to mechanisms that act through time. (b) Each data point represents a mean calculated across only time and replicates of the same community composition. Here, differences in compositional similarity across the diversity gradient cannot influence the negative relation. Only mechanisms that act through time or across replicates of the same community composition can cause the negative relation.

decreased as diversity increased, while temporal variability in the biomass of individual species increased as diversity increased (Tilman 1996). This supports the idea that diversity affects stability (of which variability is a measure; Loreau *et al.* Chapter 7) differently at different levels of ecological organization (McNaughton 1977; King and Pimm 1983; Tilman 1999a). In addition, these results also support the idea that high biodiversity insures against future environmental changes (Frost *et al.* 1995; Walker 1995; Ives *et al.* 1999; Walker *et al.* 1999; Yachi and Loreau 1999) (Loreau *et al.* Chapter 7).

Many uncertainties remain about the links between population and ecosystem-level dynamics. Theory shows that a suite of statistical effects and deterministic mechanisms either create or modify the links between population and ecosystem level dynamics (Cottingham *et al.* 2001). One effect concerns the smaller differences in composition that high diversity treatments exhibit (Fukami *et al.* 2001). That is, when diversity of replicate communities begin to approach the full diversity possible (the species pool available for experimentation), they all begin to have similar compositions. It is possible that such similarity in composition results in greater predictability among microcosms, though this cannot completely explain the greater predictability at higher diversities in McGrady-Steed *et al.* (1997). An existing analysis (McGrady-Steed *et al.* 1997) that includes only a

single community composition at each diversity level preserves the stabilizing influence of diversity on ecosystem level predictability (Fig. 11.3(b)).

Changes in species composition and diversity can also influence long-term ecosystem dynamics. Low ecosystem level variability can arise from unchanging population abundances and constant community composition or from compensatory changes in abundance and species composition (Micheli *et al.* 1999). A relatively constant trophic structure observed along productivity gradients can result from species turnover (Leibold *et al.* 1997), and suggests an important role for immigration in maintaining ecosystem functioning. Some microbes used in some aquatic microcosm experiments (e.g. Petchey *et al.* 1999) may have wide, often cosmopolitan distributions (because of extremely high dispersal rates), so that local diversity is a relatively large fraction of regional diversity (Fenchel *et al.* 1997; Finlay and Clarke 1999a,b; Finlay and Fenchel 1999) (though see Foissner 1999). Aquatic microcosm experiments with different dispersal treatments provide one route for investigating if regional scale dispersal can maintain maximal microbial functioning across sites (Finlay *et al.* 1997).

Effects of habitat isolation and fragmentation on immigration rates may threaten long-term ecosystem functioning. Experiments clearly show that dispersal or immigration affect population variability and food web structure (e.g. Luckinbill 1974; Holyoak and Lawler 1996a,b; Spencer and Warren 1996; Holyoak 2000). Similar experiments focusing on ecosystem-level properties are needed.

11.4 Diversity across multiple trophic levels

Recent experiments differ in whether diversity was manipulated within a single trophic level (e.g. Tilman *et al.* 1997a; Hobbie and Chapin 1998; Hector *et al.* 1999; Stocker *et al.* 1999) or across multiple trophic levels (e.g. McGrady-Steed *et al.* 1997; Naeem and Li 1997; Petchey *et al.* 1999; Hulot *et al.* 2000; Naeem *et al.* 2000a; Wardle *et al.* 2000a; Downing 2001). Of the 36 recent biodiversity–ecosystem functioning experiments that we surveyed, 23 manipulated diversity in a single trophic level, while 12 manipulated diversity within multiple trophic levels (Fig. 11.2(b)). Manipulating diversity across multiple trophic levels is relatively common in closed systems like microcosms where assembly and invasion is under tight control. In contrast, experiments in terrestrial systems tend to manipulate diversity in one trophic group, often plants or fungi, while higher trophic levels remain uncontrolled (often for logistic reasons) or manipulated by the addition or exclusion of multiple higher trophic groups, such as the use of pesticides to remove insects (e.g. Mulder *et al.* 1999). Consequently, in comparison to aquatic microcosms, such terrestrial experiments effectively ignore trophic diversity. Two studies illustrate how biodiversity can interact with trophic structure to determine ecosystem functioning.

11.4.1 Producer and decomposer interactions

Producers and decomposers form the base of food webs, where patterns of nutrient cycling can profoundly influence the rest of the web (e.g. Cole 1982; Vitousek and Walker 1989). Interactions between producers and decomposers can range from competition for nutrients to a mutualism based on exchange of inorganic nutrients for photosynthate (Cole 1982; Harte and Kinzig 1993; Daufresne and Loreau 2001). Naeem *et al.* (2000a) simultaneously

manipulated the number of producers and decomposer species in aquatic microbial microcosms. Each combination of 0, 1, 2, 4, or 8 species of green algae and 0, 1, 2, 4, or 8 additional species of bacteria was maintained for four weeks, sufficient time to include at least 20 generations of the manipulated organisms.

After four weeks of growth, treatments showed a positive relationship between algal diversity and algal biomass (Fig. 11.4(a)), similar to results from experiments with assembled plant communities in grasslands (Tilman *et al.* 1997a; Hector *et al.* 1999). Simultaneous variation in algal and bacterial diversity, however, caused complex changes in algal biomass, where interactions between algal and bacterial richness influenced biomass production (Fig. 11.4(a)). Thus, knowledge about diversity within a single trophic group did not adequately explain ecosystem functioning.

This co-dependency between producer and decomposer diversity raises questions about the interpretation of diversity manipulations focused within single trophic levels. Certainly, experimental manipulations of diversity within trophic levels provide useful information about how that targeted trophic level affects ecosystem functioning. Terrestrial plants may also be the dominant drivers of ecosystem functioning in their communities. Studies show, however, that below-ground community composition and diversity significantly affects above-ground processes (Laakso and Setälä 1999b; Griffiths *et al.* 2000) and that mycorrhizal diversity affects plant communities (van der Heijden *et al.* 1998b) (Raffaelli *et al.* Chapter 13). Focusing only on above-ground plant diversity may miss important complications when biodiversity changes simultaneously on several interacting trophic levels.

11.4.2 Direct and indirect effects of disturbance

A second example considers how environmental change can have both direct and indirect effects on ecosystem functioning. The direct effects result from changing metabolic rates or resource supply rates. Indirect effects of the environment on ecosystem functioning can occur through changes in the diversity, composition, and structure of interactions of a community. Petchey *et al.* (1999)

investigated direct and indirect effects by gradually warming four different microbial communities. Gradual warming had two interrelated effects on the food webs: a reduction in species richness and modified interactions between species associated with loss of higher trophic levels. Both of these changes could conceivably affect ecosystem functioning. Partitioning the net effect of warming on primary production into direct and indirect effects showed that indirect effects were consistently and logically associated with changes in community structure. Warming decreased consumer diversity, increased producer biomass, and increased primary production beyond levels expected from direct metabolic effects of temperature. The magnitude of the indirect effect peaked during the sixth week of the experiment and approached the magnitude of the direct effect (Fig. 11.4(b)) suggesting that indirect effects of warming caused by changes in trophic structure could be as important as direct effects.

These results and other lines of evidence (Raffaelli et al., Chapter 13) clearly indicate that trophic structure modulates effects of diversity on ecosystem functioning (though see Downing 2001). Recent emphasis on the role of niche complementarity, selection probability, and facilitation surely

stem in part from a focus on short-term experiments using a single trophic level, plants (e.g. Loreau and Hector 2001; Mulder et al. 2001). There are many ways that changes in trophic structure can affect ecosystem functioning, including trophic cascades, destabilizing effects of extra trophic levels, and loss or addition of keystone species (Bengtsson 1998).

11.4.3 Design issues in multi-trophic level experiments

Ideally, a factorial manipulation of diversity per trophic group and number of trophic groups could be designed. Analysis of factorial manipulations of diversity per trophic group and number of trophic groups require random assignment of species to treatments (Allison 1999). Common sense and studies of community assembly both indicate that this type of design is difficult to arrange. Obviously, treatments with predators and without prey will fail, as would treatments with herbivores and without producers. Weatherby et al. (1998) showed, furthermore, that only eight of the 63 possible combinations of six protist species persisted in the long-term. The inability to assemble some communities and the long-term instability of many communities

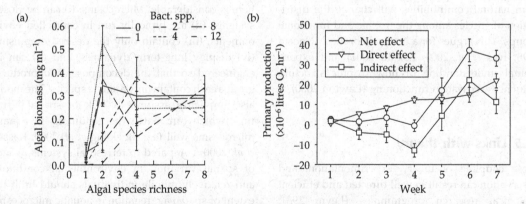

Figure 11.4 (a) Algal biomass response in microbial microcosms to variation in bacterial decomposer diversity (error bars are ±1SE). Although a generally positive relationship is observed between algal production and algal species richness, as observed in other experiments, the shape of the response curve is sensitive to decomposer diversity (see Naeem *et al.* 2000 for details of experiment). (b) Net, direct, and indirect effects of temperature on primary production during gradual warming of microbial microcosm communities (error bars are ±1SE). Changes in physiological process rates are assumed to cause direct effects of warming and were calculated by assuming a Q_{10} of 2. Indirect effects were the difference between the direct effect and the observed primary production in warmed communities and were significantly associated with changes in food-web structure.

precludes the completely random assembly of food webs (Drake 1990; Law and Blackford 1992; Holt *et al*. 1994; Leibold 1996). Designing appropriate multi-trophic level biodiversity–ecosystem functioning experiments remains challenging.

Fortunately, experiments in microcosms and lakes that do not explicitly examine biodiversity–ecosystem functioning have been common. For example, considerable effort has focused on the effects of number of trophic levels and prey heterogeneity on population, community, and ecosystem dynamics (e.g. Briand and McCauley 1978; Persson *et al*. 1992, 2001; Lawler and Morin 1993; Leibold *et al*. 1997; Kaunzinger and Morin 1998; McCann *et al*. 1998a; Bohannan and Lenski 1999; Steiner 2001). These commonplace effects of food web structure on community and ecosystem dynamics (Raffaelli *et al*., Chapter 13) suggest that changes in diversity may affect ecosystem functioning mostly through changes in the distribution of species among trophic groups. The trophic position in a food web that species are lost from may influence ecosystem functioning more than the total number of species present. Furthermore, species from some trophic, functional, and taxonomic groups suffer a greater extinction risk than others (Pimm *et al*. 1988; Cutler 1991; Tracy and George 1992; Wright and Coleman 1993; Gilbert *et al*. 1998; Petchey *et al*. 1999), suggesting that extinctions from natural communities will change the distribution of species among the trophic and functional groups. This argues for a better theoretical (Naeem 1998; Ives *et al*. 2000; Loreau 2001) and experimental understanding of how trophic structure influences ecosystem functioning (Lawton 2000).

11.5 Links with theory

Close coupling of theory, experimentation, and observation can result in well-directed and efficient ecological research programmes (Levin 1981; Kareiva 1989; Werner 1998). We have already mentioned that microcosm experiments, given their long generational duration, can test theoretical predictions about long-term dynamics. Here we show how microcosms can directly address questions raised by ecological theory. Two experiments

that focus on the influence of diversity on ecosystem level processes illustrate particularly tight links with theory (Hulot *et al*. 2000; Naeem *et al*. 2000a).

11.5.1 Producer and decomposer interactions

The first example shows that iterative feedback between experiments and theory can rapidly advance understanding. Simple models of producer–decomposer interactions describe nutrient cycling between four compartments (Harte and Kinzig 1993): primary producers (plants, algae), decomposers (bacteria, eukaryotic microorganisms), dead organic matter (litter, dead producers), and a pool of inorganic nutrients. The two living compartments, the producers and the decomposers, simultaneously compete for nutrients and exchange other nutrients in a mutualistic nature. This model makes a number of explicit assumptions: (i) The system is closed, so that rates of nutrient cycling within the system dominate dynamics. (ii) Environmental conditions are constant unless explicitly modelled otherwise. (iii) There are no other organisms to interact with; e.g., there is no predation. (iv) Steady state, long-term dynamics are considered. (v) All organisms within a trophic group are identical. The congruence between theoretical assumptions and conditions in aquatic microcosm here is considerable. Microcosms: (i) can be closed systems, (ii) are typically run in controlled environments, (iii) contain only the desired organisms, (iv) display long-term dynamics, and (v) can be engineered so that the decomposer and producer groups each contain only a single species. The model also implicitly assumes a lack of spatial structure, which can match the conditions in aquatic microcosms well (e.g. Buckling *et al*. 2000; Kassen *et al*. 2000) compared to terrestrial systems. A lack of spatial variation in environmental conditions and relatively mobile individuals should limit the extent of spatial aggregation in aquatic microcosms and result in dynamics that approximate mean field situations (Tilman and Kareiva 1997). This contrasts with terrestrial plant communities, with sedentary individual plants and greater spatial environmental heterogeneity. The six assumptions outlined above are not peculiar to Harte and

Kinzig's (1993) model; they permeate ecological theory. Consequently, conditions in microcosms often correspond well to the assumptions of simple theory.

Naeem *et al.* (2000a) examined the fifth assumption in Harte and Kinzig's (1993) model and showed the distribution of biomass between producers and decomposers depended on the number and identity of producers and decomposers (Fig. 11.4(a)). Obviously, the assumption of functionally identical producers and decomposers is unrealistic. Accordingly, Loreau (2001) developed a model including a diversity of decomposer species that differentially exploit a diversity of organic compounds derived from producers. Increasing producer diversity could reduce ecosystem process rates if it increased the likelihood that some organic compounds could not be utilized by a given set of decomposers. In contrast, increased decomposer diversity could increase ecosystem process rates if more organic compounds could be utilized. These predictions match the significant positive correlation between microbial diversity and the diversity of carbon sources decomposed described by Naeem *et al.* (2000a), but differ in other respects, such as interactions between decomposer and producer diversity. It is possible that the experiment, even though long by ecological standards, remained far from equilibrium. Consortia of microbial species may break down organic compounds in ways that are not predicted by their individual effects. These possibilities suggest that research considering transient dynamics and resource driven interactions between decomposer species would be particularly worthwhile (Loreau 2001).

11.5.2 Functional diversity and nutrient enrichment

Models of how aquatic ecosystems respond to enrichment are important because of deleterious consequences of phytoplankton blooms caused by eutrophication (Briand and McCauley 1978). Early models used many of the same assumptions as Harte and Kinzig (1993), in particular that discrete trophic levels contained functionally identical species (Oksanen *et al.* 1981). In such models, and

in experiments that strictly adhered to these assumptions (Kaunzinger and Morin 1998), enrichment increases the abundance of the highest trophic level and trophic levels located an even number of levels below the top level. The considerable natural diversity within trophic levels, however, raised questions about whether that diversity would lead to different predictions (Leibold 1989; Hunter and Price 1992; Leibold and Wilbur 1992; Hairston and Hairston 1993). Abrams (1993) predicted that multiple functional groups within trophic levels could produce very different responses than those predicted by models of simple linear food chains.

Hulot *et al.* (2000) tested this prediction by manipulating nutrient levels and fish densities in aquatic mesocosms (we use the term mesocosm only to follow Hulot *et al.* (2000) and Lacroix and Lescher-Moutoué (1991)). The mesocosms were large plastic bags suspended in a shallow lake that initially contained many different organisms. With respect to the assumptions of Harte and Kinzig (1993) this experimental system (i) is closed, (ii) is in a naturally varying environment, (iii) mostly contains only the organisms of interest, (iv) is not strongly influenced by initial conditions, and (v) contains a diversity of organisms in each trophic group. We also might expect lower levels of spatial heterogeneity than in other systems, although enclosure walls introduce some spatial heterogeneity.

Hulot *et al.* (2000) examined the importance of diversity within trophic levels for ecosystem responses to enrichment by comparing their experimental results to predictions of linear food chain models (Oksanen *et al.* 1981; Arditi and Ginzburg 1989) and a model that included functional diversity within trophic levels (as did Abrams 1993). Phytoplankton, herbivorous zooplankton, and carnivorous zooplankton were modelled either as identical species (a linear food chain), or by dividing these trophic levels into functional groups based on the size and the diet of organisms (a simplified food web). Responses of each trophic level to enrichment without and with fish were inconsistent with the linear food chain model (Fig. 11.5). For example, enrichment decreased the abundance of top trophic level, invertebrate carnivores! The more complex food-web model, however, matched experimental results in nine of ten instances, including a negative

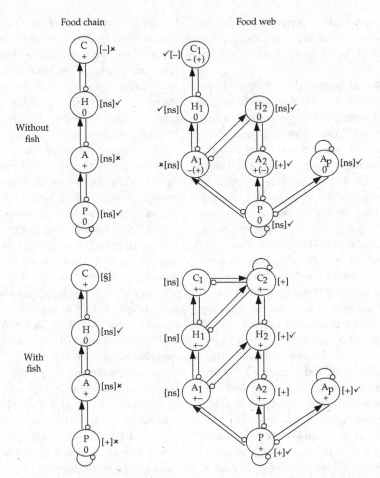

Figure 11.5 Contrasts between observed (in square brackets) and predicted (inside the circles/nodes) effects of nutrient enrichment in mesocosms without fish and with fish. Ticks indicate correct predictions and crosses indicate incorrect predictions. No tick or cross indicates when comparisons were not possible because theory made undetermined predictions or because the sum of invertebrate carnivores density and fish biomass is senseless (§). (Food chain predictions are according to a prey-dependent model.) The nodes of the food chains/webs correspond to the trophic groups. P: mineral phosphorus; A: algae; A_1: edible algae; A_2: protected algae; A_p: periphyton; H: herbivores; H_1: small herbivores; H_2: large herbivores; C: carnivores; C_1: invertebrate carnivores; C_2: fish. Links between nodes correspond to direct effects between trophic groups, determined by the coefficients of the Jacobian matrix for the dynamical system at equilibrium (Levins 1974). An arrow or a line ending with a circle from node *i* to node *j* corresponds to a positive or a negative direct effect, respectively, of trophic group *i* on trophic group *j*. Predictions and empirical results (significant at $\alpha = 0.05$ or not significant ns) are indicated by their sign: +, 0, and −, denoting a positive effect, no effect and a negative effect, respectively, of nutrient enrichment on biomass. Additional predictions were +− for undetermined outcome and +(−) or −(+) indicating a very likely positive or negative outcome, respectively. Modified from Hulot *et al.* (2000).

effect of enrichment on invertebrate carnivore abundance (Fig. 11.5). This level of agreement could not be ascribed to chance alone (sign test, $P < 0.025$). Thus, integrated use of models and experimentation showed that functional diversity within trophic levels was important in understanding the full range of ecosystem responses to nutrient enrichment and top predators.

The two studies discussed here (Hulot *et al.* 2000; Naeem *et al.* 2000a) provided data to test and modify existing theory. The microcosms conformed closely enough to model assumptions that even small increases in model complexity, which result from relaxing a particular assumption (for example, by including functional diversity) significantly increased predictive power. The mesocosm experiment (Hulot *et al.* 2000) likely meets model assumptions less well, but sufficiently so that linking the experiment to theory was profitable. The resulting models of intermediate complexity (such as in Hulot *et al.* 2000; Loreau 2001) are an important middle ground between very simple and very complex models. They contain considerable information on natural history, but are not so complex that they are completely intractable (Briggs and Godfray 1995). Combining tractable models and experimental systems that meet many model assumptions is a powerful approach for advancing ecological understanding.

11.6 Future directions

Of course, microcosms are useful model systems in their own right, and they offer important insights into the workings of well-defined ecosystems of low to moderate complexity. Whether we can extrapolate results from aquatic microcosms to other ecological situations requires an understanding of the spatial and temporal scaling of ecological processes (Lawton 1999; Peterson and Hastings 2001). The first step in solving this problem is to identify and quantify changes in scale, as we have done for the temporal scale of recent biodiversity–ecosystem functioning experiments, using non-dimensional parameters (Peterson and Hastings 2001), such as the generational duration of experiments. Changes in biological complexity also represent changes in scale, from single populations, to simple community modules in isolation, to entire communities. We have shown how the differences in trophic complexity between ecosystem-functioning experiments might explain the qualitative range of results (also see Schwartz *et al.* 2000).

An implication of the difference in duration and complexity between microcosms and many other experiments is that cross-system comparisons become difficult. Such comparisons cannot easily separate system-specific consequences (e.g. aquatic versus terrestrial), from temporal differences, from trophic complexity differences. Biodiversity manipulations of species across multiple trophic levels predominantly occur in experiments lasting more than 20 generations (Fig. 11.2(b)). Quantifying scale

at least helps us understand that any of several differences between experiments could explain the difference in the results. The next step is to record how and why the relative strength of processes changes with scale. Biodiversity–ecosystem functioning research provides an example of how the relative strength of processes change across scales. Opposite relationships between biodiversity and ecosystem functioning can occur at local and regional scales because the processes that determine functioning also change (Loreau 1998a; Lawton 2000; Levine 2000a). It should be possible to develop models of biodiversity and ecosystem functioning that include scale as an explicit parameter, rather than an implicit assumption. Similar experiments across a range of several scales (e.g. Warren 1996; Bertolo *et al.* 1999) and approaches (microcosms, mesocosms, field settings) might then parameterize the models and provide a scale-free framework for both extrapolating from small-scale experiments to larger scales, and more generally for understanding and predicting effects of scale on ecological patterns and processes.

In the rapidly expanding field of biodiversity–ecosystem functioning research, microcosms are likely to remain a principle means for exploratory research and the first proving grounds for theoretical models. There are several important directions this research can move into. Long-term experiments could reveal the effects of genetic processes on ecosystem functioning. The manipulation of immigration and emigration rates, the manipulation of volume or spatial heterogeneity to test coherency of results across different scales, longer experiments to examine long-term dynamics, and the incorporation of modern molecular methods to examine bacterial and viral community structure will permit examining pathogen/parasite and decomposer diversity at a level of detail unimaginable in larger systems. The very different spatial and temporal scales that organisms at different trophic levels operate at is currently missing from theoretical and experimental investigation. Crossing biodiversity manipulations with other co-varying global change factors, such as UV-B radiation, eutrophication, species invasions, and habitat fragmentation also represent new directions. In all cases, microcosms will help to test and guide theory, as well as provide insights into conducting more effective large-scale field experiments (e.g. whole ecosystem manipulations) and perhaps reducing the costs of such experiments by providing guidelines for effective and economic design.

O. Petchey is a NERC Postdoctoral fellow and was funded by The Biodiversity Centre at Cook College, Rutgers University and by NSF USA grant 9806427 to Peter Morin. Comments from P. Inchausti and two anonymous reviewers improved this work.

CHAPTER 12

How can marine ecology contribute to the biodiversity–ecosystem functioning debate?

M. Emmerson and M. Huxham

12.1 Introduction

Recent debate over the effects of species loss has been informed by experiments carried out largely in terrestrial ecosystems (McNaughton 1993; Tilman *et al.* 1996; Hooper and Vitousek 1997; Wardle *et al.* 1997a; Hector *et al.* 1999). A few studies have examined the effects of species loss on ecosystem processes such as biomass, density, decomposition and respiration in aquatic microbial (McGrady-Steed *et al.* 1997; Naeem and Li 1997; Petchey *et al.* 1999) and invertebrate (Hulot *et al.* 2000; Norberg 2000) ecosystems. Fewer still have explicitly addressed the effects of species richness on ecosystem processes in the marine environment (Stachowizc *et al.* 1999; Duffy *et al.* 2001; Emmerson *et al.* 2001). This contrast between terrestrial and marine systems is intriguing, not least because since the mid 1960s and 70s the field of marine ecology has encompassed the study of ecosystems, their associated communities and the interactions that take place amongst the sets of component species (Paine 1966, 1969; Hale 1975; Rowe *et al.* 1975; Pearson and Rosenberg 1978). While marine ecologists are not wholly unaware of the biodiversity–ecosystem functioning field of research (Giblin *et al.* 1995; Levinton 1995; Hutchings 1998; Snelgrove 1998, 1999; Snelgrove *et al.* 1997, 2000; Duarte 2000; Mckee and Faulkner 2000), most studies are review articles aimed at galvanizing a response from the marine community. Interestingly, these studies seem to have developed with little or no reference to the terrestrial biodiversity–ecosystem function debate. This is probably because, historically, marine ecologists have tended to examine either the effects of individual species on certain ecosystem process rates (Mayer *et al.* 1995; Kristensen and Hansen 1999) or they have ignored the contribution of individual species and concentrated on gross measures of an ecosystem process by calculating nutrient budgets and flux from whole ecosystems [e.g. Tomales Bay (Smith *et al.* 1987, 1989; Hollibaugh *et al.* 1988; Dollar *et al.* 1991)]. In fact, one of the most important marine ecosystem processes is nutrient cycling in shallow coastal waters and tidal seas, a process that is often studied independent of marine biodiversity. The flux of such nutrients fuels primary production in these habitats and hence secondary production and productivity, but such rates are mediated by both biophysical activities of benthic invertebrates and their biomass (Levinton 1995; Dame *et al.* 2001; Emmerson *et al.* 2001). Thus, explorations of the relationship between diversity and ecosystem processes in marine systems are very likely to provide valuable insights into these systems.

The extension of the biodiversity–ecosystem functioning debate to marine systems is, in fact, pertinent for several reasons. First, the consequences of human activities such as habitat modification, eutrophication and the introduction of invasive species apply as equally to marine ecosystems as they do to terrestrial ones. Second, as illustrated above for nutrient fluxes, there is already a wealth of relevant information available in the published literature. If this could be usefully analysed in a biodiversity–ecosystem functioning context, it could complement and synthesize new research. Third, certain biodiversity–ecosystem process relationships may be easier to study in marine, rather than terrestrial, systems. For example, 'natural' experiments involving anthropogenic reductions in species' richness have provided important insights into species' roles in ecosystems, but they are often confounded by parallel habitat destruction; in such cases, functioning might be affected if the system as a whole is degraded. In a biodiversity–ecosystem functioning context, the selective removal of individual species without wholesale habitat destruction is required, which is exactly the case in many third world fisheries. Fourth, keystone species and trophic cascades provide the most dramatic illustrations of how species can influence structure and function. Many of the best examples come from aquatic systems, an observation supported by Strong (1992) with reference to the highly vulnerable algal producers sustaining these systems.

In this chapter, we first attempt to address the disparity in research focus among ecologists working in marine and terrestrial habitats. The reasons for this are important because there is an impression that marine ecologists have not contributed to the mainstream biodiversity–ecosystem functioning debate in any substantial way. We then review the benthic marine literature for examples of ecosystem processes, and attempt to relate these measures back to the richness of the communities in which those measures were made, recognizing the limitations of such an approach, especially factors that might potentially be confounding. This has allowed us to synthesize the results of some experimental studies examining ecosystem process rates for marine benthic–ecosystems. We then go on to examine the suitability of certain ecosystem types for the study of biodiversity–ecosystem functioning questions and outline recent initiatives within the marine sciences that specifically address biodiversity–ecosystem functioning issues. Finally, whilst our review deals explicitly with marine benthic systems, we believe our conclusions are true for marine systems in general.

12.2 Which ecosystem processes?

Most plant-based studies to date have examined ecosystem functions that have a measurable structural component, such as biomass or cover (Naeem et al. 1995; Tilman and Downing 1994; Hector et al. 1999), which are often used as surrogates for primary production, carbon sequestration and fixation. The fixation of carbon (given increasing and urgent concerns over global warming and increased CO_2 levels) is a vitally important ecosystem service, and there are considerable advantages in using biomass as a measure of functioning. Biomass can be directly determined for each species so that individual species contributions to the net measure of function can be made; this is important for unambiguously detecting diversity effects *sensu* Hector (1998) and Loreau (1998b). The determination of individual species contributions is not always possible for ecosystem processes such as the flux of nutrients, and typically it is only possible to measure the net product or effect of all component species combined. In addition, biomass will always result in an observable effect if used as a surrogate for productivity (assuming that a species is able to grow and survive). In contrast, geochemical processes will always occur in the absence of any species, albeit at a slower rate.

Marine ecologists have tended to express ecosystem processes in terms of rates per unit area, per unit time, usually with respect to phosphate (PO_4), ammonium (NH_4), nitrate (NO_3) and nitrite (NO_2), carbon dioxide (CO_2) and oxygen (O_2), rather than using measures such as biomass (standing crop), although secondary production ($g\,cm^{-2}\,y^{-1}$) is often measured in marine systems. Interestingly, biomass production in terrestrial systems could be expressed as a rate, but in the context of the biodiversity-functioning debate typically is not.

The use of biomass as an ecosystem process has resulted in problems of pseudo-replication, the inability to separate sampling effects and niche complementarity, and the confounding problems of 'hidden' treatments (Huston 1997; Hector 1998; Loreau 1998b). We do not suggest that the examination of process rates such as flux is in any way superior to measures such as biomass, but that presently marine ecologists seem well placed to measure such processes. Marine ecologists may be tempted to simply apply biodiversity–ecosystem functioning jargon to their current approaches, but whilst describing the 'ecosystem processes' associated with particular species and locales may be of interest, this will not advance the biodiversity–ecosystem process debate if it fails to consider the difficulties in interpreting experimental findings already identified in terrestrial studies. We believe that marine ecologists urgently need to adopt the manipulative experimental approach seen in the terrestrial literature with respect to biodiversity–ecosystem functioning questions. However, cross-informing marine and terrestrial ecologists is problematic. The two disciplines have distinct funding arenas, both nationally and internationally (this is especially so within Europe), and as a result marine and terrestrial researchers are often unaware of each other's work, even during its formative stages in the peer review process. Second, marine ecologists have their own learned societies, peer reviewed journals and conferences (May 1984; Giller *et al.* 1994; Raffaelli and Burslem 2000), which are rarely accessed by terrestrial ecologists. Marine and terrestrial ecologists also have a sense of separate identity, reflecting a perception of what is and what is not relevant to their own research. This means that highly relevant work, whatever the subject area, is often overlooked on either side of the terrestrial–marine divide. To produce a true synthesis detailing the effects of biodiversity on ecosystem performance requires examination of ecosystem processes that are at least comparable amongst ecosystem types, recognizing that certain processes might behave differently in different ecosystems. Obvious candidate functions include rates of primary and secondary production and rates and mediation of nutrient flux. The ecosystem stability and resilience of these same properties could also be assessed under differing diversity regimes by measuring their temporal variance or ability to maintain process rates following some form of disturbance.

12.3 Studies of relevance

Data do exist for assessing biodiversity–ecosystem functioning relationships in the marine environment, but much of these data are not always readily accessible and will require the collaboration of original authors. The raw data are often present in 'grey' literature (i.e. governmental or agency publications) or held by individual researchers who are unaware of its relevance possibly, in some instances, because they are largely unaware of the terrestrial biodiversity–ecosystem functioning work. The original data might also be less accessible because it is presented graphically or summarized as average values bounded by error estimates, or as a range of values that have been measured seasonally. Often process rates are expressed while the composition and richness of the community are merely summarized, detailed elsewhere, or only alluded to. Here, we have synthesized the results of a few such investigations drawn from the marine literature.

12.4 Marine benthic nutrient fluxes

Asmus *et al.* (2000) provide a comprehensive review of nutrient fluxes in benthic systems. We examined 76 studies for details of both community composition and ecosystem process rate. We have included only those studies that provide details of both the species richness of the benthic assemblage and nutrient flux from the sediments containing those communities (15 studies). In some instances, details of the species richness and flux were presented in separate papers by the same author(s) and data had to be collated. We approached some authors directly for their raw data, or when raw data were not available we obtained flux measures by digitizing published graphical work to recover the data. The data we use are by no means exhaustive either in its extent or coverage of marine systems and we present it here fully aware of the potential methodological problems associated with our approach. Such examples, however, serve to formulate hypotheses regarding some of the relationships that might exist in marine systems and can help identify where future efforts may be most profitably directed. We present them here in the absence of direct experimental data.

Programmes examining the flux of nutrients have typically involved *in situ* cores, benthic chambers, bell jars or flumes (Asmus *et al.* 1998) or *in vivo* sediment cores collected from the field and returned to the laboratory. Such flux measures are usually made over relatively short time scales (6–8 h). The concentration of nutrients in the overlying seawater is determined at the start and end of the flux incubation. Usually incubations are carried out both in the light and in the dark to account for the effects of primary production (benthic and pelagic) on nutrient flux from sediments. Microscopic primary producers (microphytobenthos), such as diatoms, can have dramatic effects on the flux of nutrients from sediments, resulting in negative flux rates, i.e. nutrient concentrations decline over the course of an incubation (Asmus *et al.* 2000). Flux measures are typically determined for different habitat types or biogenic structures within intertidal or shallow water systems. These have commonly included mussel beds, oyster beds, seagrass beds, *Arenicola* flats, mudflats and macro-algal mats.

Here we have taken NH_4 flux measures made in different studies and locations and expressed them per unit area over time. Where available we have used the number of species present in the sediment for a given flux incubation to predict the flux of nutrients for that incubation. For six of the 15 sites detailed above (18 species and over) we have used published species lists by the same authors for flux measures made at that site. A synthesis of the 15 studies, shows that the production of ammonium (NH_4) from sediments appears to increase with increasing species richness (Fig. 12.1). For the 26, 27 and 31 species sites, *Nephtys-Nucula*, *Ampelisca-*, and *Mercenaria*-dominated communities within the Narragansett Bay ecosystem respectively (Fig. 12.1), the original regression analysis by Nixon *et al.* (1976) demonstrated that temperature alone explained 80% of the variability in flux rates when the data from these three community types were pooled. Thus, it is likely that covariation with seasonal temperature explains much of the scatter evident in Fig. 12.1 and might also be responsible for the average trend. Other factors such as biomass, abundance, sediment particle size and sediment organics will undoubtedly contribute to the variability evident at and between particular levels of species richness. Complete data for the temperature, biomass and abundance are limited for all sites presented here and we are unable to account for the effects of these factors on the variability

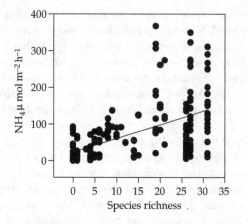

Figure 12.1 Relationship between ammonium flux (NH_4 μmol m^{-2} h^{-1}) and species richness. Line of best fit is a linear regression ($y = 3.74x + 23.48$, $R^2 = 0.27$, $P < 0.001$). NH_4 flux measures were reviewed from the literature for different species and species combinations, under both laboratory and *in situ* conditions. When detailed information regarding species richness for a given study has not been available directly for flux measures we have used published and unpublished species lists for the systems in which those measures were made (18 species and above). Data were obtained from Nixon 1976; Aller and Yingst 1978, 1985; Aller *et al.* 1983; Kristensen 1984; Asmus 1986; Andersen and Kristensen 1988; Sundbäck *et al.* 1991; Kristensen 1993; Sprung 1994; Kristensen and Hansen 1999; Asmus *et al.* 2000; Christensen *et al.* 2000; Hughes *et al.* 2000; Emmerson *et al.* 2001.

evident in Fig. 12.1. Whilst such examples of biodiversity–ecosystem process correlations are interesting, we recognize the limitations of adopting a cross-study comparison such as this for the determination of diversity effects. Combining data from different studies in a single plot is not straightforward and there are often confounding factors that are unquantified or uncontrolled. Our presentation of Fig. 12.1 here serves only to identify areas worthy of further investigation and aims to illustrate how this might proceed.

Diversity effects are commonly quantified with metrics such as over-yielding (Loreau 1998b) which require measurement of all single species (monoculture) effects on the processes. In the present analysis (Fig. 12.1), the naturally occurring diverse sites tend to feature rare or less abundant species that are typically regarded as being of less importance with respect to function. Hence these single species effects are not reported in the literature and are unavailable. If the view that diversity effects cannot be determined without such metrics is accepted then searching for patterns in existing data sets will be limited. We do not adhere to this view. In fact, marine ecologists are leading the field in attempting to reconcile pattern- and process-based approaches for the examination of ecological phenomena (Stachowizc et al., in press).

A review of studies carried out in biogeographically distinct regions, provides a range of species richness across sites, which are compositionally distinct (i.e. independent). This means that the sampling type effects described by Huston (1997) cannot influence patterns in the data. However, the lack of comprehensive covarying environmental data (e.g. temperature, biomass and abundance) represents a significant problem. Factoring out the functional diversity of communities represents an additional problem because although the functional roles of the more abundant species are well understood (Jumars and Nowell 1984; Dauwe et al. 1998; Hooper et al., Chapter 17), this is not always the case for rare species.

Because sites separated biogeographically will tend to be independent with respect to species identity, the total number of species represented in such an analysis is likely to be high. If then a mechanism for diversity effects needs to be determined and this requires knowledge of all single species effects (Huston et al. 2000), then this weakens the present approach. However, it should still be possible to assess the contribution of diversity effects relative to other variables such as temperature and biomass. Van Es (1982) demonstrated such an approach in an exploration of community respiration, primary productivity and carbon flux though the Ems–Dollard Estuary ecosystem, The Netherlands. Van Es (1982) found that temperature alone and temperature plus viable bacteria explained 50% and 70%, respectively, of the observed variation in community respiration. We take this as further evidence that at least in marine ecosystems, covarying factors seem to dominate patterns of ecosystem process rates.

12.5 Primary and secondary production

Marine ecologists have studied primary production in detail, but they have yet to relate productivity to the diversity of the intertidal or pelagic communities and ecosystems in which those measures were made. For marine systems the diversity of phyto-planktonic communities can be very high. For example, Trigueros and Orive (2001) recorded 81 diatom species and 38 dinoflagellate species over an annual cycle. Relating productivity to community composition has proved problematic in pelagic systems (Verity 1998), whilst for intertidal communities it is sometimes difficult to differentiate amongst macrophyte species and morphological types, such as *Ulva* spp. (Malta et al. 1999).

Traditionally, marine ecologists have regarded secondary production as a function of primary productivity. However, it is clear that invertebrate biomass can affect nutrient fluxes (Emmerson et al. 2001). If invertebrate production and biomass are affected by species richness then invertebrate species richness might have an indirect effect on primary productivity in marine systems. In an original analysis presented by Herman et al. (1999), secondary production (g ash free dry weight m^{-2}) was presented as a function of primary productivity. Here, rather than present secondary production as a function of primary productivity we have presented primary production as a function of invertebrate productivity (Fig. 12.2), since rates of nutrient supply to plants can be mediated by the biological activities and biomass of invertebrates living in the sediment (Emmerson and Raffaelli 2000). These data reveal an asymptotic relationship between invertebrate productivity and primary productivity. In addition there may be a relationship between invertebrate biomass and diversity (Fig. 12.3). Again comprehensive data for the covarying environmental effects is unavailable and we cannot account for this in the present analysis. It could be argued that the positive relationships documented for the two sites in Fig. 12.3 (Sylt-Rømø Bay, Germany and Ria Formosa, Portugal) might be a natural sampling effect, i.e. naturally diverse systems have a higher probability of containing larger or more productive species. However, as Emmerson et al. (2001) demonstrated, species which exert a disproportionate effect on function (a 'sampling effect') are present across the whole range of species richness in natural systems. Of course, there are competing explanations for the

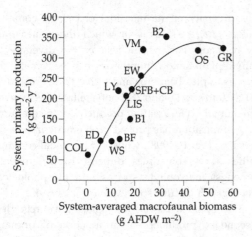

Figure 12.2 Relationship between system-averaged macrofaunal biomass (g AFDW m^{-2}) and the primary productivity of 14 shallow well-mixed estuarine systems (redrawn from Herman *et al.* 1999). Reanalysed here, system primary productivity is presented as a function of macrofaunal biomass. The best fit regression line is a quadratic polynomial $y = -0.147x^2 + 13.82x + 12.63$, $R^2 = 0.82$, $P < 0.001$. System abbreviations are OS, Oosterschelde; B1, Balgzand (Dutch Wadden Sea); B2, Balgzand (Dutch Wadden Sea); VM, Veerse Meer; ED, Ems Estuary, Inner part ('Dollard'); SFB, San Francisco Bay; LY, Lynher Estuary; WS, Westerschelde; BF, Bay of Fundy; COL, Columbia River Estuary; LIS, Long Island Sound; CB, Chesapeake Bay. See Herman *et al.* (1999) for data sources. Herman *et al.* (1999) original analysis included the Ythan Estuary. We have removed it here as an outlier.

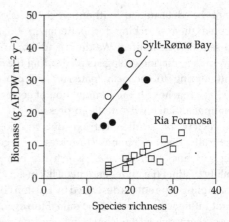

Figure 12.3 The relationship between benthic secondary production and macrofaunal species richness for two sites Sylt-Rømø Bay, Germany ($y = 1.65x - 1.91$, $R^2 = 0.54$, $P < 0.05$) and Ria Formosa, Portugal ($y = 0.53x - 4.96$, $R^2 = 0.67$, $P < 0.001$). Measurements at both sites were made for a number of different habitat types. For the Ria Formosa—sand flats, seagrass beds, mudflats and nuisance macroalgal mat impacted regions of mudflat. For the Sylt-Rømø Bay—mudflats, *Arenicola marina* flats, and seagrass beds. At Ria Formosa three mudflat data points appear to be outliers for that site (○). Open symbols. These measures were made in 'healthy' mud that had not been affected by nuisance macroalgal blooms. These data appear to be more consistent with the data for the Sylt-Rømø Bay site (where the measurements presented were made from mudflats unaffected by nuisance macroalgal mats). Consequently these mudflat data have been included in the Sylt-Rømø Bay regression. Data obtained from Asmus and Asmus 1985; Beukema 1989; Sprung 1994; Asmus *et al.* 2000.

trends shown in Figs 12.2 and 12.3. For instance, Gotelli and Colwell (2001) discuss the importance of differences in abundance and sampling effort for measures such as species richness and amongst site comparisons. It is also well documented that both species richness and biomass (and productivity) covary along environmental stress gradients (e.g. organic enrichment, salinity and hypoxia), increasing as one moves from stressed to benign conditions (Whittaker 1974). Both factors might therefore be expected to covary with both beach morphology and exposure (Raffaelli and Hawkins 1996). However, recent work suggests that these relationships may be more variable than originally supposed (Jaramillo and Lastra 2001) and the causal relations between diversity, biomass and productivity may therefore warrant further exploration.

Ricciardi and Bourget (1999) examine global patterns of benthic macroinvertebrate biomass and its distribution among functional feeding guilds for

36 rocky shores and 245 sedimentary shores. They showed that variation in biomass was related to physical variables (annual air and water temperatures, sediment grain size, intertidal slope, tide range and type, wave height and exposure) and that linear combinations of physical variables explain up to 44% and 40% of the variance in biomass on sedimentary and rocky shores, respectively. They did not use species richness as an explanatory variable in their analysis.

12.6 Suspension feeding and clearance rates

A useful surrogate for secondary production in many marine communities is the rate at which suspension feeders remove (or clear) particles from the water. Suspension feeding is one of the most ecologically significant features of aquatic environments. Whilst the ingestion of suspended particles and organisms is rare in terrestrial systems, it often represents the dominant feeding mode in marine habitats, and has major ecosystem-level significance. For example, suspension-feeding bivalves in San Francisco Bay filter a volume of water equivalent to that of the Bay every day (Cloern 1996). Such high filtration rates mean that suspension feeders often control or affect the levels of phytoplankton and other particles in the water (Petersen and Riisgård 1992; Cloern 1996). In doing so, they influence the occurrence and frequency of algal blooms, the transparency of the water and the extent of some benthic habitats, such as sea grass beds (Lemmens *et al.* 1996). Hence, clearance rates may be considered as both an ecosystem process as well as a surrogate for a different process (secondary productivity). Whilst a large body of literature exists on the feeding rates of individual organisms, relatively little work has translated this to the ecosystem level, and less still has considered the potential impact of mixed communities of suspension feeders on this important ecosystem function (Dame *et al.* 2001).

There are reasons to believe that increasing species diversity in suspension-feeding communities may lead to increased clearance rates, at least in sedimentary environments. Selective feeding on different sized particles by different species may broaden the size range of particles ingested and so increase the total feeding rate (Lesser *et al.* 1992). For example, mussels, ascidians and hydroids preferentially retain particles of 3–5, 5–15, and 30–80 μm diameters, respectively (Lesser *et al.* 1992; Armsworthy *et al.* 2001; Coma *et al.* 1995). One mechanism enhancing the filtration efficiency of multi-species communities is the likely re-ingestion of particles already emitted by other species in the community (Mook 1981). In contrast to most terrestrial and marine hard substratum environments, there is relatively little evidence for competitive exclusion in most sedimentary habitats (Huxham *et al.* 2000; Raffaelli and Hawkins 1996). Adding individuals of a new species does not, therefore, necessarily displace existing ones, and biomass and species richness may positively covary (as suggested in Fig. 12.3). Testing the proposition that

species richness affects the amount of particulate material in suspension should be easy to test.

Lemmens *et al.* (1996) conducted one of the few studies to consider the filtering capacity of whole benthic communities. They examined five sites within Cockburn Sound, Western Australia, sampling from within seagrass and bare sediment habitats at each site. Plotting filtration rates (based on the mean biomass they report for each taxon at each site) against reported taxon richness, produces a significant positive relationship (Fig. 12.4).

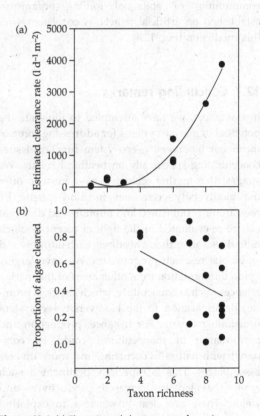

Figure 12.4 (a) The estimated clearance rates for each taxon are scaled by their respective biomass in the system (data from Lemmens *et al.* 1996), and the relationship between taxon richness and clearance rate is plotted. The figure conflates data from seagrass and open mudflat habitats in Cockburn Sound, Australia. The trendline shows a power regression $y = 23.542x^{2.2}$ $R^2 = 0.92$ and demonstrates a significant positive relationship. (b) Laboratory-measured clearance rates (measured as the proportion of initial particles removed after 2 h) plotted as a function of taxon richness for epifaunal fouling communities from the Forth Estuary, Scotland. Despite a negative slope, the linear regression is non-significant $R^2 = 0.17$. Data provided by H. Lindsay (Napier University, unpublished).

In contrast, hard-substratum communities, which commonly experience intense competition for space, may show a different pattern. Here, large species with high clearance rates, such as mussels and tunicates often exclude smaller species during settlement and reduce species richness (Paine 1966). Such exclusion by dominant species would lead to a negative relationship between taxon richness and clearance rates through a natural sampling effect. Unpublished data (Hazel Lindsay, Napier University) from a study examining feeding rates in mixed communities of epifaunal fouling organisms, established on artificial panels, is consistent with this prediction (Fig. 12.4).

12.7 Concluding remarks

In this paper, we have attempted to illustrate the potential of marine systems for addressing some of the major biodiversity–ecosystem function issues concentrating specifically on benthic habitats. We suggest that marine ecologists have most to offer the biodiversity–ecosystem function debate by re-examining published and unpublished data that can be re-evaluated in the light of recent research initiatives, rather than a headlong rush into new and ambitious research programmes. We have largely ignored information from other areas of the marine sciences, such as aquaculture, which has great promise for application to the biodiversity–ecosystem functioning debate. For instance, production and performance in monocultures could be compared with naturally occurring and more diverse assemblages. This is especially pertinent if such approaches led to increased productivity and yields. There are clear advantages to exploring biodiversity–ecosystem function questions in shallow and intertidal water benthic habitats, including the relative ease of *in situ* and *in vivo* manipulative experiments, the short time scales involved in measuring ecosystem process rates, the methodological techniques for measuring ecosystem processes and a good understanding of the effects of the biota on these processes. It is also obvious that marine ecologists have long been involved with research programmes that are germane to biodiversity–ecosystem functioning questions and that considerable data already exists which could be analysed in a biodiversity–ecosystem process context. It is clear that so far marine ecologists have not contributed substantially to the biodiversity–ecosystem process debate. If marine ecologists are to play a more considered role in biodiversity–ecosystem functioning research, then it is important that they and terrestrial ecologists communicate more effectively to ensure that the appropriate questions are asked by both types of ecologists and that the lessons learnt by one group can be applied by the other.

We thank D. Raffaelli for his contribution to the manuscript and M. Solan for useful comments. We thank S. Nixon for provision of species lists for the Narragansett Bay ecosystem, E. Kristensen, H. Asmus and R. Asmus for the kind supply flux and species list data, H. Lindsay for the provision of filtration rate data. To T. Hollibaugh and S. Dollar our thanks for their help in trying to track down the Tomales Bay species lists. We thank P. Inchausti, M. Loreau, S. Naeem and P. Morin for constructive comments over the manuscript. We also thank those others that were kind enough to respond to our requests for data.

CHAPTER 13

Multi-trophic dynamics and ecosystem processes

D. Raffaelli, W. H. van der Putten, L. Persson, D. A. Wardle,
O. L. Petchey, J. Koricheva, M. van der Heijden, J. Mikola,
and T. Kennedy

13.1 Introduction

It is axiomatic in ecology that the dynamics of one set of species cannot be understood without reference to the dynamics of other species and processes. For instance, biodiversity loss within a trophic level is likely to impact on species at other levels, whether directly through changes in competitive and consumer–resource interactions, or indirectly via changes in ecosystem processes (Fig. 13.1). Such interactions imply that predicting the consequences of biodiversity loss for ecosystem processes will be difficult, especially when, as is often the case, there are simultaneous reductions in diversity at more than one trophic level. Given the complexity of real ecological systems, the task of producing a predictive framework for providing advice to managers and policy makers on the consequences of biodiversity loss might seem daunting. However, there exists in the literature much theory, as well as observational and experimental data, which could potentially provide the basis for such a framework, although the orientation of the original studies may have been different from that of the present 'biodiversity–ecosystem function' issue (henceforth termed 'ecosystem process'). In this chapter, we explore how our present knowledge and understanding of interactions between species and between different trophic levels can be used to inform the biodiversity–ecosystem function debate. Specifically, we consider:

- the empirical evidence for the kinds of interactions shown in Fig. 13.1, and the utility of experiments where biodiversity within a single species set (usually trophic level) is manipulated;
- the feasibility of the controlled experimental approach for examining the effects of simultaneous changes in biodiversity at several trophic levels;
- the potential of our current theoretical and modelling base for addressing the effects of biodiversity change for ecosystem process;
- future directions where an approach which considers several trophic levels might usefully inform the current debate.

13.2 The need to account for multi-trophic effects

Figure 13.1 shows the likely feedbacks between a species set (e.g. terrestrial plants), the ecosystem process of interest (e.g. primary production) and biota at other trophic levels (see also Fig. 13.1 in Mulder *et al.* 1999). In addition to material flows from consumers to resources, top-down and bottom-up effects on community organisation (e.g. diversity) are also likely between trophic levels, which in turn will affect system process.

There is now considerable evidence for each of the feedbacks shown in Fig. 13.1. Much of this comes from the growing literature on multi-trophic interactions (e.g. Gange and Brown 1997; Van der Putten *et al.* 2001; Tscharntke and Hawkins, in press), which has demonstrated often complex and unexpected feedbacks between the elements of bi- and tri-trophic systems, with implications for the structure and functioning of the entire food web. In this

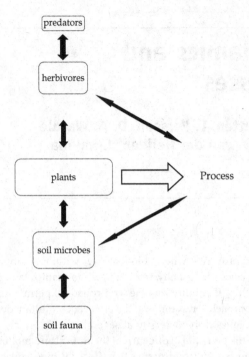

Figure 13.1 Multi-level feedbacks on function and biodiversity. Feedbacks may be both positive and negative. These kinds of interactions are of course not limited to plant communities.

section, we deal explicitly with the empirical evidence from a number of such studies that interactions between different trophic levels affect both biodiversity and ecosystem process. Most of the recent experimental work on the biodiversity–ecosystem process issue has focussed on terrestrial systems, especially on plants, and it is appropriate therefore to first examine the evidence for multitrophic effects from terrestrial studies. We then extend our review to aquatic systems, where similar processes occur, but where the biodiversity–ecosystem process issue has been approached differently (see Emmerson and Huxham, Chapter 12; Raffaelli 2000; Persson 1999 for commentaries).

In terrestrial systems, plant species richness and functional diversity have been shown to affect both herbivore abundance and diversity (Andow 1991; Siemann *et al.* 1998; Koricheva *et al.* 2000). Herbivores in turn may affect plant diversity and ecosystem processes (e.g. primary productivity and nitrogen cycling) by selectively removing plant species, creating refuges for competitively inferior species or affecting carbon and nitrogen cycles

directly (e.g. Brown and Gange 1989, 1992; Olf and Ritchie 1998; Pacala and Crawley 1992; Mulder *et al.* 1999; Knops *et al.* 2001). Other studies have indicated effects of both the biodiversity of terrestrial plant assemblages and ecosystem processes (net primary production NPP) on below ground processes. These have been discussed at length by Wardle and Van der Putten (Chapter 14) and we provide a summary here.

The potential effects of plant richness have been investigated for root-associated fungal and bacterial abundance (Christie *et al.* 1974), arbuscular mycorrhizal fungi (Helgason *et al.* 1998; Cuenca *et al.* 1998), plant pathogens, soil microbial biomass (Wardle and Nicholson 1996) and litter decomposition (Hector *et al.* 2000; Mulder *et al.* 1999), although plant functional group effects may be more important than richness *per se* (Hooper and Vitousek 1998; Wardle *et al.* 1997, 1999; Spehn *et al.* 2000a). Similarly, there are different effects of litter from mono- and multi-species plant assemblages on soil biota, decomposition and mineralization (Chapman *et al.* 1988; Kaneko and Salamanca 1999; Hansen 2000; Blair *et al.* 1990; Wardle *et al.* 1997; Hector *et al.* 2000a; Bardgett and Shine 1999; Seastedt 1984), but these are variable in direction and magnitude and dependent on location and plant species identity.

Similarly, it is likely that the ecosystem process measured in several plant species richness experiments (net primary production, NPP) has demonstrable effects on terrestrial soil systems (e.g. Mikola and Setälä 1998c; Scheu and Schaeter 1998; Chen and Wise 1999; Myrold *et al.* 1989; Zak *et al.* 1994; Groffman *et al.* 1996; Jonasson *et al.* 1996; Vance and Chapin 2001; Bääth *et al.* 1978). However, in many investigations surrogates for NPP (e.g. glucose) were the treatment variables, rather than NPP itself. Here again, the effects of the treatment vary greatly between studies and seem similarly context-dependant, perhaps because the soil food web is complex and characterized by a mixture of bottom-up and top-down processes, the relative influence of each depending on the energy channel most affected (*sensu* Moore and Hunt 1988; de Ruiter and Moore 1997) and trophic position (de Ruiter *et al.* 1995, 96, Chapter 9). The precise effects of terrestrial NPP on below-ground systems are therefore difficult to predict (Wardle and Yeates 1993).

There are also well-documented interactions between terrestrial plants and mutualistic, parasitic and pathogenic soil microbes, which greatly complicate the overall multi-trophic picture for terrestrial systems. For instance, mycorrhizal fungi have been shown to modulate the relationship between plant diversity and ecosystem processes (Grime 1987; Klironomos et al. 2000; van der Heijden et al. 1998b, Chapter 16) by re-allocating available resources between co-existing plants and promoting the growth of specific plants. These fungi can also alter ecosystem processes such as nutrient and carbon cycling, soil structure, decomposition and productivity (Smith and Read 1997; van der Heijden and Sanders 2001). Similarly, plant soil pathogens have the potential to affect co-existence of above-ground herbivores by inducing chemical defenses that affect herbivores differentially (see Van der Putten et al. 2001 for a review).

Whilst it is not hard to find parallel interactions between species and trophic levels in aquatic systems, there are some fundamental differences between aquatic and terrestrial systems. In contrast to terrestrial plants, the major primary producers in aquatic systems are algal and not intimately associated with sediments. The only truly marine angiosperms are the sea grasses; salt marsh and mangrove taxa are confined to the marine–terrestrial ecotone. This makes it difficult to draw detailed comparisons between aquatic and terrestrial systems for the multi-trophic interactions described above. Neither does there seem to be an ecological equivalent of mycorrhizal fungi for marine algae, probably reflecting the different nutrient uptake routes by plants. Whilst effective chemical defences have been documented for marine algae (reviewed in Raffaelli and Hawkins 1996), their inducibility by pathogens and subsequent role in mediating herbivore communities (so-called 'crosstalk'; Felton and Korth 2000; Van der Putten et al. 2001) has not been explored.

Nevertheless, many parallels do exist. For instance, the compartmentation of terrestrial soil food webs described as energy channels by Moore and Hunt (1998) and de Ruiter and Moore (1997) have their equivalents in marine and aquatic pelagic systems: in addition to the classical grazing food chain from phytoplankton to piscivorous fish,

there exists a microbial loop, within which significant carbon flows occur (Stockner and Porter 1988) and both channels appear linked (Rieman et al. 1986; Rieman and Christoffersen 1993; Wickham 1995; Pace et al. 1998). Furthermore, the amount of carbon flowing through each of these channels is a function of the biomass of piscivorous fish, at least in freshwater systems. Clearly, the coupling between different branches of aquatic food webs will need to be addressed in future biodiversity studies.

Compensatory interactions are known to occur within a trophic level in aquatic systems (Osenberg and Mittelbach 1996; Polis and Strong 1996; Persson 1999). This is thought to reflect a trade-off between efficiency of resource utilization and ability to avoid predators (Rosemond 1996; Leibold et al. 1997). Thus, in an experiment by Leibold and Wilbur (1992), a planktonic herbivore (Daphnia) enhanced periphytic growth, whilst a surface grazer (Rana) enhanced phytoplankton growth. Similar facilitation between different taxa of litter processors in stream systems is well documented (e.g. Johnsson and Malmqvist 2000). Wherever heterogeneity occurs within a trophic level, compensatory responses are likely to occur in response to biodiversity changes elsewhere in the food web and will have effects on ecosystem process.

Interestingly, some of the strongest evidence for feedbacks between trophic levels comes from aquatic systems, in particular the occurrence of trophic cascades in pelagic food chains in lakes (see Persson 1999; Carpenter et al. 2001 for reviews) and in benthic food chains in lacustrine, riverine and marine intertidal systems (e.g. Martin et al. 1992; Paine 1981; Brömark et al. 1991, 1997; Power 1990, 1992; Wootton 1995; Diehl and Kornijów 1997). The common feature of all these cascades is the extent to which carnivores limit the negative impacts of herbivores on plants, and thereby have positive effects on plant production. Such cascades might also be common in terrestrial systems (Pace et al. 1999; Schmitz et al. 2000). However, cascades are unlikely to be easily revealed in terrestrial plant biodiversity experiments, where the focus is exclusively on a single trophic level and/or when other trophic levels are rigorously excluded or otherwise controlled for. Also, the time scales over which cascading effects are detectable are quite different for terrestrial and aquatic systems, due to the relative generation times of the main primary producers in terrestrial (decades to centuries) and aquatic (weeks to months) systems (Huntly 1991; Persson 1999), implying that terrestrial field experiments should run for longer, relevant time scales (Raffaelli and Moller 2000; but see also Leibold *et al.* 1997).

In addition to recognizing that significant interactions and feedbacks occur between different trophic groups and between these groups and ecosystem processes, there are many other compelling reasons why a multi-trophic approach is required in investigations of biodiversity change on ecosystem processes. For instance, at the largest geographical scale and with climate change, the immediate cues for migration used by migratory birds may become different from those of their prey, so that the birds arrive on their spring territories at an inappropriate time (Both and Visser 2001). There is growing appreciation that trophic cascades are not restricted to aquatic systems (Strong 1999; Schmitz *et al.* 2000), so that failure to take a multi-trophic approach in biodiversity studies may well underestimate the impacts of environmental change on processes. Similarly, habitat generalist species are likely to become more dominant and specialists likely to decline through habitat loss (Warren *et al.* 2001), with consequences for tropho-dynamics and hence ecosystem processes. Lastly, habitat fragmentation not only leads to a decline in biodiversity (Hanski 1988, 1999), but may also change the relative influence of top-down and bottom-up forces in the food web as predators find it harder to locate prey patches (Kruess and Tscharntke 1994). The effects of biodiversity loss in the real world may therefore be to dislocate food chains or change trophodynamics in subtle and unexpected ways. In this respect, it is worth noting that the highly focussed debate on the relative importance of complementarity versus sampling effects for explaining the results of previous biodiversity experiments (e.g. Huston 1997; Tilman *et al.* 1997b; Aarssen 1997) will seem largely irrelevant for multi-trophic systems, where the mechanisms will prove much more complex.

13.3 Approaches incorporating several trophic levels

From the evidence presented above, it is clear that significant feedbacks occur between trophic levels which may regulate the process of interest in biodiversity experiments. Just how this might affect experiments which focus on a single trophic level (e.g. plants) has been elegantly illustrated by Mulder *et al.* (1999) comparison of the outcome of subsets of the BIODEPTH experiments (Hector *et al.* 1999) in the presence or absence of insect herbivores: when insects were present, the relationship between plant richness and biomass was much weaker than when insects were excluded (Mulder *et al.* 1999). Experiments clearly need to incorporate multi-level effects and several soil and freshwater microcosm studies have already done so (e.g. Laakso and Setälä 1999b; de Ruiter *et al.* 1995, 1996, Chapter 9; Mikola *et al.*, Chapter 15; Wardle and Van der Putten, Chapter 14; Petchey *et al.*, Chapter 11). Whilst all of these studies have examined the effects of species richness on food web wide biota and processes, only a handful (Wardle *et al.* 2000a; Laakso and Setala 1999b; de Ruiter *et al.*, Chapter 9; Petchey *et al.*, Chapter 11) manipulated biodiversity at more than one trophic level as main factors in the design. In the majority of the studies, the non-manipulated species are best seen as response variables.

The controlled manipulation of species richness at several trophic levels is not a trivial task. For many systems, there will be problems associated with the different temporal and spatial scales over which different trophic levels naturally operate. For instance, the rapid responses of above-ground herbivores to changes in plant biodiversity, is in marked contrast to those of soil organisms and processes which are much slower (Wardle *et al.* 1999). Thus, in the CLUE-Project (Van der Putten *et al.* 2000), the short-term response of the soil community to plant diversity was weak, suggesting a marked lag behind above-ground changes, and higher trophic levels might respond even more slowly. This decoupling of the temporal dynamics of different trophic levels in biodiversity experiments has spatial parallels. Above-ground herbivores and their predators often have foraging extents which are orders of magnitude larger than the scale of experimental plant diversity plots, and these extents tend to increase dramatically with increasing trophic level. Similarly, the larger consumers which operate at higher trophic levels in aquatic systems effectively couple quite different major habitat types, such as pelagic, benthic and littoral systems (Vanni 1996; Schindler 1996), and, in the case of migratory fish, such as salmonids, may significantly subsidize freshwater and terrestrial systems (Polis et al. 1997). Similar examples could be provided for terrestrial consumers. Issues of scale are dealt with in detail elsewhere (Bengtsson et al., Chapter 18), and it clear that it will be extremely difficult to design multi-trophic level experiments which can accommodate the necessary ranges of temporal and spatial scales for all species.

There are also scale-dependent taxonomic issues. Whilst terrestrial plants and most higher trophic level organisms are relatively easily identified and enumerated, the taxonomic resolution and enumeration of many lower trophic levels or smaller metazoan consumers can be extremely difficult. This is especially true for the biota of terrestrial soils and freshwater and marine sediments. Here, microbial diversity is huge (e.g. several thousand bacterial species per g of soil (Torsvik et al. 1994)) and molecular techniques are often necessary for species identification. Indeed, the majority of soil and sediment microbes have yet to be named and many cannot be cultured. Even the larger soil and sediment biota (nematodes, mites, tardigardes, turbellarians, harpacticoid copepods, kinorhynchs, etc.) require expert taxonomists, and can usually only be grouped into feeding guilds.

These logistic difficulties constrain the range of approaches available in soils and sediments, which have either employed selective soil sterilization techniques (e.g. Degens 1999; de Ruiter et al., Chapter 9) or the assembly of soil decomposer (Mikola et al., Chapter 15) or marine benthic invertebrates communities in microcosms (Emmerson et al. 2001). The microcosm approach has many advantages for assessing the effects of biodiversity on ecosystem process, especially in aquatic systems. Communities can be assembled successfully to provide a range of trophic levels with varying patterns of connectance; the experiments can be run for many generations of

the species concerned; replication of treatments can be high; a wide variety of system properties and processes can be measured (Petchey et al., Chapter 11). These features also permit close links between empirical observation and developing theory. Petchey et al. (Chapter 11) discuss this approach in detail and here we provide only two examples to illustrate the potential of this approach for incorporating multi-trophic level treatments in experiments. In their microcosms, Naeem and Li (1998) and Naeem et al. (2000a) manipulated primary producer and decomposer richness as well as the number of species per functional group. Their results show that more diverse decomposer assemblages made more efficient use of the carbon substrate, in turn probably enhancing nutrient supply to the primary producers. Similarly, Petchey et al. (1999) constructed food webs with five trophic groups, each having a range of richnesses represented, and subjected them to disturbance (warming). Extinctions were more frequent in the disturbed treatments and were trophic-level dependent, with changes in food web structure having likely effects on ecosystem processes (primary production).

The potential of microcosms for addressing questions about biodiversity and ecosystem process looks promising, but there are limitations to this approach. By definition, microcosms are best used to support assemblages of microorganisms, usually protists or small metazoans. Investigations of the interactions between larger species and ecosystem processes require larger-scale versions, often termed 'mesocosms', (although the distinction is rather arbitrary; Lawler 1998). Mesocosms have been built to accommodate larger-sized species, such as macroinvertebrates and fish (Gamble 1991), but, with the exception of Hulot et al. (2000) and some of the Ecotron experiments, large-scale enclosed systems have not been used to specifically address the biodiversity-process issue. The difficulties of designing large-scale enclosed 'mesocosm' experiments which capture the dynamics of larger-scale systems are similar to those for any large-plot manipulative field experiment (Raffaelli and Moller 2000). Financial and physical resource constraints limit the degree of replication that can be achieved, so that there is usually a trade-off between enclosure size and number of replicates. Simply maintaining large enclosures is prohibitively expensive, but the potential has been demonstrated for lakes (Carpenter *et al.* 1995, 2001; Persson 1999), terrestrial systems (Krebs *et al.* 1995; Erhlinge 1987; Brown 1999), marine pelagic systems (Gamble 1991; Lalli 1991) and islands (Vitousek *et al.* 1995; Wardle *et al.* 1997a). Whilst many of the lake studies used whole lakes or natural boundaries for enclosure, a very real problem with marine pelagic mesocosms is the removal of turbulent structure within the water column, a primary driver of ecosystem processes (Gamble 1991).

Finally, a feature of most microcosm experiments is that they have required the assembly of ecological systems. Whilst the ability to rigorously control and replicate the composition and structure of the resulting food web must be viewed as a major advantage of the microcosm approach, it is also a potential weakness for addressing questions about the effects of biodiversity loss from intact systems. Assembled species-poor systems are not necessarily equivalent to systems made depauperate through species loss: their dynamics (and hence process) may be quite different in the two scenarios. The selective removal of species from established species-rich microcosms (or of *in situ* primary producers from terrestrial field sites) is logistically much more difficult than species assembly, but may be the most productive way forward.

13.4 Theory and models

It is clear from the above that empirical and experimental investigations of the effects of biodiversity change on ecosystem process need to encompass several trophic levels. Similarly, modelling and theoretical approaches need to include a multi-species–multi-level approach. At first sight, existing mainstream ecosystem and food web theory ought to be able to inform our understanding of the consequences of biodiversity loss. However, these fields have been more concerned with the relationship between diversity and stability, rather than with diversity and ecosystem process *per se* (Loreau *et al.*, Chapter 7). Nevertheless, there are clear lessons to be learned from how diversity–stability debate developed, with respect to how we might best

proceed with the exploration of the relationship between biodiversity and ecosystem processes. The paradigm that diversity generated stability (Odum 1953; Elton 1958; MacArthur 1955) was challenged by competing theory (May 1972, 1973), and field investigations to explore these contrary propositions rapidly followed, most focussing on a single trophic level (e.g. McNaughton 1985). As McCann (2000) has pointed out, the degree to which single trophic level approaches could have hoped to inform the diversity–stability issue was always debatable, given the multi-level perspective of the theory and the prevalence of top-down and bottom-up regulatory interactions in nature. The parallels with the present biodiversity–ecosystem process debate are clear.

Most of the theoretical exploration of the effects of diversity on ecosystem dynamics has focused largely on single trophic levels (Tilman *et al.* 1997b; Loreau 2000a), perhaps because empirical research in terrestrial systems has itself dealt mainly with single trophic levels (see also Petchey *et al.*, Chapter 11). Theoretical analyses of multi-trophic dynamics in relation to species diversity include plant–herbivore interactions (Loreau 1996; Holt and Loreau 2002) and decomposer food webs (Zheng *et al.* 1997; Loreau 2001). Studies by Ives *et al.* (2000) and Johnson (2000) also show how species diversity may affect the stability of ecosystem processes in food webs. Overall, these theoretical studies confirm our contention that an understanding of the effects of diversity on ecosystem processes in most cases necessitates a multi-trophic approach (see also Petchey *et al.*, Chapter 11).

An increased understanding of the effects of diversity on ecosystem processes also means that a closer link between organism-oriented and element flux-oriented research needs to be developed. This need was highlighted several years ago at two international meetings in the 1990s which evaluated the potential contribution of a multi-trophic level approach to predicting the consequences of biodiversity change (Hochberg *et al.* 1996; Polis and Winemiller 1996). Both these meetings concluded that a better framework for linking ecosystem processes to trophic structure and dynamics needed to be developed. There has been some progress towards incorporating non-living components, such as nutrients, along with species dynamics in models (e.g. De Angelis 1992), but developments in this area have remained relatively limited. Exceptions include de Ruiter *et al.* (1995, 1996, 1997, Chapter 9) stability analyses of below-ground food webs, and Andersen's (1997) analyses of the promotion of multiple equilibria in herbivore–plant systems when nutrients are included. Individual based and/or ecosystem type models based on nutrient or energy flows also offer promise (e.g. Gurney *et al.* 1995; Sterner 1995; Wedin 1995; Andersen 1997; Loreau 1996, 1998c) because their building blocks and currencies have immediate relevance to biogeochemical fluxes and other ecosystem processes and services in which society is interested.

Historically, food web ecology has been dominated by approaches where highly complex, species-based information in real food webs has been summarized in order to reveal dominant patterns more clearly, e.g. the grouping of species into trophic levels. More recently various food web statistics, such as the number of species, connectance, food chain length, and proportions of functionally different groups of species (Cohen *et al.* 1990) have been used in an attempt to facilitate comparisons of different ecosystems, an approach which has become known as 'food web theory'. If systems have similar statistics, then this would imply dynamic constraints on the way species interact within natural systems and provide insights into the mechanisms behind these constraints, as well as providing realistic parameters for multi-species models and/or permitting tests of those models (Cohen *et al.* 1990).

Analysis of statistics derived from food web diagrams (depictions of webs in the literature) did indeed suggest recurrent patterns in nature (e.g. short food chain lengths, lower connectance when richness is high, a low prevalence of omnivores in webs), many of which are consistent with the outcomes of conceptual and formal models (Cohen *et al.* 1990; Pimm *et al.* 1991). However, the meaningfulness of these analyses and the patterns which they reveal has been critically questioned (Moore *et al.* 1989; Polis 1991; Paine 1992; Hall and Raffaelli 1993, 1997), mainly on the basis that the majority of web diagrams from which the statistics are derived

are entirely inappropriate for such analyses. Often these diagrams are summaries of much larger food webs or parts of larger webs, and most were never intended for this kind of analysis by the original authors. A particular shortcoming of all but a few food web diagrams is the absence of pathogens and parasites (Marcogliese and Cone 1997). The addition of only a few parasite species to a food web diagram can fundamentally alter the web's statistics (Huxham *et al.* 1993), whilst pathogens are much more interactive with their host populations than are the host's predators, and can have major impacts on community organization and ecosystem processes (e.g. rinderpest: Sinclair 1974; Dublin *et al.* 1990, and myxomatosis: Sumption and Flowerdew 1985). Whilst we agree with Pimm *et al.* (1991) and May (1997) that such shortcomings should serve as a spur to refining data sets and theory, the promises claimed for the use of food web theory for tackling problems of multi-species fisheries, integrated pest management and climate change (Pimm *et al.* 1991) have yet to be realized. It is not at all clear how the 'food web theory' approach can be usefully applied to the issue of biodiversity loss and ecosystem process.

An alternative approach to food web theory for the analysis of multi-trophic systems is the simplification of the interacting elements to trophic types (e.g. Gange and Brown 1997; Van der Putten *et al.* 2001; Tscharntke and Hawkins, in press), which has major advantages for the development of models of these interactions (e.g. Holt 1997). Such studies have demonstrated that multi-trophic models can generate more complex community patterns, such as alternative equilibria as a result of omnivory (Diehl and Feissel 2000; Myllius *et al.* 2001) or prey size refuges (Chase 1999). Although these studies have largely been carried out without explicit reference to biodiversity issues, they reinforce the need for a multi-trophic perspective. Approaches based on community modules (*sensu* Holt 1997) also permits analysis of the dynamics when the system is not close to equilibrium.

13.5 Conclusions

Whilst recognizing the need to take a multi-level perspective on biodiversity issues, it is clear that many areas of trophic ecology have a long way to go before they can be used to develop predictive tools for managers. Large-scale, highly controlled experimental tests of hypotheses of the relationships between biodiversity and ecosystem process will be logistically difficult at the spatio-temporal scales appropriate for those ecological systems which include vertebrates and other taxa with large foraging ranges. However, general experimental tests of the effects of specific disturbances on biodiversity change and ecosystem process will continue to provide significant insights and should be encouraged.

At the theoretical level, multi-trophic level models are beginning to appear. It is important that these models to a larger extent address the biodiversity-process issue at the whole-system scale. The most relevant theory and modelling studies have focussed on the relationship between diversity and stability (see Loreau *et al.*, Chapter 7; McCann 2000). We have also pointed out that here is an urgent need to integrate population oriented models with models focusing on biogeochemical cycling, an integration that lags behind the empirical work described here, despite earlier appeals (Polis and Winemiller 1996). Most of the modelling approaches which embrace fluxes and flows of materials are ecosystem based and it is not immediately obvious how species richness can be easily incorporated. The high hopes that analysis of food web patterns might be useful for addressing biodiversity change

issues seem unlikely to come to fruition given the paucity of good data on real food webs and the undermined foundations of earlier work. Instead, approaches based on simplified 'trophic groups' may prove more useful in exploring the full dynamical properties of ecological systems. Such lumped approaches have, however, to be carefully balanced against the fact that indirect effects penetrating long distances through complex food webs could have substantial effects on overall food web dynamics.

Our review has highlighted many gaps and short-comings in our present knowledge and understanding of complex multi-trophic interactions. The microcosm approach shows promise for controlled experimental tests, the development and forming of theory and modelling, and in linking biodiversity change to process at spatio-temporal scales appropriate for complete food webs (see Petchey *et al.*, Chapter 11; McCann 2000). By developing and testing relevant theory, microcosm-based research offers much scope for informing and directing the work urgently required on larger scale, real ecological systems. Such work must include whole ecosystem manipulations, controlled field experiments and observational studies. Only through such an integrated approach will ecologists be able to deliver the advice demanded by policy makers and managers on the consequences of biodiversity loss.

CHAPTER 14

Biodiversity, ecosystem functioning and above-ground–below-ground linkages

D. A. Wardle and W. H. van der Putten

14.1 Introduction

All terrestrial ecosystems consist of an explicit producer subsystem and a decomposer subsystem. While traditionally these components have usually been considered in isolation from one another, they are obligatory dependent upon each other. Producers provide the organic carbon sources that drive the decomposer community and decomposer activity is in turn responsible for mineralizing nutrients required for maintaining growth of the producers. Mutualists, herbivores, pathogens and parasites affect producer–decomposer interactions both by directing changes in the flow of energy and resources, and by imposing selective forces that lead to evolutionary changes in individual producer and decomposer populations. The direct and indirect interactions between above-ground and below-ground communities therefore have the potential to operate as major drivers of population-, community- and ecosystem-level processes (Hooper et al. 2000; Van der Putten et al. 2001; Wardle 2002).

A major focus of recent activity in assessing how biodiversity may influence ecosystem functioning has been the issue of how plant diversity affects plant productivity (e.g. Naeem et al. 1994a; Tilman et al. 1996; Hooper 1998; Hector et al. 1999). A growing number of studies have also considered how soil biodiversity may affect decomposer processes (e.g. Andrén et al. 1995; Mikola and Setälä 1998a; Griffiths et al. 2000), and how plant species diversity may affect above-ground consumer diversity (e.g. Southwood et al. 1979; Siemann 1998; Koricheva et al. 2000; Mortimer et al. 2001), soil diversity and

soil processes (e.g. Wardle et al. 1999; Malý et al. 2000; Spehn et al. 2000a; Korthals et al. 2001). However, to develop a more complete understanding of how diversity affects ecosystem functioning requires a combined above-ground–below-ground approach, because of the likelihood that feedbacks between the two components have a role in governing ecosystem functioning. In this article, we will evaluate how plant diversity may influence below-ground organisms and processes, and how soil biodiversity may in turn affect primary production and plant diversity. We will also discuss how above-ground plant-related biota may interact with soil biota as mediated by plant community processes. In doing this, we assess the utility of considering the above-ground and below-ground subsystems in tandem for enhancing our understanding of how biodiversity influences ecosystem properties, processes, and stability.

14.2 Plant diversity effects on decomposers

Diversity of plant species or functional groups potentially influences decomposer processes by affecting either the quantity or quality of resources entering the soil. Such effects could potentially occur either through the effects of live plants or dead plant parts, each of which will now be considered.

14.2.1 Effects of live plant diversity

The relatively few studies that have investigated the effects of increasing plant species richness on soil organisms and processes, have claimed results that are idiosyncratic (Chapman *et al.* 1988; Wardle and Nicholson 1996), inconsistent (Mulder *et al.* 1999; Spehn *et al.* 2000a; Wardle *et al.* 2000a; Knops *et al.* 2001), weakly positive (Hooper and Vitousek 1998; Hector *et al.* 2000), more strongly positive (Spehn *et al.* 2000a; Stephan *et al.* 2000), or not detectable (Malý *et al.* 2000; Korthals *et al.* 2001) (Table 14.1). However, those studies which have found stronger positive effects of plant diversity on soil organisms or processes involve situations in which multiple species treatments were composed of random draws of species from the total species pool, including unnaturally low species diversities, and it is possible that these results can be explained in part by 'sampling effect' in which plant species with disproportionate effects on soil processes have a higher probability of occurring in those treatments which have more species present (see Aarssen 1997; Huston 1997). However, Hooper and Vitousek (1998) did find that microbial immobilization of nitrogen was greater in a treatment with four functional groups of plants present than in treatments consisting of monocultures of each of the four groups. Some other studies suggest that both positive and negative effects of diversity on decomposition are possible even within the same experiment (Chapman *et al.* 1988; Wardle and Nicholson 1996; Mulder *et al.* 1999), and the direction of these effects appears to be dependent upon what combinations of plant species are present (Chapman *et al.* 1988; Wardle and Nicholson 1996).

Adequate assessment of how plant diversity affects decomposer processes requires explicit consideration of the mechanisms involved; possible mechanisms are depicted in Fig. 14.1. One possible mechanism by which plant diversity may affect decomposition is through their effects on net primary production (NPP) and hence the amounts of resources entering the decomposer subsystem. If increasing plant diversity is to predictably affect decomposition processes through this route then there are three requirements:

1. That increasing plant diversity predictably affects NPP. This topic continues to be debated in the recent literature (see e.g. Grime 1997; Huston 1997; Tilman *et al.* 1997; Wardle *et al.* 1997a; Hector

et al. 1999; Huston *et al.* 2000; Van der Putten 2000; Lepš *et al.* 2001) and will not be discussed here. Although there is evidence that plant species richness can positively influence NPP through resource use complementarity at low levels of diversity (e.g. two or three versus one species), the issue of whether it influences NPP at higher diversity levels remains unresolved.

2. That increased NPP predictably influences the levels (biomasses or densities) of soil organisms. Data sets exist in which increasing NPP has been shown to have effects on the soil microbial biomass (the primary decomposers in the soil food web) that are positive (e.g. Myrold *et al.* 1989; Zak *et al.* 1994), neutral (e.g. Groffman *et al.* 1996) or negative (e.g. Wardle *et al.* 1995). Variable responses of organisms in higher trophic levels of soil food webs to increasing NPP are also a feature of the literature (reviewed by Wardle 2002). There are two reasons for idiosyncratic responses of decomposers to NPP. Firstly, the relative importance of top down and bottom up forces in regulating soil food web components may be context dependent. Although the relative importance of these two forces may vary predictably across gradients of NPP for above-ground food webs (Hairston *et al.* 1960; Oksanen *et al.* 1981), the issue appears to be more complex (and therefore more difficult to generalize) for below-ground food webs. Secondly, plants not only provide carbon resources for microbes but also compete with them for nutrients (Okano *et al.* 1991; Kaye and Hart 1997). Therefore the direction of effects of increasing NPP on decomposer organisms may be governed by which of two opposing effects (stimulation of microbes by carbon addition, inhibition of microbes by resource depletion) dominates.

3. That increased levels of soil organisms have predictable effects on decomposer processes. Greater rates of decomposition are often associated with greater levels of microbial biomass (Flanagan and van Cleve 1983; Beare *et al.* 1991). However, soil animals that feed directly on microbes can either promote or reduce rates of carbon mineralization depending upon nutrient conditions and faunal density (e.g. Seastedt and Crossley 1980; Hanlon 1981). Higher trophic level consumers that induce trophic cascades in soil food webs have also been shown to have effects on carbon mineralization,

Table 14.1 Key results of studies that have investigated the effects of experimentally varied diversity of live plants, or of dead plant parts, on decomposer organisms or processes. Note that this list is representative, not exhaustive

	Reference	Ecosystem type	Species richness — Species in replicated treatments	Monocultures?	Effects of increased diversity as claimed by authors
Live plants	Chapman et al. (1988)	Forest (north-west England)	1, 2	Yes	Idiosyncratic effects on litter respiration, and densities of earthworms and enchytraeids
	Wardle and Nicholson (1996)	Grassland (glasshouse)	0, 1, 2	Yes	Idiosyncratic effects on litter decomposition, and microbial biomass, respiration and metabolic quotient
	Hooper and Vitousek (1998)	Serpentine grassland (California, USA)	FG = 0, 1, 2, 3, 4[a]	Yes	Weak positive effects on microbial N immobilization
	Mulder et al. (1999)	Grassland (Umeå, Sweden)	1, 2, 4, 8, 12	Yes	No consistent effect on litter decomposition rate
	Hector et al. (2000)	Grassland (Ascot, UK)	0, 1, 2, 4, 8, 11	Yes	Weak increase in decomposition rate
	Malý et al. (2000)	Grassland (Planken Wambuis, The Netherlands)	4, 15	No	No effect on nitrogen mineralization, nitrification, ammonification, and soil microflora
	Spehn et al. (2000)	Grassland (Lupsingen, Switzerland)	0, 1, 2, 4, 8, 32	No	No effect on decomposition rate or microbial metabolic quotient, weak positive effect on microbial biomass, strong positive effect on earthworm densities
	Stephan et al. (2000)	Grassland (Lupsingen, Switzerland)	0, 1, 2, 4, 8, 32	No	Positive effect on abundance and catabolic activity of soil bacteria
	Wardle et al. (2000)	Grassland (glasshouse)	0, 1, 2, 4	Yes	No consistent effect on microbial biomass or activity, decomposition rates or the resistance or resilience of these properties to experimental disturbance
	Korthals et al. (2001)	Grassland (Planken Wambuis, The Netherlands)	4, 15	No	No effect on major nematode groups
	Knops et al. (2001)	Prairie (Minnesota, USA)	1, 2, 4, 6, 8, 12, 24	No	No consistent effect on litter decomposition or N loss
Dead plant parts	Blair et al. (1990)	Deciduous forest (North Carolina, USA)	1, 2, 3	Yes	Enhanced N release, but no effect on decomposition rates; idiosyncratic effects on microflora and fauna
	Wardle et al. (1997a)	Varied vegetation (Hamilton, New Zealand)	1, 2, 4, 5, 8	Yes	Idiosyncratic effect on decomposition rates, litter nitrogen release, microbial biomass
	Bardgett and Shine (1999)	Grassland species (laboratory)	1, 2, 3, 4, 5, 6	Yes	Generally positive effects on litter mass loss and microbial biomass
	Kaneko and Salamanca (1999)	Mixed forest (Matsue, Japan)	1, 2, 3	Yes	Generally positive effects on decomposition, and abundance and diversity of mesofauna
	Nilsson et al. (1999)	Forest species (mesocosm study; Umeå, Sweden)	1, 2, 3	Yes	Idiosyncratic or null effect on decomposition, microbial biomass and activity, and tree seedling growth
	Hansen (2000)	Deciduous forest (North Carolina, USA)	1, 2, 3, 7	Partial[b]	Reduced litter mass loss and greater population densities of microarthropods
	Hector et al. (2000)	Grassland (Ascot, UK)	1, 2, 4, 8, 11	Yes	Weak stimulation of decomposition rates

[a] Functional group diversity manipulated; [b] All species in 1, 2, and 3 species treatment appear in replicated monocultures.

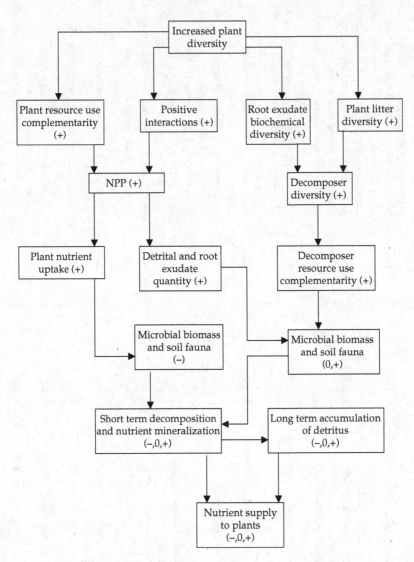

Figure 14.1 Hypothesized mechanisms by which increasing plant species richness may influence decomposition and nutrient mineralization. +, 0, and − indicates positive, neutral and negative effects, respectively.

which are positive (e.g. Hedlund and Öhrn 2000) or negative (e.g. Santos *et al*. 1981). Further, increased microbial biomass levels can also either promote net mineralization or immobilization of nutrients (notably nitrogen) depending upon available nutrient levels and in particular the carbon to nutrient ratios of organic matter inputs and of decomposing microbes.

It is therefore to be expected that if enriched plant diversity indeed promotes NPP, then the effect on both decomposer organisms and decomposer processes may be either positive or negative, depending upon site properties such as soil fertility, and the functional attributes of the dominant plant species present. In this light, a variety of responses of decomposer processes to plant diversity gain or loss, both between and within studies, could be expected; this would create difficulties in predicting even the direction of effect. This would in turn result in the sorts of idiosyncratic effects that are characteristic of the published literature.

Another mechanism by which increasing species richness of live plants could potentially influence decomposer activity is through promoting the release of a greater heterogeneity of carbon sources to the soil in the plant-rooting zone (Fig. 14.1). Here, a greater diversity of plant root exudates could result in an increased functional diversity of microbes, and this leads to a greater biomass of microbes through enhanced microbial resource use complementarity. However, those studies in which mixtures and monocultures of plant species have been compared find inconsistent effects of growing plants in mixtures on densities of soil microorganisms, with both positive and negative effects being possible (Aberdeen 1956; Christie *et al.* 1974; Wardle and Nicholson 1996). Further, if increased heterogeneity of exudates does indeed promote greater levels of decomposer organisms then this would be expected to cause idiosyncratic or context-dependent effects on the processes that those organisms carry out, for the reasons described earlier.

14.2.2 Effects of diversity of dead plant parts

Most of the organic matter that a plant produces is not consumed by herbivores, but is instead returned to the soil as plant litter where it has important 'afterlife effects' (*sensu* Findlay *et al.* 1996). The proportion of NPP that is not consumed varies from 50% in highly productive ecosystems to >99% in highly unproductive ecosystems (McNaughton *et al.* 1989; see also Cebrian 1999). Therefore, if increasing plant species richness was to influence decomposition rates, it is reasonable to expect that interactions among dead plant parts should contribute to such an effect. A growing number of studies have investigated the effects of species diversity of plant litter on decomposer organisms and processes, through the use of 'litter mixing' experiments in which the performance of multiple species litter mixes is compared with what would be expected based on the performance of litter from each component species in monoculture. Increased diversity of plant litter has been found through this approach to exert effects on decomposer organisms which are either idiosyncratic (e.g. Blair *et al.* 1990; Wardle *et al.* 1997a; Nilsson *et al.* 1999) or positive (e.g.

Bardgett and Shine 1999; Kaneko and Salamanca 1999; Hansen 2000) (Table 14.1). Further, effects of increasing litter diversity on the rates of decomposition processes have been found to be negative (e.g. Hansen 2000), neutral (e.g. Blair *et al.* 1990), idiosyncratic (e.g. Wardle *et al.* 1997a; Nilsson *et al.* 1999) or weakly positive (e.g. Bardgett and Shine 1999; Kaneko and Salamanca 1999; Hector *et al.* 2000). Recent theoretical work (Loreau 2001) predicts that greater diversity of organic compounds should lead to either negative or neutral effects on soil processes, although this is consistent with the results of only a subset of those 'litter mixing' experiments which have been performed to date; the question remains, however, to what extent greater species diversity of plant litter leads to a greater diversity of organic compounds.

If increasing diversity of litter types is to promote decomposition rates, then it is first necessary for litter diversity to influence the biomasses or activities of those decomposer organisms inhabiting the litter. This can occur through increased litter diversity promoting greater diversity of decomposers and hence greater resource use complementarity among these organisms (Fig. 14.1). However, while Kaneko and Salamanca (1999) did find that litter mixing promoted diversity of saprophagous mites, only some studies provide evidence that litter diversity promotes biomasses or densities of decomposers (Table 14.1). Further, even if increasing litter diversity was to promote greater levels of decomposer organisms, this should be expected to induce idiosyncratic effects on plant litter decomposition and nutrient mineralization for the reasons described earlier. Consistent with this, while Kaneko and Salamanca (1999) and Hansen (2000) both found densities of saprophagous mites to be promoted by increasing litter species diversity, the former study found the rate of decomposition to be enhanced by litter mixing while the latter study found it to be reduced. Varied responses of decomposer-mediated process rates to litter mixing, such as are characteristic of the recent literature, may arise because high quality litter may accelerate the decomposition of associated low quality litter, while litter of poorer quality (e.g. that with high concentrations of phenolics) may have the reverse effect (Seastedt 1984).

14.2.3 The relative importance of plant diversity effects

It is important to place the effects of species diversity on decomposer organisms and processes in the context of other factors which can also influence decomposition such as climate, disturbance, species composition and soil fertility. If species diversity of plants is to operate as a driver of the decomposer subsystem, then this first requires that diversity has important effects relative to these other drivers of the decomposer subsystem. It has long been recognized that community composition of plants and the ecophysiological traits of dominant plant species are major drivers of ecosystem functioning, both above- and below-ground. This is apparent, e.g. through the mull and mor theory of Muller (1884) and through the pioneering work of Handley (1954) showing that plants which produce high levels of phenolics have different effects on nitrogen availability to those that do not. It is also being increasingly recognized that plants that grow slowly, invest carbon into secondary compounds, and exhibit a certain suite of ecophysiological attributes (e.g. long leaf life span, low leaf nitrogen concentrations, low specific leaf area) tend to produce slower decomposing leaf litter and promote accumulation of organic matter relative to faster growing plants that do not invest in secondary compounds and possess a different suite of ecophysiological attributes (Grime 1979; Chapin *et al.* 1993; Aerts and Chapin 2000). The question which therefore emerges is whether species richness of plants exerts effects which are important relative to the effects of vegetation composition and traits of dominant plant species in driving the decomposer subsystem.

Although this issue has been addressed with regard to plant biomass and productivity (e.g. Grime 1973b; Waide *et al.* 1999) it has been seldom addressed with regard to the decomposer subsystem. However, in a study involving 50 lake islands in northern Sweden, Wardle *et al.* (1997b) assessed above-ground and below-ground properties and processes for each island, and found that the least plant-diverse (larger) islands supported greater plant biomass, rates of decomposition and nutrient mineralization, and soil microbial biomass and activity than that of the most plant-diverse islands. This is because the larger islands supported an earlier-successional faster growing vegetation dominated by plant species which invested carbon into growth rather than defence compounds and therefore produced high quality litter. The dominant plant species, which promoted these rapid process rates, were also responsible for maintaining low vegetation diversity, probably through competitive exclusion of subordinate plant species. Meanwhile, the smaller islands supported slower growing vegetation known to produce litter with high concentrations of phenolics, which retarded process rates both above and below ground. However, the diminished success of highly competitive species on these smaller islands enabled a greater diversity of plant species to coexist. These results suggest that any positive effects of diversity on ecosystem functioning that might occur through resource use complementarity among species are overridden by the key biotic drivers of the ecosystem such as the traits of the dominant plant species present. It is recognized that spatial scale in itself is likely to be a major determinant of relationships between plant diversity or composition and ecosystem processes, and that the ecosystem drivers that dominate at one spatial scale may be different to those that dominate at another. While the study of Wardle *et al.* (1997a) provides evidence that species richness is not a key ecosystem driver at the spatial scale of between islands, current experiments are evaluating whether it might have a more important role at the within-island spatial scale.

Another approach which enables insights into the importance of plant diversity relative to that of community composition as an ecosystem driver involves 'removal experiments', such as have been widely used by plant ecologists for studying the role of competition in plant communities (reviewed by Aarssen and Epp 1990). Here, subsets of the plant community (e.g. particular species or functional groups) are manually removed and excluded from the community through the duration of the experiment. This approach enables assessment of the question of what happens to ecosystem functioning as a result of losses of species (such as occurs through local extinctions) more directly than is possible through experiments in which species richness is varied as an experimental treatment (Lamont 1995; Wardle *et al.* 1999; Symstad and Tilman 2001). There have been few attempts to investigate the influence of plant species removals on the decomposer subsystem in real ecosystems. Wardle *et al.* (1999) conducted an experiment in a New Zealand perennial grassland in which different plant functional groups were permanently excluded from different field plots over a three year period. Loss of functional groups sometimes had negative effects on the rates of ecosystem processes, as well as on the biomasses or populations of decomposer microorganisms and soil fauna. However, these effects were largely determined by which functional groups were removed, rather than how many species were lost. This study therefore points to the significance of the functional attributes of the principal species in the plant community as the key drivers of the decomposer community and the ecosystem processes that it carries out. It is recognized that removal experiments can have limitations because of physical disturbances created by mass removal effects (Grime 2001), although in the study of Wardle *et al.* (1999) the experiment was initiated on vegetation-free plots, and all removed plants were weeded out as emerging seedlings.

14.3 Decomposer diversity effects on plants

It is well known that soil organisms influence plant growth, usually positively, through increasing rates of supply of plant available nutrients (see Mikola *et al.*, Chapter 15). In particular, several experimental studies have shown that manipulations of components of the soil food web can have significant effects on plant growth; this has been shown for microfauna (e.g. Alphei *et al.* 1996; Bardgett and Chan 1999), mesofauna (Setälä and Huhta 1991) and macrofauna (e.g. Lavelle *et al.* 1994; Brussaard *et al.* 1996). Further, soil food web composition can be important in influencing plant nutrient uptake and different soil fauna occupying similar trophic levels may differ in their effects on plant productivity (Ingham *et al.* 1985; Alphei *et al.* 1996;

Bonkowski *et al.* 2000). However, the impact of decomposer diversity per se on plant nutrient uptake and NPP remains little explored. In pot experiments with *Betula pendula* seedlings and constructed soil communities of different trophic levels and different numbers of species within trophic levels, Laakso and Setälä (1999a) concluded that primary production of plants is mainly controlled by organisms at low trophic positions in the decomposer food web but not by their species richness (see Mikola *et al.*, Chapter 15). Further, Laakso *et al.* (2000) found through the use of mini-ecosystems that the decomposer food web structure at the functional group level affected the N uptake and growth of plants, although the impact of a given trophic group was also dependent of the species composition of that group. The limited available evidence suggests that the impact of decomposer diversity in affecting NPP through influencing plant available nutrient supply operates more at the functional group level than at the species within functional group level, although there are instances in which species composition within functional groups can be important.

14.4 Nitrogen fixation

Nitrogen fixing bacteria are important keystone organisms in many terrestrial ecosystems; they represent only a minute proportion of the total heterotrophic bacterial biomass and perform a unique function through the addition of aerial nitrogen into compounds that can be used by primary producers and ultimately consumers. These effects are apparent when ecosystems are invaded by plants with effective nitrogen fixing symbioses; invasion of the actinorhizal shrub *Myrica faya* into native ecosystems in Hawaii which lacked symbiotic nitrogen fixing plant species has resulted in a greater than four fold increase of input of nitrogen to the system (Vitousek *et al.* 1987). Further, the inclusion of the plant functional group consisting of nitrogen fixing legumes has been shown to exert disproportionately large effects on plant productivity in plant diversity experiments (Tilman *et al.*

1997a; Hooper and Vitousek 1997; Hector *et al.* 1999), a phenomenon which has long been recognized by agronomists.

Most of the Leguminosae and one genus of the Ulmaceae are able to form associations with nitrogen fixing bacteria of the Rhizobiaceae. The actinomycetes of the Frankiaceae have the same capacity, but these are known to be associated with a much more diverse range of at least 137 plant species within seven plant families (Squartini 2001). The diversity of strains of nitrogen fixing bacteria associated with host plant species can vary considerably. For example, in a study of 96 sites, each containing a single *Myrica* species, *Myrica pensylvanica* had a diverse reservoir of Frankia strains, whereas *Myrica gale* had only a few dominant sequences (Clawson and Benson 1999). This suggests that Frankia diversity may be strongly plant species-dependent. There are no studies that have explicitly investigated the effects of nitrogen fixer diversity or composition on plant productivity. However, there is evidence that heavy metal toxicity reduces Rhizobium diversity by killing strains that are effective at fixing nitrogen, resulting in nodulation by only ineffective strains (Giller *et al.* 1998). Further, the specificity of the relationship between plants and strains of symbiotic bacteria suggests that the diversity of symbiotic nitrogen fixing microorganisms may influence plant species diversity, because although the host range may be wide, the effectiveness of nitrogen fixation is often highly dependent on the specific plant–microorganism combination.

In most undisturbed, natural terrestrial ecosystems, root-associated or free living nitrogen fixing bacteria are the main source of nitrogen input, and nitrogen fixation is carried out by a wide range of bacterial phyla, from Archaebacteria to Eubacteria (Poly *et al.* 2001). As such, there is a high diversity of these bacteria in most ecosystems, although there is considerable spatial variability in their diversity both across ecosystems and at the microscale (Bagwell and Lovell 2001; Poly *et al.* 2001). Although the above-ground effects of diversity of associative or free living nitrogen fixing bacteria remains unexplored, their relatively high diversity suggests a greater amplitude for functional redundancy of specific strains than might be expected for nodulating nitrogen fixing bacteria.

14.5 Mycorrhizas

Mycorrhizal fungi may drive community and ecosystem processes through enabling greater uptake and accessibility of nutrients and water by plants (Smith and Read 1997), or through affecting plant susceptibility to pathogens (Newsham *et al.* 1995a) and insect herbivores (Gange and West 1994). However, there is little evidence that plant species diversity *per se* has important effects on mycorrhizal fungal colonization or species diversity, at least at the across-ecosystem scale (Allen *et al.* 1995). For example, in some tropical forests, there are 1000 plant species and fewer than 25 arbuscular mycorrhizal fungal (AMF) species (Allen *et al.* 1995). Meanwhile, some grassland may contain over 30 AMF species (Bever *et al.* 1996). Further, some forest types can contain few dominant ectomycorrhizal plant species and over 1000 ectomycorrhizal fungal species (Allen *et al.* 1995). Agricultural fields with low botanical diversity often support a lower diversity of AMF than do plant species rich grasslands, but in such fields low plant diversity is maintained by sowing a single plant species and using herbicides to suppress the rest, while mycorrhizal networks (and probably AM fungal diversity) are adversely affected by ploughing and suppressed by fertilization (Helgason *et al.* 1998).

Mycorrhizal fungi can influence the competitive balance among coexisting plant species and thus influence plant community diversity both positively (Grime *et al.* 1987; Van der Heijden *et al.* 1998a) and negatively (Moora and Zobel 1996; Smith *et al.* 1999). Much remains unknown, however, about how mycorrhizal fungal diversity affects plant productivity and diversity. There is experimental evidence pointing to increased AMF species diversity promoting plant diversity and productivity (Van der Heijden *et al.* 1998b), although whether this is due to sampling effect or biological mechanisms remains unresolved (Wardle 1999; Van der Heijden *et al.* 1999; see also discussion by Van der Heijden and Cornelissen, Chapter 16). Further, Jonsson *et al.* (2001) found that ectomycorrhizal fungal species richness did significantly influence the growth of tree seedlings, but both positive and negative effects were possible and the nature of effects was dependent upon both tree species and soil fertility.

14.6 Root pathogens and herbivores

There is a wide range of studies that show effects of root pathogens and herbivores on plant productivity and plant community processes (e.g. Augspurger and Kelly 1984; Stanton 1988; Brown and Gange 1989a,b, 1992; Van der Putten et al. 1993; Bever 1994; Olff et al. 2000; Packer and Clay 2000). Examples range from tropical to temperate forest and from subtropical savannah to temperate old-field grasslands. Bever et al. (1997) presented a conceptual plant–soil feedback model that predicts that individual plant species can selectively attract or promote both organisms that enhance plant performance and those that reduce it. This model predicts that root herbivory or soil pathogen infestation may decrease plant competition and therefore change the relative dominance of different plant species in the community.

Effects of plant species diversity on the diversity and abundance of root pathogens and root herbivores has received little attention. However, in a plant removal experiment in a perennial grassland, Wardle et al. (1999) found that only the removal of all plants reduced root herbivore density and diversity. In an experiment with later-succession plant species sown in abandoned arable land, Korthals et al. (2001) found little effect of plant species diversity or identity on abundance or diversity of root herbivorous nematodes. Further, both these studies suggest that delayed response, dispersal constraints and generalist feeding behaviour may make root pathogens and root herbivores less responsive to plant diversity changes than is the case for above-ground herbivores (e.g. Southwood et al. 1979; Siemann 1998; Koricheva et al. 2000; Mortimer et al. 2001).

While a number of studies in natural ecosystems have investigated how root pathogens or root herbivores may affect plant productivity, few have experimentally varied the community composition of root pathogens or root herbivores. De Rooij-Van der Goes (1995), in an attempt to identify potential pathogens of the coastal foredune grass Ammophila arenaria, investigated the effects of different combinations of natural soil fungi alone and with nematodes. All combinations of soil fungal species were able to reduce plant production to varying extents, suggesting some degree of functional similarity between species within one trophic level. It is not known if this type of functional similarity is a more general pattern for root pathogens. However, other studies, such as by Augspurger and Kelly (1984), Holah and Alexander (1999), Mills and Bever (1998), Mihail et al. (1999) and Packer and Clay (2000) all point to the capacity of a limited number of fungal species (mostly öomycete fungi such as Pythium and Phytophthora) to reduce NPP or increase seedling mortality.

The application of soil insecticides (Brown and Gange 1989a) and soil nematicides (see Stanton 1988) can affect the relative abundance of different plant species in a community, as well as NPP. For example, Brown and Gange (1992) found for a grassland community that reduction of root herbivores through the application of an insecticide favoured domination by dicots while the application of foliar insecticides favoured grasses. While these types of experiments may be confounded by non-target effects, they nevertheless suggest that primary production and plant species diversity can be affected by soil biota, including root herbivores. Further, root herbivores may partition resources both spatially and temporally (Brown and Gange 1990; Yeates et al. 1985; Bongers and Bongers 1998). As a result, plants may be under constant root herbivore pressure and the loss of root herbivorous species could conceivably result in a reduction in net root herbivory. However, this issue, as well as the effects of the influence of root herbivore community composition in general, remains largely unexplored.

14.7 Interactions involving above-ground consumers

Above-ground consumers of plant material are an important driving force of relations between plants and soil biota. Foliar herbivores can exert important effects on decomposer communities and processes (and ultimately plant-available nutrient supply) through a variety of mechanisms which operate over a range of spatial scales, and these can

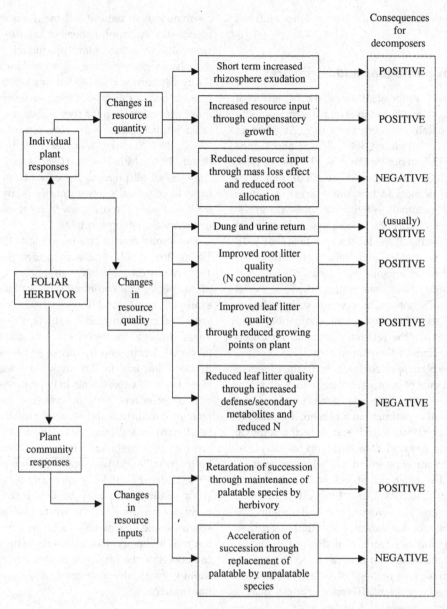

Figure 14.2 Mechanisms by which foliar herbivores may influence soil decomposer organisms, both positively and negatively.

be either positive or negative (Fig. 14.2). These processes operate both at the individual plant level and at the level of the plant community, and can involve either positive or negative effects on both the quantity and quality of resources that enter the decomposer subsystem (reviewed by Bardgett *et al.* 1998; Wardle 2002). Consistent with this variety of possible effects, those studies that have used fenced

exclosure plots to investigate the ecological consequences of browsing by mammalian herbivores have found overall effects on decomposer organisms that can be positive (e.g. Bardgett *et al.* 1997; Stark *et al.* 2000) or negative (e.g. Suominen 1999; Suominen *et al.* 1999), effects on carbon mineralization that are positive (e.g. Bardgett *et al.* 1997; Kielland *et al.* 1997) or negative (e.g. Pastor *et al.*

1988; Stark *et al.* 2000), and effects on net nitrogen mineralization that are positive (e.g. Frank and Groffman 1998; Tracy and Frank 1998) or negative (e.g. McNaughton *et al.* 1988; Stark *et al.* 2000). Further, in a study of 30 long term (13–36 year old) fenced plots aimed at excluding introduced deer and goats in a range of New Zealand rain forests, Wardle *et al.* (2001) found that while these mammals consistently negatively affected plant biomass and diversity in the browse layer, their effects on soil microflora, microfauna, carbon mineralization, and carbon and nitrogen sequestration were highly idiosyncratic, with roughly equal numbers of sites showing strong positive and strong negative effects. However, larger soil animals showed consistently negative responses, probably as an effect of physical effects by the browsers such as trampling.

There have been studies that describe the effects of removal of subsets of the herbivore community on plant species diversity (Bowers 1993) and on plant spacing and its consequences for infestation by fungal pathogens (Ericson and Wennström 1997). These point to the ecological importance of herbivore community composition. However, no study to date has investigated the effects of terrestrial foliar herbivore diversity beyond two species on ecosystem processes, either above- or below-ground. We could predict that even if enhanced resource use complementarity resulting from increased herbivore diversity was to increase standing herbivore biomass and thus herbivore effects on vegetation, the flow through effects to the decomposer subsystem would be idiosyncratic given the complex and varied ways in which herbivores influence the quantity and quality of materials returned to the soil (Fig. 14.2). Further, the effects of herbivore diversity on the decomposer subsystem may be context-dependent, for two reasons. Firstly, the direction of foliar herbivore effects on plant productivity and diversity appears to differ between high fertility and low fertility situations (Proulx and Mazumder 1998; Olff and Ritchie 1998). Secondly, the relative importance of top-down and bottom-up forces in regulating herbivore densities varies in relation to environmental factors such as total plant productivity (Oksanen *et al.* 1981; Oksanen and Oksanen 2000), and the effects of herbivory (and therefore, potentially, effects of herbivore diversity)

on ecosystem functioning are likely to be of diminished importance when top-down mechanisms regulate herbivore densities.

Although herbivores may exert important effects on the functioning of the below-ground subsystem, they are also in turn influenced by soil organisms, and consideration of indirect feedbacks between herbivores and soil biota as mediated by plants may be an important consideration in better understanding linkages between diversity and ecosystem functioning. At the individual plant level, there is a vast literature showing that decomposer, organisms, and organisms which interact with plant roots either as mutualists or as antagonists, can influence the quantity and quality of foliage that the plants produce; this may in turn exert large effects on those organisms that consume this foliage. A number of studies have shown that indicators of success of invertebrate herbivores (e.g. biomass production, fecundity) can be indirectly influenced by the presence of mycorrhizal fungi (e.g. Gange and West 1994; Goverde *et al.* 2000), root herbivores (e.g. Masters and Brown 1992; Masters 1995) and decomposer fauna (Scheu *et al.* 1999) (Table 14.2). A recent study also points to the indirect effects of root herbivores on above-ground secondary consumers, i.e. parasitoids of insect seed predators (Masters *et al.* 2001). At the plant community level, and at longer time scales, effects of herbivores on decomposer organisms and nutrient cycling can have significant effects on above-ground NPP; this may occur for systems driven by both vertebrate herbivores (see McNaughton 1979; De Mazancourt and Loreau 2000) and invertebrate herbivores (Belovsky and Slade 2000), and is likely to represent an important feedback ultimately affecting food availability and quality for the herbivores.

Above-ground–below-ground interactions are not just driven by primary plant compounds (carbohydrates, nutrients), but may also be influenced by secondary plant compounds. Many anti-herbivore compounds are synthesized in the roots, whereas most examples of their activity are known from above-ground studies (Karban and Baldwin 1997). Plant defence inductions following herbivore attack may also affect plant–pathogen interactions (Pieterse and van Loon 2000; Paul *et al.* 2000). While several studies have looked

Table 14.2 Examples of how soil organisms can indirectly influence above-ground herbivores via plants. Note that this list is representative, not exhaustive

Type of below-ground consumer	Below-ground consumer (BC)	Plant species	Above-ground consumer (AC) response variable	Effect of BC on AC (% of control)[a]	Reference
Mycorrhizal fungi	Arbuscular mycorrhizae	*Plantago lanceolata*	*Arctia caja* (Lepidoptera: Arctidae) larval growth rate	41.1 (*P*<0.05)	Gange and West (1994)
			A. caja food consumption	76.5 (*P*<0.05)	
			Myzus persicae (Homoptera: Aphididae) adult weight	177.2 (*P*=0.05)	
			M. persicae embryo production	300.0 (*P*<0.01)	
	Arbuscular mycorrhizae[b]	*Lotus corniculatus*	*Polyommatus icarus* (Lepidoptera: Lycaenidae) larval survival	119.0 (*P*<0.05)	Goverde *et al.* (2000)
			P. icarus larval weight	200.0 (*P*<0.05)	
Root herbivores	*Phyllopertha horticola* (Coleoptera: Scarabidae)	*Sonchus oleraceus*	*Chromatomyia synergesiae* (Diptera: Agromyzidae) pupal weight	112.7 (*P*<0.05)	Masters and Brown (1992)
			C. synergesiae leaf mine frequency	115.3 (NS)	
	P. horticola	*S. oleraceus*	*M. persicae* adult weight	120.8 (*P*<0.05)	Masters (1995)
			M. persicae growth rate	111.4 (NS)	
			M. persicae fecundity	108.2 (NS)	
Decomposer fauna	Springtails	*Poa annua*	*M. persicae* juvenile production	300.0 (*P*<0.05)	Scheu *et al.* (1999)[c]
		Trifolium repens	*M. persicae* juvenile production	63.1 (*P*<0.05)	
	Earthworms	*P. annua*	*M. persicae* juvenile production	253.3 (*P*<0.05)	
		T. repens	*M. persicae* juvenile production	173.9 (*P*<0.05)	

[a] *P* values correspond to the level of significance of the effect of below-ground biota on above-ground herbivores, i.e. whether the effect ratio differs significantly from 100%. NS = non-significant.

[b] Data averaged for all mycorrhizal treatments.

[c] Data presented for the 16 week measurement only; 'juveniles produced' = numbers of juvenile aphids produced per eight days. Collembola species used are *Heteromurus nitidus* and *Onychiurus scotarius*; earthworm species are *Aporrectodea caliginosa* and *Octolasion tyrtaeum*.

at above-ground–below-ground linkages as being driven by fluxes of major nutrients and carbohydrates, few studies have considered the role of plant defence mechanisms and secondary defence compounds as possible drivers of these sorts of interactions (Van der Putten *et al.* 2001). To determine how these complex interactions may affect, and may be affected by, diversity in the first and higher trophic levels is one of the major challenges, since this will provide information on possible cascading effects of biodiversity loss through both the above-ground and below-ground components of food webs.

14.8 Ecosystem stability

Terrestrial ecosystems are not static entities, and even if species richness is unimportant as an ecosystem driver at a given point of time, the issue

remains as to whether diversity is important under changing environmental conditions through buffering ecosystem properties and process rates against disturbance (see Loreau 2000; Cottingham *et al.* 2001; Loreau *et al.*, Chapter 7). The question of how the loss of diversity of plants influences the stability of below-ground organisms or processes remains little explored. In the study of Wardle *et al.* (1999), experimentally induced loss of functional groups from a total plant community was found to have possible positive, negative or neutral effects on the stability (measured as temporal variability) of NPP, microbial biomass and activity, and the main soil animal groups. The direction of these effects was related to plant functional group identity. In a complementary glasshouse experiment with imposed disturbance (drought), Wardle *et al.* (2000) found that the species composition of the plant community, and the identity of those species which were present, often emerged as being the drivers of the response to disturbance of below-ground properties. This is consistent with the study of MacGillivray *et al.* (1995) which suggested that ecosystem responses to disturbance are governed to a large extent by the ecophysiological traits of the dominant plant species present, and with the results of analyses pointing to the types of vegetation present as drivers of long term temporal variability of above-ground processes such as NPP (e.g. Knapp and Smith 2001). Idiosyncratic responses to plant diversity and species identity of the stability of soil organisms and processes may also be influenced by the nature of interactions between the types of plants and soil organisms under consideration, including their degree of specificity (Van der Putten *et al.* 2001).

Although there have been theoretical studies considering the stability of soil food webs and processes (De Ruiter *et al.*, Chapter 9), there have been few empirical investigations of this. Griffiths *et al.* (2000) (see also De Ruiter *et al.*, Chapter 9) provided evidence that manipulation of microbial community composition through chloroform fumigation affected the resistance of litter decomposition rates to a second disturbance. In their study, fumigation reduced microbial diversity, although whether the reduced stability in the fumigated system was due to changes in diversity or composition of the microbial community cannot be determined. Nevertheless, their study provides evidence that the microbial community may influence the stability of those soil processes that are directly relevant for regulating the rates of supply of plant available nutrients. However, in real ecosystems most disturbances that are likely to influence the soil community are also likely to directly influence the plant community, meaning that the stability of processes above-ground and below-ground will often be inextricably linked.

In order to understand better the effects of loss of plant species diversity on stability of ecosystem processes, there is also a need for studies that adequately take into account the functional attributes of plant species that are lost. For example, subordinate plant species which are functionally unimportant at a particular point of time may confer an element of stability on above-ground and below-ground processes following a change in environmental conditions that adversely affects the dominant species relative to the subordinate species (cf. Grime 1998). This issue has not received much attention, but may lead to insights regarding the relationship between ecological complexity and the stability of ecosystem processes.

14.9 Conclusions

Although there has been increasing recent activity regarding how diversity of plants may influence organisms and processes in the soil *vice versa*, few consistent trends have emerged and many aspects of this issue remain unexplored. Specific plant–soil organism relationships may have positive, negative and neutral effects on species diversity and plant productivity. Firstly, the impact of both live plant diversity and diversity of plant litter input on decomposer processes and organisms shows a variety of patterns across different studies (see Table 14.1). Secondly, consumer organisms have been shown to exert effects on processes on the other side of the soil surface interface that operate in different directions dependent upon context. These varied patterns are due to the range of mechanisms by which plants can affect soil organisms and *vice versa*, and because these different mechanisms frequently have effects that operate in opposite directions. The challenge now is to develop general principles that may enable us to at least predict the direction of response in a given set of circumstances. Plant–soil organism relationships that are based on host specificity (e.g. some root–pathogen or root–herbivore interactions) or involve relatively few key species which are uniquely capable of carrying out a specific process (e.g. symbiotic nitrogen fixation) are more likely to yield positive relationships of diversity between plants and soil organisms. Many of the plant—soil organism interactions are, however, based on differential susceptibility of plant species to certain mutualists, pathogens or herbivores in their root environment (rather than on absolute specificity). These interactions may be of particular significance in driving the species composition of vegetation.

There are recognized major challenges in studying the regulation and functional significance of biodiversity in relation to above-ground–below-ground feedbacks. These include the opaque nature of soil, the sheer diversity of soil organisms, and the fact that most small organisms (microfauna, microflora) remain undescribed and are unculturable. For this reason, conventional species richness concepts that are readily applied to plants and larger animals cannot be adapted for much of the soil biota and other approaches based on functional diversity and composition become increasingly necessary. Moreover, to better understand how diversity of either plants or soil organisms influences other organisms (including those on the other side of the soil surface interface) and ecosystem processes requires an explicit consideration of the mechanistic basis of such interactions. Such mechanisms include, e.g. the nature of trophic interactions (e.g. the relative importance of top-down and bottom-up interactions, mutualisms, and indirect interactions), the role and interaction of plant primary and secondary chemistry (e.g. the role of induced defences and changes in allocation patterns in response to herbivory), relationships between soil biota and plant nutrient acquisition, and the roles of the different types of feedbacks that might be involved.

We thank P. Inchausti, M. Loreau and S. Naeem for organizing a stimulating workshop which represented a balanced spectrum of points of view on the so-called 'diversity debate', F. de Souza for discussing plant mutualisms, and A. Hector and an anonymous reviewer for thorough comments on the submitted version of this chapter.

CHAPTER 15

Biodiversity, ecosystem functioning and soil decomposer food webs

J. Mikola, R. D. Bardgett, and K. Hedlund

15.1 Introduction

Decomposer organisms are essential for the functioning of terrestrial ecosystems; they mineralize carbon and nutrients that are bound to dead organic matter and provide resources for primary production. However, linking soil decomposer diversity with ecosystem functioning is difficult as processes in soil are not controlled by a uniform group of organisms but an array of interacting bacteria, fungi and fauna (Beare *et al.* 1995; Ekschmitt and Griffiths 1998). Further, the considerable lack of knowledge about the roles of individual species (especially within bacteria, fungi, microfauna and mesofauna) hinders our ability to identify the unique functions that decomposer species might perform.

In this chapter, we will firstly demonstrate how decomposer organisms and their interactions affect nutrient mineralization and primary production. We will then examine the evidence for declining decomposer diversity in terrestrial ecosystems, concentrating on the effects of agricultural land management. We will also investigate the potential for species diversity effects among decomposers and discuss the existing empirical evidence for the effects of decomposer diversity on ecosystem processes. Lastly, we will examine covarying diversity factors in existing decomposer studies and raise topics that merit further investigation.

In addition to decomposers, soils provide a habitat for other organisms that impact on ecosystem functioning; of these organisms mycorrhizal fungi are covered by van der Heijden and Cornelissen (Chapter 16) and symbiotic nitrogen fixers, root pathogens and root herbivores by Wardle and van der Putten (Chapter 14).

15.2 Control of ecosystem processes by decomposer food webs

Basal resources of decomposer food webs are mainly composed of plant residues and organic compounds released from plant roots. Bacteria and fungi, often collectively called microbes, are the primary decomposers of these resources because of their almost unique capability of directly breaking down complex organic compounds. Despite their irreplaceable role in the decomposition of dead organic matter, a clear picture of who is doing what among microbes remains poorly understood (Tunlid 1999) and most of the species are still to be discovered (Øvreås 2000). Decomposer fauna can influence ecosystem processes either indirectly by modifying the biomass, composition and activity of soil microbial communities or directly by consuming detritus and releasing inorganic nutrients. The influence of animals on microbial biomass often varies between negative, neutral and positive even within trophic groups (Table 15.1), and it seems that soil fauna is rather inefficient in controlling microbial biomass. However, animals can still modify the structure of microbial communities; for instance, microarthropods (mites and collembolans feeding on microbes and/or detritus) can decrease the ratio of fungal biomass to bacterial biomass in leaf litter (Hanlon and Anderson 1979) and affect fungal competition through selective grazing (Parkinson *et al.* 1979; Newell 1984a,b). Similarly, detritivorous macroarthropods (such as isopods and diplopods) and earthworms can reduce the proportion of fungal biomass within the total microbial biomass (Kayang *et al.* 1994; Saetre 1998)

Table 15.1 Effects of trophic groups of soil fauna on soil microbes and ecosystem processes (the list of references is not complete but a representative sample of existing literature)

Faunal group	Effects on microbes		Effects on ecosystem processes	
	Microbial biomass	Microbial activity	C and/or N mineralization	Plant growth
Microfauna				
Protists				
Woods *et al.* 1982	−[a]		+	
Kuikman *et al.* 1990	−	+	+	+
Bonkowski *et al.* 2000b	0		+	+
Nematodes				
Woods *et al.* 1982	−		+	
Ingham *et al.* 1985	+/−		+	0/+
Mikola and Setälä 1998c	−/0	+	+	
Setälä *et al.* 1999				+
Mesofauna				
Microarthropods				
Hanlon and Anderson 1979	−	+/−		
Hedlund *et al.* 1991	+	+		
Beare *et al.* 1992	−		+	
Bardgett and Chan 1999	+/0		+	−
Setälä 2000				0/+
Enchytraeids				
Williams and Griffiths 1989	0		+	
Sulkava *et al.* 1996			+	
Cole *et al.* 2000	0		0/+	
Macrofauna				
Macroarthropods				
Gunnarsson and Tunlid 1986	+			
Hassall *et al.* 1987	+	+	+	
Teuben and Roelofsma 1990		+/−	+/−	
Van Wensem *et al.* 1993		−/0	0/+	
Earthworms				
Haimi *et al.* 1992				+
Saetre 1998	−		+	
Wolter and Scheu 1999	0/+			
Alphei *et al.* 1996	−	0	+	−

[a] Symbols −, 0 and + indicate that fauna has been found to reduce, have no effect or increase microbial biomass and activity or ecosystem process rate.

and alter species composition in microbial communities (Gunnarsson and Tunlid 1986).

The effects of microbial-feeding microfauna (such as protists and nematodes) on microbial activity (which mostly determines the rate of C mineralization), nutrient mineralization and primary production are generally positive (Table 15.1) and the mechanisms behind these effects are relatively well

known. Enhanced C mineralization results from increased turn-over rate, activity and respiration of grazed microbial populations (Anderson *et al.* 1981; Kuikman *et al.* 1990; Mikola and Setälä 1998c), whereas enhanced N mineralization is mainly due to direct animal excretion of excess N (Woods *et al.* 1982). Nutrients released from microbial biomass by microbial grazers in turn enhance plant growth

(Clarholm 1985) since soil microbes compete with plants for inorganic nutrients (Kaye and Hart 1997). Moreover, microfauna may stimulate plant growth by modulating the composition of microbial communities in plant rhizosphere (Jentschke et al. 1995; Alphei et al. 1996; Griffiths et al. 1999). The effects of meso- and macrofauna on microbial activity and ecosystem processes are also mostly positive (Table 15.1) but depend more on specific circumstances. For instance, micro- and macroarthropods may either reduce or enhance microbial activity and mineralization depending on the season (Teuben 1991), nutrient content of microbial resources (Hanlon 1981), abundance of grazers (Hanlon and Anderson 1979; Hanlon 1981) and litter degradation stage (Van Wensem et al. 1993). Similarly, the effects of enchytraeids on ecosystem processes may depend on their abundance and microbial growth conditions (Wolters 1988). The mechanisms by which meso- and macrofauna affect ecosystem processes include fragmentation and consumption of litter, consumption of detritus and microbial biomass and alteration of microbial communities, but the relative importance of these mechanisms is not well known.

Faunal groups also prey on each other, which complicates interpretation of how soil fauna affects ecosystem processes. For instance, several species of bacterial-feeding nematodes also feed on protists (Elliott et al. 1980; Woods et al. 1982), and depending on particular circumstances this may lead to enhanced (Elliott et al. 1980) or reduced mineralization (Woods et al. 1982). Similarly, the reduced abundance of microbial-feeding nematodes by predation may lead to effects on mineralization and plant growth that are either positive (Allen-Morley and Coleman 1989), neutral (Laakso and Setälä 1999a) or negative (Bouwman et al. 1994; Mikola and Setälä 1998b,c; Setälä et al. 1999; Laakso and Setälä 1999a). The limited data from other faunal groups also point to several possible outcomes; predation on collembolans has been shown to reduce litter mass loss (Lawrence and Wise 2000) and enhance C mineralization (Hedlund and Sjögren Öhrn 2000), while predation on fungal-feeding and detritivorous mesofauna was found to have no effects on plant growth despite significant reductions in prey populations (Laakso and Setälä 1999b).

These examples illustrate how trophic groups and their *interactions* in decomposer food webs significantly influence ecosystem functioning, thus warranting a food-web approach when studying the diversity-functioning relationship in soil (see also Beare et al. 1995; Giller et al. 1997; Ekschmitt and Griffiths 1998; de Ruiter et al., Chapter 9).

15.3 Evidence for declining decomposer diversity in terrestrial ecosystems—the case of agricultural land management

Compared to what is known about organisms that live above-ground there is very little information available on changes in decomposer diversity, especially at the species level. Available information is also inevitably patchy with respect to the ecosystems that are studied, the stresses that they are subjected to, and the group of soil organisms considered. The main reasons for this are the historical neglect of decomposers by conservation inventories (Lawton et al. 1996), the lack of methodologies that can identify and quantify the diversity of decomposers, especially the unculturable microbes (Tiedje et al. 1999), and the fact that taxonomic work on soil organisms is extremely time-consuming (Bloemers et al. 1997). As a result, there is insufficient data available from specific locations to allow rigorous testing of general ecological theories, such as those that predict that maximum diversity occurs at intermediate levels of disturbance (Grime 1973b; Connell 1978) and productivity (Al-Mufti et al. 1977; Tilman 1982). It has been suggested that there is a unimodal relationship between soil decomposer diversity and agricultural intensification (Giller et al. 1997; Bardgett and Cook 1998), but convincing field data on this relationship is still lacking.

Terrestrial ecosystems are subject to many disturbances, but most studies on soil biodiversity focus on agricultural land management which is likely to affect decomposer diversity by altering: (1) the quality and quantity of detritus and non-detritus inputs through changes in plant community production and composition (see Wardle and van der Putten, Chapter 14); and (2) the soil microhabitat stability and complexity through chemical inputs and changes in soil physical structure (Bardgett and Cook 1998). The general pattern that emerges from these studies is that the conversion of natural ecosystems to agriculture reduces the diversity of decomposers. For example, the diversity of soil macrofauna (Lavelle and Pashanasi 1989) and nematodes (Bloemers *et al.* 1997) are dramatically reduced by the conversion of primary tropical forest to agriculture. Studies in Mexico, Peru and India reveal that earthworm communities in agroecosystems have lower species richness and a lower number of native species than those in undisturbed ecosystems (Fragoso *et al.* 1997). The diversity of collembola is higher in native prairie than in prairie that has been influenced by agriculture (Brand and Dunn 1998) and the diversity of soil nematodes has been shown to decrease with agricultural improvement of native pastures in New England Tablelands (Yeates and King 1997). However, in contrast to most studies which report that agriculture degrades nematode diversity (Bongers 1990; Freckman and Ettema 1993), Boag and Yeates (1998) found in their literature review that species richness of nematodes was higher in cultivated soils than in many natural ecosystems. The high diversity in cultivated soils was thought to be due to the dynamic nature of agricultural habitats in which a wide range of soil nematode species can survive and multiply.

Intensification of agricultural management may also reduce the diversity of decomposers. For instance, tillage of soils can reduce the abundance and diversity of earthworms (Springett 1992), although the effects appear to depend on soil type, climate and tillage operation (Chan 2001). Wardle (1995) found in a literature review that the diversity of macrofaunal groups could either be substantially elevated or reduced by tillage while the diversity of microfauna was little affected. Fertiliser application to grassland has been shown to reduce the diversity of nematodes and increase the proportion of bacterial-feeding groups in the nematode community (Yeates *et al.* 1997). Similarly, the density and species diversity of soil microarthropods is much lower in fertilized, high-input grasslands than in unfertilized, low-input sites (Siepel 1996). Furthermore, fungal-feeding grazers often dominate the microarthropod community in the low-input sites, whereas in the high-input sites these organisms may be replaced by opportunistic bacterial-feeders (Siepel 1996). Studies that have used molecular techniques to characterize soil microbial communities point to similar shifts in microbial community composition. For example, more intensive management of grazed pastures was shown to increase the proportion of bacteria relative to fungi in the soil microbial community (Bardgett *et al.* 1999; Grayston *et al.* 2001), and the abandonment of management increased the growth of fungi relative to bacteria in alpine meadows (Zeller *et al.* 2001). Information on associated changes in microbial diversity is scarce, but Yang *et al.* (2000) showed that DNA diversity declined with fertilizer use and Clegg *et al.* (1998) found a reduced microbial diversity with increased pasture management as determined by DNA reassociation and cross-hybridization techniques. In contrast, McCaig *et al.* (1999) used 16S ribosomal DNA sequence analysis to show that intensifying management of temperate pasture did not have wide effects on bacterial diversity.

The loss of decomposer species with agricultural intensification may not be a random process. For instance, Siepel (1996) showed that declines in the diversity of soil microarthropods with intensifying agriculture were accompanied by clear shifts in the life-history characteristics and feeding guilds of the community. The loss of diversity in low-input agriculture was explained by the disappearance of drought intolerant species; low-input grasslands are cut in summer thereby increasing the chance of drought in the litter layer. Meanwhile, the loss of species in high-input grassland was explained by the elimination of fungal-feeding grazers that were replaced by opportunistic bacterial-feeders. Moreover, abandoned high-input sites still lacked fungal-feeding mites after twenty years of management for nature conservation due to the low population growth and dispersal rate of these species (Siepel 1996). Similarly, Scheu and Shultz (1996) found a very slow change in the oribatid mite community following cessation of cultivation, despite dramatic changes in the plant community and rapid recolonization of uncultivated soils by the more mobile macrofauna. Earthworms, by virtue of their limited dispersal ability, also slowly recolonize favourable habitats that result from land-use change as illustrated by the absence of earthworms in the Dutch polders reclaimed between 1942 and 1975 (Brussaard *et al.* 1996).

To summarize the response of decomposers to agricultural land management, two patterns seem to emerge: (1) natural or unmanaged ecosystems have more diverse, fungal-dominated decomposer food webs than agricultural systems (Bardgett and Cook 1998); and (2) changes in diversity are not necessarily reversible in the short term (Brussaard et al. 1996). We still do suggest caution in making these broad conclusions since the diversity of feeding habits and habitat preferences of decomposers implies that responses to management will be complex and may be limited to specific groups of organisms only.

15.4 The potential for species diversity effects among decomposers

As species richness in decomposer food webs is high, it has been speculated that most decomposer species are probably redundant, i.e. replaceable by other species without general soil functions such as nutrient and C mineralization being impaired (see Andrén et al. 1995; Lawton et al. 1996; Groffman and Bohlen 1999). On the other hand, it is recognized that the existence of redundancy among decomposers is largely assumed without evidence (Behan-Pelletier and Newton 1999), and some authors support the view that within trophic groups differences among species may be sufficient to produce species-specific impacts on ecosystem functioning (Wall and Virginia 1999). Species diversity can only matter for ecosystem functioning if species within trophic groups differ markedly with regard to their resource use, response to abiotic factors and direct effects on ecosystem processes. We now examine whether there is such potential for species diversity effects in decomposer food webs.

Some microbial-feeding animals appear to be selective in their feeding; for instance, fungal-feeding mites and collembolans forage selectively among species of saprophytic (Verhoef et al. 1988; Maraun et al. 1998), arbuscular mycorrhizal (Klironomos and Kendrick 1996) and ectomycorrhizal fungi (Schultz 1991). Moreover, Hedlund et al. (1995) showed that collembolan species prefer fungal species from their own soil layer, which suggests that collembolans are adapted to feeding on fungal species growing in their own habitat or

choose a habitat with most palatable fungi. Although most top predators seem to be generalists (e.g. gamasid mites that capture collembolans, mites and nematodes (Sardar and Murphy 1987)), some species show morphological and behavioural adaptations for catching special prey (e.g. mites specialized in catching nematodes (Koehler 1997) and carabid beetles specialized in feeding on collembolans (Bauer 1982)). Therefore, it appears that soil fauna includes specialized consumers that may not be easily replaced by other species. Preferential feeding by consumers may in turn affect the structure of prey communities. For instance, selective feeding by a species of collembola influenced competition between fungal species and subsequent litter decomposition (Newell 1984a,b), as well as the replacement of primary saprophytic fungi by secondary saprophytes in decomposer succession (Klironomos et al. 1992). Among detritivorous macrofauna, animals may also have species-specific effects on microbes, such as the effects of gut passage of different earthworm species on microbial communities (Brown 1995).

Species may also differ within trophic groups in their response to predation, competition and abiotic factors. Mikola and Setälä (1998a) found in a microcosm experiment that species of microbial-feeding nematodes differed in their vulnerability to predation and interspecific competition. Soil moisture and drought tolerance of microarthropod species in turn appear to determine the species-specific spatial distribution of this fauna in soil (Verhoef and Witteven 1980; Verhoef and van Selm 1983; Vegter et al. 1988) and the combination of drought stress and food availability can explain species-specific distributions in several soil animal groups, such as earthworms, collembolans and oribatid mites (Vegter 1983; Hassall et al. 1986; Scheu and Schultz 1996). Bongers and Bongers (1998) have further summarized evidence of substantial differences in population growth rates and disturbance tolerances for nematode genera within trophic groups. In addition to spatial distribution, responses to abiotic factors may determine the *temporal* distribution of species (Sohlenius 1985; Holmstrup 2001); for instance, Sohlenius (1985) showed that a combination of physical factors may cause species-specific differences in the distribution of nematodes in time. Similarly, differences in the thermal adaptation of metabolic activity of species of bacterial-feeding nematodes may explain their temporal distribution in the field (Ferris *et al.* 1995).

Finally, species within trophic groups are known to differ markedly in their direct effects on ecosystem processes. Weekers *et al.* (1993) showed that although species of bacterial-feeding amoebae all preferred the same type of bacteria, population densities and the parallel production of ammonium differed ten-fold among the species, and similar results have been obtained using species of fungal-feeding nematodes (Chen and Ferris 2000). Among macrofauna, earthworms may have species-specific effects on plant performance (Blakemore 1997) and often more than one species of earthworms is needed for maintaining soil structure (Shaw and Pawluk 1986; Blanchart *et al.* 1997), which shows that species are not replaceable in terms of their effects on soil formation.

It appears that within trophic groups soil faunal species: (1) can be specialized in food intake (i.e. are not necessarily generalists that could easily replace each other), which may have impacts on the structure and succession of microbial communities, decomposition rate and release of nutrients to plants; (2) may respond differently to competition, predation and abiotic factors, such as soil moisture; and (3) may have different direct effects on ecosystem functioning. Moreover, when such differences have not been observed, it is possible that the techniques that are used to describe, for instance, changes in microbial communities are not fine-tuned enough to detect the differences (Tunlid 1999). Therefore, it seems that there is no a priori reason to assume that decomposer fauna is likely to be exceptionally redundant with regard to effects on ecosystem processes. In the next section, we will examine whether this potential for species diversity effects is realized in decomposer food webs.

15.5 Empirical evidence for diversity effects in decomposer food webs

Although the evidence that is available to test the above-mentioned contrasting hypotheses—widespread redundancy (Andrén *et al.* 1995; Lawton *et al.* 1996; Groffman and Bohlen 1999) versus a

significant role for diversity (Wall and Virginia 1999)—is still limited, a few studies exist that have manipulated the diversity of soil microbes and fauna (Table 15.2). For instance, Salonius (1981) diluted suspensions of field soil to produce a decreasing gradient of microbial diversity in sterile soil microcosms. Total microbial activity decreased with the supposedly decreasing microbial diversity even though microbial abundance increased with decreasing diversity; this was taken to provide evidence for the importance of species diversity rather than biomass in determining total metabolic activity in soil microbial communities. In contrast to the dilution technique, Hedlund and Sjögren Öhrn (2000) established simple decomposer food webs in soil microcosms using three known species of fungi. They manipulated the species composition of the fungal community (each of three fungal species alone versus all species together) and food chain length by adding a collembolan species and a top predatory mite, and found that C mineralization rate and fungal hyphal length in the three-species fungal mixture did not exceed corresponding values in the best performing monoculture. However, both variables differed between fungal communities, which suggests that fungal biomass and C mineralization rate were determined by species-specific characteristics of fungi rather than fungal species richness. Moreover, fungal species composition did not modify trophic interactions within the community as the influence of fungal grazers and top predators on C mineralization was equally positive in each system. Using similar approach, Robinson *et al.* (1993) compared carbon and nutrient mineralization from plant litter in the presence of four fungal species and their pair-wise combinations. They found that C mineralization was in species combinations constantly higher than predicted from the mineralization in monocultures, which was taken to suggest a positive species richness effect. Finally, Dauber and Wolters (2000) compared microbial communities living in mounds of different ant species and found that C mineralization rate and the diversity of carbon sources utilized by the microbial communities were unrelated (the diversity of utilized carbon sources was assumed to represent microbial functional diversity). However, both variables clearly differed

Table 15.2 Effects of biodiversity, species identity and species composition on ecosystem functioning in soil decomposer studies (sorted by study method)

Reference	Study method	Diversity gradient	Diversity effect	Species identity or composition effect
Faber and Verhoef 1991	Community construction	1–3[a]	No	Yes
Robinson et al. 1993	Community construction	1–2[a]	Yes (+)[d]	Yes
Mikola and Setälä 1998a	Community construction	2–6[a]	No	Yes
Laakso and Setälä 1999b	Community construction	1–5[a]	No	Yes
Hedlund and Sjögren Öhrn 2000	Community construction	1–3[a]	No	Yes
Cragg and Bardgett 2001	Community construction	1–3[a]	No	Yes
Liiri et al. 2002	Community construction	1–41[a]	Yes (+)	Not examined
Degens 1998	Destruction by fumigation	8% reduction–natural[b]	No	Not examined
Griffiths et al. 2000	Destruction by fumigation	60% reduction–natural[c]	Yes (−/+)	Not examined
Salonius 1981	Dilution of soil suspension	Unknown	Yes (+)	Not examined
Griffiths et al. 2001b	Dilution of soil suspension	50% reduction–natural[c]	Yes (−/+)	Not examined
Heneghan et al. 1999	Correlative field data	13–29 and 40–60[a]	Yes (+)	Yes
Dauber and Wolters 2000	Correlative field data	±5% of control (natural)[b]	No	Yes

[a] Species richness of fungi or fauna.

[b] Functional diversity of microbes (measured as a diversity of utilized carbon sources).

[c] Overall taxonomical diversity of microbes and microfauna.

[d] (+) and (−) indicate positive and negative diversity effects on ecosystem processes, (−/+) indicates that both effects were observed.

between the mounds thus indicating a species composition effect. Supporting these results, Degens (1998) found that decreased functional diversity of microbes in fumigated soil does not necessarily reduce the ability of microbes to decompose plant residues.

A few studies have also manipulated soil faunal diversity. Laakso and Setälä (1999b) compared the effects of two trophic groups (fungal-feeders and detritivores either singly or in combination) on microbial and top predator biomass, nutrient mineralization and plant growth. Within each trophic group treatment they manipulated the species composition and richness (one versus five species) of animals. They found that soil microbial biomass was not affected by faunal treatments whereas the biomass of fungal-feeders, detritivores and top predators was determined by species composition within trophic groups. Ammonium concentration in soil, plant nitrogen uptake and primary production were likewise affected by species composition within trophic groups, the effects being mainly driven by one enchytraeid species. Trophic group identity, trophic group richness and species richness within trophic groups did not significantly

affect either nutrient mineralization or plant growth. The results support the view that decomposer species within trophic groups, and even trophic groups themselves, are replaceable in relation to mineralization and plant growth. On the other hand, the dominant role of one species demonstrates how species composition may still greatly influence ecosystem processes.

In another microcosm experiment that included microbes, microbial-feeding nematodes and predatory nematodes, species composition of microbial grazers also determined the biomass of microbes, grazers and the top predator, whereas species richness (two versus six microbial grazers) did not have a consistent effect (Mikola and Setälä 1998a). The effect of varying species composition on biomasses at three trophic levels was attributed to species-specific characteristics of microbial grazers; the six species used differed with regard to resource use efficiency and responses to predation and competition. However, the effects of species composition on trophic interactions were not simply linked to subsequent changes in C and N mineralization: net N mineralization was not affected and C mineralization rate appeared to be positively related to the number of microbial grazers rather than species composition. The importance of species identity and composition relative to species richness is further supported by a microcosm study by Cragg and Bardgett (2001) in which the effects of a collembolan community (including one, two or three species in different combinations) on litter mass loss, microbial activity and N mineralization were mostly determined by one dominating species. In a comparable field study, organic matter loss and N mineralization were also differently affected by three collembolan species and were not higher in systems that had all three species present than in the best performing one-species system (Faber and Verhoef 1991). In contrast to above-mentioned studies with soil fauna, Liiri et al. (2002) provide evidence for a weak species diversity effect. In their study, plant production and total N uptake increased asymptotically with increasing richness of microarthropod species. However, microarthropod diversity did not alter trophic interactions since no effects were found on fungal diversity or microbial biomass. The results also indicated that microarthropod species richness did not modify the effect of an experimentally imposed disturbance (drought) on birch growth. The authors suggested that the most probable mechanism to mediate the effect of microarthropod diversity on plant growth was the increased N availability with increasing species richness.

Unlike other studies that have manipulated species and trophic group richness of specific decomposer groups, Griffiths et al. reduced the biodiversity of both microbes and microfauna using either lengthening periods of chloroform fumigation (Griffiths et al. 2000) or inoculation of sterile soil with serially diluted soil suspensions (Griffiths et al. 2001b). In both experiments, the effect of declining biodiversity on ecosystem processes depended on the specific process in question (Griffiths et al. 2000, 2001b). For instance, in the fumigation experiment (Griffiths et al. 2000), net N mineralization increased in soil with increasing fumigation time, whereas C mineralization and decomposition rate of added plant residues were unaffected. Similarly, decreasing biodiversity did not affect net N mineralization in the dilution experiment but slowed down plant residue decomposition

(Griffiths et al. 2001b). Moreover, while declining diversity had negative effects on the stability of soil processes in the fumigation experiment, no effects were found in the dilution experiment (these results and the general relationship between soil biodiversity and stability are covered in depth by de Ruiter et al., Chapter 9).

It has been suggested that decomposition rate of plant litter can be predicted without a knowledge of decomposer food web structure because abiotic factors, such as temperature and moisture, control the activities of soil organisms that in turn limit decomposition (Andrén et al. 1999). Recently, Heneghan et al. (1999) presented data that illustrates the relative importance of abiotic and biotic factors (in this case the soil microarthropod community) on litter decomposition. Their data showed that the rate of decomposition of oak litter was higher in two tropical sites than in a temperate site, demonstrating the influence of climate. Second, the contribution of microarthropods to litter mass loss differed between the tropical sites, suggesting that litter decomposition depended on the different microarthropod species composition at the two sites. Finally, within two of three sites, species richness of microarthropods was positively correlated with the contribution of fauna to litter mass loss. It thus appears that although the climate drives decomposition rate at a large scale, species composition and richness of decomposers may be important at a local scale. This is analogous to the situation for primary production, which is driven mainly by abiotic factors at a large scale (Huston, Chapter 5) but is also affected by species composition (Wardle and van der Putten, Chapter 14) and species richness (Tilman, Chapter 3) at a small scale.

15.6 Disentangling covarying diversity factors in soil diversity experiments

Experimental studies of biodiversity are prone to factors that covary with diversity treatments (covarying diversity factors) and confound the interpretation of data (e.g. Huston 1997). The choice of an experimental design is therefore important, and even more so as different designs provide different powers for finding diversity effects (Allison 1999).

There are two main types of design that can be used when constructing communities for diversity experiments; both designs involve the establishment of an experimental gradient of species richness, but differ in the way that replication is realized within each level of diversity. In the first type of design, richness levels contain replicated monocultures of constituent species or several replicated multiple-species communities (see, e.g. Wardle *et al.* 1997a). Four of the decomposer studies that have been performed to date (Faber and Verhoef 1991; Mikola and Setälä 1998a; Hedlund and Sjögren Öhrn 2000; Cragg and Bardgett 2001) can be placed into this category, even though most of them include only one multiple-species system. Although covarying diversity factors do not confound results in these kinds of experiments, the experiments do not allow for a strong test for species richness effects due to the absence of wide richness gradient (Table 15.2). The number of species used is also very low relative to natural communities, which may partly explain the idiosyncratic results obtained in these studies (Mikola and Setälä 1998a).

In the second type of design, each replicate at each richness level is chosen by a separate random draw from the total pool of species (see, e.g. Tilman *et al.* 1996). The disadvantage of this design is that both complementary resource use (due to niche differentiation) and sampling effect (the increasing probability of selecting species of excessive characteristics in randomly selected samples of increasing species richness) may induce a positive relationship between species richness and ecosystem functioning (Huston 1997; Aarssen 1997; Tilman *et al.* 1997b), and the only way to effectively separate these two mechanisms is to compare the performance of species mixtures with that of monocultures (Garnier *et al.* 1997; Loreau 1998b; Huston *et al.* 2000). Of the decomposer studies, the experiment by Liiri *et al.* (2002) resembles this type of design. The experiment contained randomly selected microarthropod communities and since it did not involve monoculture treatments of microarthropod species, the sampling effect cannot be totally excluded as a potential cause for the positive relationship between microarthropod species richness and plant growth. Moreover, in this study the interpretation of results and mechanisms is

hampered by the fact that microarthropod biomass correlated with species richness from the beginning of the experiment, and microarthropod biomass rather than the number of species may have influenced plant production.

The advantage of dilution and destruction approaches used by Salonius (1981), Degens (1998) and Griffiths *et al.* (2000, 2001b) is that they produce species richness levels that are comparable to natural levels. However, these approaches do not eliminate the problem of covarying diversity factors; e.g. dilution selects for originally abundant species (Griffiths *et al.* 2001b) and cannot separate species composition effects from species richness effects. The removal of microbial and faunal species by increasing fumigation time is not random either, and it is likely that fumigation selects for organisms with particular physiological characteristics (Griffiths *et al.* 2000). Fumigation can also release nutrients, in which case soil nutrient availability and decomposer diversity may covary (Degens 1998).

It appears that most experimental studies that have reported diversity effects on ecosystem processes, i.e. Salonius (1981), Griffiths *et al.* (2000, 2001b) and Liiri *et al.* (2002), are susceptible to factors whose effects on measured functional variables cannot be separated from diversity effects. It thus seems that although there is evidence for species composition effects in decomposer food webs (Table 15.2), convincing data for significant species diversity effects is still mostly lacking.

15.7 Why diversity effects are rare and inconsistent in decomposer food webs

In contrast to the ample evidence for positive effects of plant diversity on primary production in grasslands (Hector *et al.*, Chapter 4; Tilman *et al.*, Chapter 3; Schmid *et al.* 2002), significant effects of decomposer diversity are rare (Table 15.2): only in six of 13 studies was decomposer diversity found to influence ecosystem processes, and even in these studies the effects were often weak (Liiri *et al.* 2002) or inconsistent (Griffiths *et al.* 2000, 2001b). We examine three possible reasons for this apparent difference between plant and decomposer studies: (1) decomposers are generally redundant;

(2) diversity gradients are narrow in decomposer studies; and (3) soil processes are controlled by multi-trophic interactions, which complicate the diversity–ecosystem functioning relationship.

Species redundancy would be the most logical and simplest explanation for the absence of diversity effects in decomposer systems. However, the role of redundancy is not supported by the fact that species identity and species composition of microbes and fauna had significant effects on ecosystem processes in each study in which their role was examined (Table 15.2). Thus, as we suggested, there seems to be a potential for diversity effects in decomposer food webs although it is seldom realized. Instead, diversity gradients are narrow in several experiments (Table 15.2), and it seems that diversity effects have not been observed in studies in which the gradient is below five species (except in the study by Robinson et al. 1993), whereas studies with wider gradients have often reported significant diversity effects (Heneghan et al. 1999; Griffiths et al. 2000, 2001b; Liiri et al. 2002). Finally, Mikola and Setälä (1998a) have argued that if a community consists of several trophic groups (like decomposer communities do), ecosystem processes may decrease as well as remain unchanged or increase with decreasing species diversity depending on the nature of trophic interactions and the group from which species disappear. This is because: (1) different trophic groups may have different (sometimes opposite) effects on ecosystem processes (Table 15.1); (2) trophic group biomass (e.g. micro-arthropod abundance), which is likely to change when species are lost, and process rates (e.g. mineralization rate) may have parabolic, rather than linear, relationships (Hanlon 1981); and (3) species disappearance can produce indeterminate and unexpected changes in the abundance of remaining species due to the numerous indirect interactions within food webs (Yodzis 1988). In that case, the inconsistent effects of diversity on soil processes observed by Griffiths et al. (2000, 2001b) may in fact be the most likely scenario in multi-trophic decomposer systems.

To summarize, it seems that the usage of narrow diversity gradients is one possible reason for the general rarity of diversity effects in decomposer studies, whereas the inconsistent, but significant effects of diversity on different soil processes may be best explained by the complexity of food webs that govern these processes.

15.8 Future challenges

Rigorous experimentation and hypothesis testing are essential when examining the relationship between soil biodiversity and ecosystem functioning (Freckman et al. 1997; Swift et al. 1998), especially because positive effects of diversity are not the only reasonable hypothesis in multi-trophic systems (Mikola and Setälä 1998a). We also recommend that future studies should: (1) evaluate several aspects of community structure to clarify the *relative* role of biodiversity in ecosystem processes; (2) select appropriate experimental designs to avoid covarying diversity factors; (3) investigate diversity effects on the aspects of *stability* of ecosystem functioning; and (4) include explicit contemplation of attributes that characterize soils, such as patchy distribution of resources, the diversity of basal resources and soil porosity.

Above-ground studies have revealed the importance of clarifying the relative roles of trophic group composition, trophic group richness, species composition (within trophic groups) and species richness (within trophic groups) when aiming to fully understand the role of biodiversity in ecosystem functioning. Until now, the effects of trophic group richness of soil fauna on mineralization and plant growth have received little attention (but see Laakso and Setälä 1999b) and studies manipulating trophic group composition are also few. However, there is evidence that trophic group composition of fauna is important in determining nutrient mineralization and primary production. Moreover, trophic groups seem to modify each other's influence, suggesting that the influence of trophic group mixtures cannot be predicted based on the influence of each group in isolation. For instance, Alphei et al. (1996) examined the effects of protists, nematodes and earthworms on plant growth, and found, amongst other things, that while protists and nematodes generally enhanced and earthworms reduced primary production, the effect of protists depended on the presence of earthworms. Alphei *et al.* (1996) did not explicitly test the effect of trophic group richness on mineralization and plant growth but their data suggests that the influence of trophic group mixtures never exceeded (or fell short of) the influence of the best (or worst) performing component group in isolation, suggesting a non-significant richness effect. Similarly, Laakso and Setälä (1999b, see above) did not find a significant effect of trophic group richness, and in the data presented by Setälä (2000) plant growth was not higher in the presence of a combination of mites, collembolans and enchytraeids than in the presence of either collembolans or enchytraeids alone. However, some data suggest that trophic group richness might matter; Huhta and Viberg (1999) reported that net nitrogen mineralization in soil was higher when both enchytraeids and earthworms were present than in the presence of either group alone.

Experimental designs used in diversity studies usually test the effect of random disappearance of species or trophic groups on ecosystem functioning. However, when the disappearance of species is known to be non-random, as in those systems described by Siepel (1996), these experimental designs are not appropriate. The extensive literature that is available on pitfalls in experimental designs in which diversity is manipulated as a treatment (see Huston 1997; Loreau 1998b; Wardle 1999; Huston *et al.* 2000; Hector *et al.* 2000b) will help to avoid the problems raised by covarying factors in future research. However, it may be impossible to disentangle the effects of diversity and covarying factors in decomposer studies if achieving of natural levels of species richness is a priority (Griffiths *et al.* 2000). Constructive experiments using known species may be able to exclude covarying factors but they are unlikely to be tractable for examining natural levels of diversity in soil systems. On the other hand, destructive methods (like chloroform fumigation) allow for natural levels of diversity but have their own limitations as discussed above. The solution for the apparent trade-off between achieving natural diversity and accurate treatments may be to apply several approaches and critically evaluate the combined results (Griffiths *et al.* 2000).

Even if soil biodiversity would be unimportant in driving ecosystem functioning under static conditions, it may significantly affect the stability of functioning in response to disturbance (Bongers and Bongers 1998; Loreau *et al.*, Chapter 7). Moreover, many soil organisms are active only during short periods over the year, which suggests that the diversity of organisms has a potential to affect the temporal variability of functioning. We suggest that diversity ought to be manipulated and functioning measured: (1) in constant abiotic environments (to find the innate effect of diversity on process variability); (2) in naturally varying environments (e.g. under fluctuating temperature and moisture); and (3) in response to large-scale disturbances, such as pollution and extreme drought. Those studies that have investigated the diversity–stability relationship among decomposers suggest that diversity has neutral (Griffiths *et al.* 2001b; Liiri *et al.* 2002) and positive (Griffiths *et al.* 2000) effects on the resilience and resistance of ecosystem processes to disturbances, and no effects on the variability of processes in constant environment (Griffiths *et al.* 2001b). However, when measuring the resistance and resilience of soil processes to perturbations, it should be noted that these variables are greatly affected by soil biophysical properties, such as the amount of organic matter, cation exchange capacity and soil structure, which have developed with the help of soil organism activities over a long period of time (Bardgett *et al.* 2001). Short-term changes in soil biodiversity may therefore not manifest their effects on the stability of soil processes for a long time after organisms are lost, suggesting that delayed responses may be an important legacy of changes in soil biodiversity.

The effects of soil characteristics on the relationship between diversity and ecosystem processes provide the final challenge for future biodiversity studies. Soils are typically heterogeneous environments consisting of patches of resources in space and time (Griffiths 1994; Giller 1996), which affects the distribution and abundance of decomposers, the nature of their interactions and decomposition rate (Anderson 1978, 1988; Sulkava and Huhta 1998; Mikola and Sulkava 2001). Even though we cannot hypothesize the mechanisms by which patchiness, diversity and ecosystem functioning might be linked, there appear to be good reasons to believe that patchiness can modify the interaction between the other two. Besides being heterogeneously distributed in soil, composition of organic resources shows different degrees of heterogeneity, for instance due to plant community structure (Wardle and van der Putten, Chapter 14). It can be hypothesized that soils with more heterogeneous resources allow a higher number of species to coexist, and on the other hand, need a higher diversity of consumers to become more fully utilized.

Another feature that is typical for soils is that soil pore size controls the movement of many micro- and mesofaunal groups and limits the existence of interactions between microbes and fauna. For instance, Elliott *et al.* (1980) showed that amoebae and nematode numbers are higher in coarse- than in fine-textured soil because a higher proportion of bacteria can be reached by the grazers in coarse-textured soil, and that in fine-textured soil amoebae can enter a greater proportion of soil pores than nematodes due to their smaller size. This has two important consequences: first, in fine-textured soil the species and trophic group composition of microbial feeders may determine their ability to reduce microbial biomass and affect mineralization, and second, soil texture is able to modify the diversity-functioning relationship of microbial grazers by determining the size range of grazers that are needed to fully exploit microbial biomass.

15.9 Conclusions

Diversity research in decomposer food webs is clearly warranted due to land-use changes that decrease decomposer diversity in terrestrial ecosystems and the absence of a priori reasons for assuming non-significant diversity effects among decomposers. Existing empirical results suggest that species identity and composition of decomposers have significant impacts on ecosystem processes while the effects of decomposer diversity are less common and often weak and inconsistent. Moreover, most of those studies that have reported significant diversity effects on nutrient mineralization and primary production are susceptible to covarying diversity factors that confound the interpretation of results. The rarity of diversity effects in decomposer studies, relative to their occurrence in plant studies, may be a consequence of narrow diversity gradients used in many studies, whereas the inconsistent but significant effects of diversity on different soil processes may be a manifestation of complex trophic interactions that govern these processes. In the future, aims of priority are the assessment of the role of biodiversity in relation to other aspects of decomposer food web structure, investigation of diversity effects on the stability of ecosystem functioning, and more explicit contemplation of attributes that characterize soils.

CHAPTER 16

The critical role of plant–microbe interactions on biodiversity and ecosystem functioning: arbuscular mycorrhizal associations as an example

M. G. A. van der Heijden and J. H. C. Cornelissen

16.1 Introduction

The relationship between biodiversity and ecosystem functioning has been a main focus of ecological research in recent years as is summarized in this book. Studies that investigated this relationship mainly concentrated on plants and how their diversity influences productivity and stability of terrestrial ecosystems (e.g. Hector *et al.*, Chapter 4; Tilman *et al.*, Chapter 3). Plants, as primary producers, are the basal component of most terrestrial ecosystems, which justifies this approach. However, most ecosystems harbour diverse communities of vertebrate and invertebrate animals, and many microorganisms. These organisms interact with plants and hence influence the relationship between biodiversity and ecosystem functioning. That such multitrophic interactions are an integral part of ecosystem functioning is increasingly being recognized (e.g. see de Ruiter *et al.*, Chapter 9; Mikola *et al.*, Chapter 15; Raffaelli *et al.*, Chapter 13; Wardle and van der Putten, Chapter 14). One group of organisms that is often neglected, are symbiotic microorganisms that associate with plants. These associations are ubiquitous and an essential part of terrestrial ecosystem functioning. Up to 80% of all plant species form intimate symbiotic associations with soil fungi and the number of plants that harbour symbiotic bacteria may be much higher than we currently know. Nutrient acquisition by plants, a process essential for the functioning of all terrestrial ecosystems is, in many ecosystems, driven by symbiotic microorganisms. Moreover, vegetation productivity, often seen as one of the Key parameters that characterizes ecosystem functioning, is largely determined by nutrient availability and, thus, influenced by symbiotic microorganisms. Here we will show that symbiotic interactions between plants and microorganisms contribute greatly to plant diversity and to the functioning of terrestrial ecosystems.

We will only discuss those symbiotic associations where each symbiont benefits from the presence of the other. After a brief discussion of several of such mutualistic plant–microbe symbioses we will focus on associations between plants and mycorrhizal fungi. We will argue that a robust understanding of biodiversity–ecosystem functioning relationships in terrestrial ecosystems requires an understanding of how plants are affected by mycorrhizal fungal abundance and their diversity of taxa and functional types.

16.2 Overview of plant–microbe symbioses and their contribution to biodiversity and ecosystem functioning

16.2.1 Non-mycorrhizal associations

Intimate symbiotic associations are widespread and very old. Endosymbiotic associations between ancient bacteria are thought to have led to the origin of the eukaryotic cell (Margulis 1993). These early associations contributed to the evolution and diversification of life on Earth. Plant–microbe symbioses vary considerably including associations with prokaryotes and fungi that range from loose to very close associations. Close symbiotic associations with prokaryotes include the well-known associations between nitrogen fixing Rhizobium bacteria and leguminous plants (e.g. Werner 1992; Spaink *et al.* 1998). These and other nitrogen fixing associations are an essential part of the global nitrogen cycle (e.g. Sprent and Sprent 1990). It is not surprising that this symbiosis exists. Plants are not capable of biological nitrogen fixation (that trait is restricted to organisms with a prokaryotic cell structure; Marschner 1995), while they depend on nitrogen since it is one of the major elements that limits their growth (Aerts and Chapin 2000). Many of the 16,000–19,000 species of legumes (Allen and Allen 1980) host nitrogen fixing bacteria from the Rhizobiaceae family (de Faria *et al.* 1989). These Rhizobium bacteria fix atmospheric nitrogen into ammonium in special root organs: the root nodules. Often, a large part of nitrogen in legumes is symbiotically fixed (Sprent and Sprent 1990) and rhizobia probably contribute to the survival of legumes in many natural communities.

Nitrogen fixing bacteria have a huge impact on ecosystem functioning. That follows from the indirect observation that their hosts, the legumes, are a key functional group in grasslands and in many other ecosystems. Indeed several studies have shown that legumes are key determinants of nitrogen acquisition and productivity (Hector *et al.* 1999; Tilman *et al.* 1997a; Loreau and Hector 2001). Studies with two species mixtures have, in addition to this, shown that the presence of legumes (including their bacterial symbionts) increases the relative yield by approximately 20% compared to mixtures of non-legumes (Trenbath 1976). Natural abundances of the nitrogen isotope 15 N can be used to estimate the proportion of nitrogen in the vegetation that is symbiotically fixed (e.g. Boddy *et al.* 2000; Robinson 2001). Namely, plants that derive nitrogen from their nitrogen fixing symbionts contain different delta 15 N levels compared to those that acquire organic or inorganic forms of soil

nitrogen. For example, Jacot *et al.* (2000) estimated that nitrogen fixers contributed 8–15% of total above ground nitrogen of alpine grasslands. This number varied depending on the relative proportion of legumes in the total biomass.

Biological nitrogen fixation is not restricted to the legumes and their bacterial symbionts. A number of other plant species, including the water fern Azolla, members from the ancient cycad plant family and species of the angiosperm genus Gunnera acquire additional amounts of nitrogen through association with nitrogen fixing cyanobacteria (e.g. Paracer and Ahmadjian 2000). Moreover, approximately 140 plant species belonging to seven plant families form actinorhizal symbiosis with nitrogen fixing bacteria of the genus Frankia (Squartini 2001).

Recent molecular and physiological studies suggest that the symbiosis between plants and bacteria may be more widespread than previously thought. Bacteria, such as Azospirillum, Burkholderia and Acetobacter, that possess nitrogen fixing capabilities have been identified in the leaves of several grasses (Reinhold-Hurek and Hurek 1998) and even in hyphae of symbiotic mycorrhizal fungi (Minerdi *et al.* 2001). These bacteria can have positive effects on plant growth as has been shown in the case of sugar cane (Boddey *et al.* 1995). The contribution of such endophytic bacteria to nitrogen capture and productivity of natural or semi-natural ecosystems is not known. A combination of sensitive molecular identification tools together with ecological experiments may help to overcome problems that previously could not be solved. Besides, free-living diazotrophic bacteria such as Azospirillum, Clostridium and Klebsiella fix nitrogen and form loose associations with roots and even leaves (e.g. see Marschner 1995 for references).

Nitrogen fixation is also an important feature of lichens in many biomes of the world, particularly in boreal, arctic and tropical zones. They can, indirectly, provide significant nitrogen input into the soil, as was shown, for instance, for terrestrial lichens in a sub-arctic woodland (Crittenden 1983) and for epiphytic canopy lichens in a tropical forest (Forman 1975). These associations may significantly contribute to nitrogen nutrition in nitrogen-limited ecosystems. Worldwide, an estimated

90–130 Tg N/year $(1\,\mathrm{Tg} = 10^{12}\,\mathrm{g})$ is biologically fixed by nitrogen fixing bacteria, a figure comparable to anthropogenic nitrogen fixation (Vitousek *et al.* 1997a). These symbiotic bacteria, thus, contribute to the annual nitrogen requirement of natural communities pointing to their important role in terrestrial ecosystems. Plants form a number of loose associations with many other bacteria. It is questionable whether these can be classified as 'symbiotic' since often there is no direct association, although reciprocal benefit occurs. Many rhizosphere colonizing bacteria such as Bacillus and Pseudomonas produce hormone like substances that stimulate plant growth or that inhibit root pathogens (Glick 1995). Agricultural plants benefit from the presence of such bacteria (e.g. Schippers 1992). Extrapolation of these results to natural plant communities suggests that such bacteria might influence ecosystem functioning as well. Experimental evidence for this is, to our knowledge, absent. Moreover, bacteria and fungi are the primary decomposers of any ecosystem (Mikola *et al.*, Chapter 15) providing the minerals that are necessary for plant productivity. Again this can be viewed as a loose 'symbiotic' association that is necessary for ecosystem functioning.

Another group of proposed mutualistic microorganisms are endophytic fungi. Endophytic fungi are ubiquitous, symptomless colonizers of leaves and stems which have been isolated from virtually all plant taxa, from hepatics to almost every gymnosperm and angiosperm (see Wilson and Faeth 2001 for references). Endophytic fungi have been shown to protect plants against herbivores through the production of alkaloids (Omacini *et al.* 2001) and one study showed that a host specific grass endophyte reduced plant diversity (Clay and Holah 1999). The presence of the endophyte, Neotyphodium coenophialum, stimulated the relative dominance of the endophyte infected grass and it decreased species number in that study. A number of potential mechanisms have been suggested, including: increased drought tolerance, improved herbivore resistance and a higher seedling establishment of the endophyte infected grass. These mechanisms are all related to the enhanced competitive ability of the endophyte-infected grass.

From the above overview it is clear that, even without mycorrhizas (see below), symbiotic relationships between microorganisms and plants are widespread and very diverse in taxonomic identity, form and habitat choice. Although empirical data are still lacking, it appears most likely from the above (partly anecdotal) evidence, that the functioning of many terrestrial ecosystems is affected by the availability of a diverse range of functional types and taxa of symbiotic microorganisms and their hosts.

16.2.2 Mycorrhizal associations: overview

The mutualistic symbiosis between plants and mycorrhizal fungi is widespread and very abundant in most ecosystems (see below). Mycorrhizal fungi provide an extensive hyphal network that forages for soil nutrients much more effectively than plant roots. Nutrients such as nitrogen, phosphorus, zinc and copper, that are often strongly bound to soil particles or that are sequestered in organic carbon sources, are supplied by the fungus (Smith and Read 1997). It has also been proposed that mycorrhizal fungi are involved in mineral weathering and this can act as an additional nutrient source (Landeweert *et al.* 2001). Plants typically show enhanced growth when colonized by these fungi, especially when nutrient availability is low. The heterotrophic mycorrhizal fungi in turn benefit from assimilates that they receive from the autotrophic plant (Smith and Read 1997). Mycorrhizal fungi are multifunctional: they also provide other 'services' such as protection against soil pathogens, enhanced access to water and they contribute to soil aggregation and soil structure (Newsham *et al.* 1995; Miller and Jastrow 2000). Recent studies suggest that mycorrhizal fungi are a key group in many ecosystems that contribute to important ecosystem functions such as nutrient acquisition, nutrient cycling, productivity and plant diversity (see Section 1.3). Decomposition processes are also under influence of mycorrhizal fungi (Hodge *et al.* 2001; Leake *et al.* 2002). The ecological significance of mycorrhizal fungi is highlighted in a new book on mycorrhizal ecology (van der Heijden and Sanders 2002).

16.2.3 Functions and functional groupings: three major types of mycorrhizal associations

Six different types of mycorrhizal associations exist, arbuscular mycorrhizal associations, ecto-mycorrhizal associations, ericoid mycorrhizal associations, orchid mycorrhizal associations, arbutoid mycorrhizal associations and monotropoid associations (see Smith and Read 1997 for an extensive description of each type). The first three types are important on a global scale (Read 1991). These three associations can be classified in different functional groups as will be discussed in the following section. First however, we will give a short overview of the general biology of the three major types.

The most abundant type is the symbiosis between arbuscular mycorrhizal fungi (AMF) and many grasses, forbs shrubs and trees. It is particularly abundant in grasslands and (sub) tropical ecosystems. AMF are ancient zygomycetous fungi belonging to the order Glomales (Remy *et al.* 1994; Redecker *et al.* 2000). AMF are probably not host specific, indicating that each AMF species is potentially able to colonize most of the approximately 1,50,000 terrestrial plant species that are estimated to associate with AMF (Trappe 1987). These associations are characterized by arbuscules, microscopic, tree-like structures that are often formed by the fungus within cortical root cells. Arbuscular mycorrhizal associations are also distinguished from other mycorrhizal types by the formation of extensive amounts of fungal hyphae that run parallel to the endodermis inside the root cortex (Fig. 16.1(a)).

Another abundant type of mycorrhizal association is that between ecto-mycorrhizal fungi (EMF) and most trees from boreal and temperate ecosystems. Dense mycelial webs of ecto-mycorrhizal fungi occupy the floor of most northern forests. EMF belong to the class of Basidiomyceteous and Ascomyceteous fungi. The roots of ecto-mycorrhizal trees are often completely surrounded by a fungal mantle and the largest part of the fungus remains outside the root, hence the name ecto-mycorrhizal (Simard *et al.* 2002; Fig. 16.1(b)). Nutrients must pass this fungal mantle before they can enter the root thus pointing to the importance of EMF for mineral nutrition. Only a few EMF hyphae enter the root

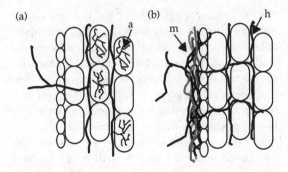

Figure 16.1 Schematic presentation of the structural characteristics of the association between plants roots with AMF (a) and ecto-mycorrhizal fungi (b). A small hypothetical subsection of a root is taken where (a) stands for an arbuscule inside a root cortical cell, (m) fungal mantle surrounding the root and (h) for the Hartig net between root cells.

and form the so-called Hartignet (Smith and Read 1997). Many EMF show some degree of host specialization, e.g. the Basidiomycete genus Suillus with Pinaceae trees (Molina *et al.* 1992). Other EMF have a broader host range (e.g. Laccaria). Ecto-mycorrhizal fungi, like many wood decomposer fungi, excrete a range of extra-cellular enzymes that are able to break down complex organic substances (Leake *et al.* 2002). These nutrients subsequently can be translocated to their hosts and hence improve their growth. This ability is especially important in forests where most available nutrients are embedded in organic substances in plant litter or root exudates (Aerts 2002).

The third type, ericoid mycorrhizal associations, are common in heath lands and tundra ecosystems that are dominated by species of the Ericaceae family. In this case, Ascomyceteous fungi colonize single cells of the root epidermis leading to the formation of structures that are typical for ericoid mycorrhizal associations. Ericoid mycorrhizal fungi are even more specialized in uptake of organically bound nutrients compared to EMF (Read 1991).

16.2.4 Impacts on plants

Mycorrhizal fungi often have a positive impact on plant growth. For instance, 19 of the 28 plant species typical for nutrient poor European calcareous grassland showed enhanced growth when colonized by mycorrhizal fungi, compared to un-colonized plants (van der Heijden 2002). A similar study showed that 56 out of 95 plant species characteristic of nutrient poor American tall grass prairie benefit from AMF (Wilson and Hartnett 1998). AMF stimulate the growth of such mycorrhiza-dependent plant species through the supply of nutrients, particularly phosphorus.

AMF are not only beneficial to plants. The growth of several mycorrhizal plant species is reduced by AMF and several non-hosts are inhibited by AMF (e.g. Francis and Read 1995). The responsiveness of plants to AMF decreases when nutrient availability increases as has been shown in many studies (reviewed in Koide 1991). Ecto-mycorrhizal fungi are also beneficial for tree growth and sapling survival (Marx 1991). Trees such as Quercus or Pinus are often heavily colonized by ecto-mycorrhizal fungi and these species often acquire a large part of their below ground resources from their fungal associates.

16.2.5 Plant–mycorrhizal functional groupings

It is important to note that the functionality of these different mycorrhizal types goes beyond the transfer of phosphorus, nitrogen and/or carbon between plant and fungus. Different types of mycorrhizal fungi may, by affecting their hosts, exert an important role on nutrient and carbon cycling traits of ecosystems (Fitter 1990; Read 1991; Kielland 1994; Rygiewicz and Anderson 1994; Michelsen et al. 1996). Recently, Cornelissen et al. (2001) compared 83 diverse British plant species for important traits with respect to ecosystem functions such as productivity and decomposition. The 83 species were also classified according to their known predominant mycorrhizal association type. It was consequently tested whether plant species could be classified in different plant functional groups according to the type of their mycorrhizal associate. We used data, obtained from standardized tests, for the potential seedling relative growth rate (RGR) and the relative leaf litter decomposability of all these 83 species. Inherent RGR's of the component species of ecosystems are, together, good indicators of ecosystem

CO_2 capture and productivity (Grime and Hunt 1975), while the combined litter decomposabilities of component species have strong repercussions for litter breakdown and CO_2 release from ecosystems (Pastor et al. 1984; Hobbie 1996; Wardle et al. 1997b). Cornelissen et al. (2001) revealed large, significant differences among the different plant mycorrhizal types, with essentially similar patterns for inherent RGR and leaf litter decomposability (Fig. 16.2). Ericoid and EMF plant species were inherently slower growing and had more poorly decomposable litter than AMF plant species. These findings provided strong quantitative support for Read's (1991) hypothesis that mycorrhizal type is an important component of a plant's strategy in the context of nutrient availability in ecosystems.

Ericoid and EMF plant species generally predominate in ecosystems of low nitrogen availability. These plants tend to produce well-defended, long-lived leaves, which turn into recalcitrant litter that inhibits nitrogen mineralization and suppresses inorganic nitrogen availability to plant roots (van der Krift and Berendse 2001). The latter would reduce the establishment, survival and/or competitiveness of potentially fast-growing AMF and non-mycorrhizal plant species. It would also benefit EMF and ericoid plants since these plants have, due to their association with ericoid and ecto-mycorrhizal fungi, access to organic nutrients, a nutrient source of which AMF plants are mainly deprived. Thus, ericoid and EMF plants probably benefit indirectly from the recalcitrant litter they themselves produce (Berendse 1994). However, structural and secondary chemistry of foliar defences by many ericoid and ecto-mycorrhizal plants tends to reduce the assimilation capacity of their leaves, thereby reducing potential growth rates (Coley 1988; Poorter and Bergkotte 1992; Cornelissen et al. 1998). This would put these plants in a disadvantaged position in ecosystems of high inorganic nitrogen availability, where fast-growing plants, notably non-mycorrhizal and AMF plant species, are likely to out-compete them and maintain nitrogen availability by producing nutritious, easily decomposable litter (REF). In this way, interspecific variation in plant mycorrhizal type seem prominent in a hypothesized positive feedback between nutrient availability, vegetation productivity and

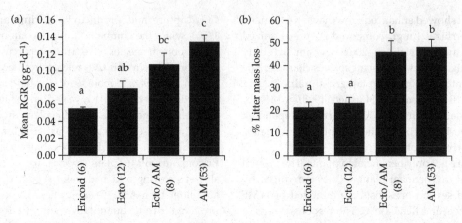

Figure 16.2 Mean values (+1 SE) for plant traits of four plant functional types in terms of their mycorrhizal association. Plants that associate with ericoid mycorrhizal fungi (Ericoid), ecto-mycorrhizal fungi (Ecto), arbuscular mycorrhizal fungi (AMF) or which form a dual symbiosis with ecto-mycorrhizal fungi and arbuscular mycorrhizal fungi (Ecto/AMF) are shown. (a) Inherent seedling mean RGR ($F_{7.44}$, $P < 0.001$); (b) Litter mass loss percentage during 20 weeks of winter/spring incubation in outdoor decomposition bed ($F_{11.1}$, $P < 0.001$). Means which share the same letter were not significantly different in *post-hoc* Games Howell tests. Numbers of species in each group are in parentheses. For further details see Cornelissen *et al.* (2001). This graph shows that plants can be classified in different plant functional groups according to their fungal partner (reprinted with permission, *Oecologia*, Springer Verlag).

litter turnover rate of ecosystems (Pastor *et al.* 1984; Reich *et al.* 1992; Van Breemen 1993; Berendse 1994; Cornelissen *et al.* 1999; Aerts 2002), and thus plays a key role in carbon cycling.

16.2.6 Ubiquity of the mycorrhizal symbiosis

Most ecosystems are dominated by plants having mycorrhizal associations (Read 1991). Over three-quarters of the Dicotyledones naturally occurring in the United Kingdom are likely to have mycorrhizal associations (Harley and Harley 1987). Examples from an European calcareous grassland, an American tall grass prairie, several temperate deciduous forests, a tropical rain forest and shrub land of the threatened South African Cape region show that most plants in these ecosystems are able to associate with mycorrhizal fungi (van der Heijden and Sanders 2002). In contrast, disturbed habitats with many ruderals, wetlands and alpine regions contain fewer mycorrhizal plant species (Smith and Read 1997). Plants in disturbed habitats often harbour many non-mycorrhizal plant species perhaps because nutrient availability is high in such environments or because they do not suffer

antagonistic effects there from mycorrhizal fungi (Francis and Read 1995).

Mycorrhizal fungi are therefore common to most habitats and, not surprisingly, they are most likely an important part of biodiversity–ecosystem functioning experiments using experimental plots. We estimated the number of mycorrhizal plants in the Biodepth Biodiversity experiments (as discussed by Hector *et al.*, Chapter 4). It hosted 110 plant species, of which 69 normally associate with AMF, while six plant species are non-mycorrhizal (Fig. 16.3). The mycorrhizal status of the remaining 35 plant species is still unknown (the question mark in Fig. 16.3). These observations suggest, assuming that the proportion of mycorrhizal to non-mycorrhizal species is equal for those remaining species, that 90 percent of all the plant species in the Biodepth Experiments are mycorrhizal. In the Cedar Creek Biodiversity (Tilman, Chapter 3) only one out of the 40 plant species is supposedly non-mycorrhizal (Fig. 16.3).

Given the prominence of plants with known mycorrhizal associates in these experiments and the fact that dominant organisms are often responsible for the major impacts on functioning and structuring of communities in such communities (Grime 1987), mycorrhizal fungi may be playing a key role

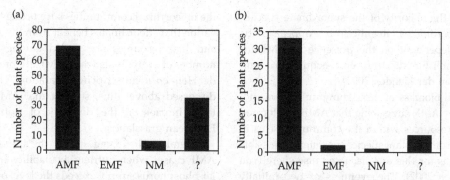

Figure 16.3 Relative abundance of mycorrhizal plant species in the experimental grasslands of the Biodepth experiments (a) or in the Cedar Creek experiment (b). The number of plant species that normally form an association with AMF or EMF are shown. Some plant species form no mycorrhizal associations (NM), and for some species it is not known whether they form a symbiosis with mycorrhizal fungi (?). Plant species composition of the Biodepth Experiments after Hector *et al.* (1999); data kindly provided by A. Hector. Plant species composition of the Cedar Creek experiment after Tilman *et al.* 1996 (and on: www.lter.umn.edu/research/exper/e120/species.html). Mycorrhizal status of the plants in the Biodepth experiment after Harley and Harley (1987) and of the Cedar Creek Diversity experiment after Wilson and Hartnett (1998) and N.C. Johnson (personal communication).

in the Biodepth and Cedar Creek biodiversity experiments. Both the studies at Cedar Creek and at the Biodepth sites focused on plants and revealed a positive relationship between plant diversity and productivity (Tilman *et al.* 1996; Hector *et al.* 1999). The contribution of mycorrhizal fungi to the relationship between plant diversity and productivity needs to be considered in such experiments for two reasons. First, mycorrhizal fungi affect plant diversity and can alter relationships between plant diversity and ecosystem functioning (see Section 1.3; Klironomos *et al.* 2000). Second, soil preparations during the set up of such experiments might have eliminated or altered the abundance and composition of the mycorrhizal community. These changes could impair ecosystem functioning, especially when nutrient availability is low (see Section 1.3).

16.3 Plant–microbe interactions and their impacts on ecosystem functioning: Arbuscular mycorrhizal fungi as a well-studied example

16.3.1 Overview

We will now concentrate on the impact of AMF on plant diversity and ecosystem functioning. Arbus-

cular mycorrhizal associations are typical for grassland ecosystems, which have been a focus of study within the biodiversity–ecosystem functioning debate (e.g. see Hector *et al.*, Chapter 4; Tilman, Chapter 3). Many other factors have been proposed that contribute to plant species diversity and ecosystem functioning. These include climate, environmental heterogeneity, disturbances, herbivory, mutualism, trophic interactions, ecological history and opportunities for seedling established (Huston, Chapter 5; Wardle and van der Putten, Chapter 14; Zobel 1992). The relative importance of each of these factors is unknown and may vary between ecosystems. The importance of AMF in the hierarchy of these factors is also unknown, but could be large, in view of the evidence presented below.

16.3.2 AMF as intrinsic determinants of plant diversity

There is accumulating evidence that AMF influence plant species diversity in temperate communities. Several field and greenhouse experiments revealed, for instance, that plant diversity of nutrient poor European grasslands is considerably higher when AMF are present (Grime *et al.* 1987; Gange *et al.* 1990, 1993; van der Heijden *et al.* 1998b). AMF enhanced plant diversity in such grasslands by stimulating growth of the majority of the subordinate species. These subordinates, mainly forbs, appear to be strongly dependent on the presence of AMF for their establishment, nutrient acquisition and growth (van der Heijden 2002).

The total biomass of these communities was not changed by AMF, suggesting that AMF re-allocated available resources within the community. This in turn can promote plant species coexistence. Possible mechanisms for this are discussed elsewhere (van der Heijden 2002). The evenness of two initially identical plant communities that were inoculated either with a mixture of four different AMF species or with an autoclaved mixture of these four AMF species was enhanced when AMF were present (Fig. 16.4). Both the mycorrhizal and the non-mycorrhizal communities were dominated by one grass species, Bromus erectus, (species 1 in Fig. 16.4) which contributes to approximately 70% of the total biomass. However, the presence of AMF led to a more even distribution of the biomass of the remaining species as can be derived from the longer tail of the mycorrhizal community. This finding follows also from the observation that eight of the 11 plant species in these communities are almost completely dependent on AMF. Moreover, seedling survival and seedling establishment was higher in the mycorrhizal communities supporting the observation that subordinates benefit from the symbiosis and that resources are shared among a higher number of individuals when AMF are present (van der Heijden, manuscript in preparation). The studies discussed above, thus, suggest that AMF have a large impact on the diversity of nutrient-poor European grasslands.

Several studies suggest that the importance of AMF ceases when nutrient availability, in particular phosphorus supply exceeds the level of demand (Koide 1991). The importance of AMF in productive fertile grasslands is, therefore, likely to be lower. The reduced abundance and activity of AMF which is observed in productive grasslands points also to this (I. Kotorova, unpublished results). These observations suggest that AMF are important for unproductive, nutrient-poor, species-rich grasslands, while they are expected to be relatively unimportant in productive, nutrient-rich, species-poor grasslands (see also van der Heijden 2002 for a discussion).

It is not only nutrient availability that determines the extent to which AMF affect diversity and ecosystem functioning. Plant species composition is important as well. Intuitively one expects that those plant communities that only comprise non-mycorrhizal plant species or plants with a low dependency on AMF, such as communities at ruderal sites, are hardly affected by AMF. In contrast, those communities that only harbour mycorrhizal dependent plant species are likely to have a greater productivity when AMF are present. Moreover, AMF can also reduce plant diversity as has been observed by Hartnett and Wilson (1999). They observed that the C_4 tall grasses that dominated the investigated tall grass prairie plots were heavily dependent on AMF while many of the subordinate species had a low dependency on AMF. These subordinates benefited from the absence of AMF, likely due to reduced competition intensity, and this can explain why AMF reduced plant diversity in this study. The comparison of this study with European calcareous grasslands, where AMF promote growth of subordinate species, showed that the outcome of AMF effects on plant diversity were strongly dependent on plant species or functional type composition.

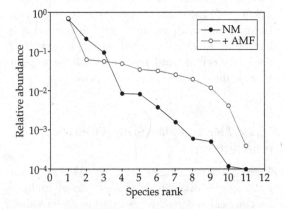

Figure 16.4 Rank-abundance diagram for a mycorrhizal (+AMF) and a non-mycorrhizal (NM) plant community. The rank number and relative abundance of the 11 plant species in this diagram are based on their biomass in microcosms simulating calcareous grassland (after van der Heijden *et al.* 1998). This graph shows that AMF enhance evenness in experimental calcareous grassland communities.

16.3.3 Carbon transfer between mycorrhizal plants

The same mycorrhizal fungus can link the roots of different plants (due to the absence of host specificity). Nutrients and carbon can consequently move from one plant to another via hyphal bridges connecting the roots of different plants (reviewed by Newman 1988). For instance, Simard et al. (1997) observed that shaded tree seedlings acquired substantial amounts of carbon from neighbouring plants via mycorrhizal webs. The carbon gain of the shaded seedlings was on average equivalent to 6% of carbon uptake through photosynthesis. This might be sufficient to contribute to the survival of shaded tree seedlings in the forest. Moreover, Grime et al. (1987) observed inter-plant carbon transfer in experimental grassland communities. In that experiment, the canopy dominant Festuca ovina was labelled with radioactive carbon and this carbon was consequently detected in neighbouring subordinate species that were colonized by AMF. This additional carbon could help the subordinates (assuming that they were carbon limited) and this has been proposed as a mechanism by which AMF could enhance plant diversity. In contrast, Fitter et al. (1998) measured inter-plant carbon transfer and found that AMF distribute nutrients within their hyphal network but that the net transport from one plant to another is almost zero. The ecological significance of such inter-plant nutrient and carbon transport is unclear and needs further investigation. Future studies should test whether transfer occurs between pairs of plant species with different strategies and life histories. The existence of a wood wide mycorrhizal web in terrestrial plant communities is exciting since it points to the existence of a communication pathway between different plant individuals. Such a mycelial web might, in theory at least, promote the overall stability and persistence of plant communities by facilitating individuals that are temporarily less vigorous.

16.4 Mycorrhizal fungal diversity, plant diversity and ecosystem functioning

So far we have not considered the fact that communities of AMF occur that vary in species composition and diversity. This could be important in view of the observation that different plant species benefit, in term of biomass and nutrient acquisition, from different AMF species (Streitwolf-Engel et al. 1997; van der Heijden et al. 1998a; Klironomos 2000). Up to 25 different AMF species have been recorded within plant species-rich grassland communities (Johnson et al. 1991; Bever et al. 1996) while other communities such as intensively used agricultural fields contain only a few AMF species (Cuenca et al. 1998; Helgason et al. 1998). Moreover, a recent study investigated AMF species richness in the Cedar Creek plots where plant diversity was experimentally manipulated (Burrows and Pfleger in press). The highest number of AMF species occurred in plots that contained the greatest plant species richness suggesting that plant diversity promotes AMF diversity or visa versa.

We have investigated whether AMF diversity and AMF species composition affect plant diversity by manipulating the number of AMF species in experimental grasslands simulating European calcareous grassland and North-American old-field systems (van der Heijden et al. 1998b). The composition of calcareous grasslands changed when different AMF species were present. This was caused by differential growth responses of plant species when growing in experimental ecosystems that were inoculated with different AMF species. In addition, the lowest plant diversity in the assembled American old fields was found in plots that contained only one AMF species, while plant species richness was twice as high when 14 different AMF species were present (Fig. 16.5). Trends in plant productivity followed that of AMF diversity and the lowest productivity was observed when only a few AMF species were present.

It is not only plants that are influenced by the species composition of AMF communities. Recently, we observed that a herbivore that feeds on plants colonized by different AMF species was also differentially affected by different AMF species (Goverde et al. 2000). These observations, thus, indicate that plant species composition, plant diversity and ecosystem functioning depend on the identity and the number of fungal symbionts that occur within ecosystems. A recent study by Jonsson et al. (2001) also showed that the diversity of EMF

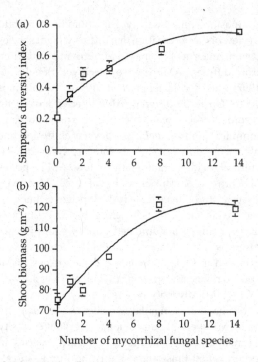

Figure 16.5 The effect of AMF species richness on (a) plant diversity (fitted curve is $y = 0.271 + 0.077x - 0.003x^2$; $r^2 = 0.63$; $P < 0.0001$), and (b) shoot biomass ($y = -0.334x^2 + 8.129x + 72.754$; $r^2 = 0.69$; $P < 0.0001$). Experiment was performed using macrocosms simulating an old-field ecosystem. Squares represent means (±SE) (reprinted from van der Heijden *et al.* 1998a; reprinted with permission, *Nature*, MacMillan Magazines).

might play a role in seedling establishment of boreal forest trees. They demonstrated that the productivity of birch (*Betula pendula*) seedlings growing on nutrient poor sterilized soil increased with more than 200% when the number of EMF increased from one to eight species. They also showed that the effects of EMF diversity on seedling productivity depend on nutrient availability.

In conclusion, the results discussed in this section indicate that below-ground biological diversity is one of the drivers of the diversity we see above ground. Studies with other below-ground organisms show variable results and there may be positive, negative or neutral effects on plant diversity (Wardle and van der Putten, Chapter 14). Moreover, studies that test effects of below-ground diversity need to mimic the highly variable temporal and spatial conditions that prevail below-ground in the soil. In this way, soil biota may occur in niches they

occupy in nature and that could show that microbial diversity is not so redundant as sometimes has been suggested.

16.5 Mechanisms

There is much debate concerning the mechanisms of observed diversity effects. Many of the mechanisms proposed are likely to apply to mycorrhizal fungi. Below we list five hypothetical mechanisms that may explain effects of AMF diversity on plant species composition and diversity:

1. Association complementarity effects: Complementarity occurs when co-occurring species increase resource capture or primary productivity compared to species growing alone. The mycorrhizal symbiosis is a typical example of complementarity since the plant and fungus show enhanced growth by association. Increasing mycorrhizal fungal diversity increases the number of different kinds of beneficial associations possible in an ecosystem. This is due to the fact that functional compatibility and optimal matching occurs between specific plant–fungal pairs (Ravnskov and Jakobsen 1995; van der Heijden *et al.* 1998a; Klironomos 2000). Thus, the addition of new species increases the possibility that additional beneficial associations establish. Increasing plant diversity would also enhance these effects.

2. Exploitative complementarity effects: Complementarity also occurs when different mycorrhizal fungi acquire mineral resources from different localities within the same soil volume. It has indeed been observed that there are spatial differences in the acquision of soil phosphate among different coexisting AMF species (Smith *et al.* 2000) indicating that niche separation occurs among AMF species. This suggests that the amount of phosphorus extracted by the fungi from a specific soil volume increases when the number of AMF species increases as has been observed in experimental old field communities inoculated with different numbers of AMF species (van der Heijden *et al.* 1998b). Moreover, it is well known that ecto-mycorrhizal fungi differ in their ability to degrade organic substances, and hence have a variable 'feeding' niche (e.g. Taylor *et al.* 2000). This may explain why increases in ecto-mycorrhizal species richness lead to increased plant growth and resource capture under nutrient limited conditions (Baxter and Dighton 2001; Jonsson *et al.* 2001).

3. Spatial coexistence effects: Several studies predict that spatial variations in abiotic resources explain plant species coexistence and diversity (e.g. Tilman *et al.* 1997b; Loreau 1998a). Such spatial variations also change the distribution of different AMF species since AMF species have specific niche requirements (e.g. Stahl and Smith 1984; Hayman and Tavares 1985) and the addition of new species of mycorrhizal fungi increase the probability that more spatial niches are occupied.

4. Sampling/selection effects: The sampling effect occurs when productivity or other ecosystem functions increase with diversity since larger species pools are more likely to contain more productive species (Huston 1997; Wardle 1999). This might in some cases be a natural phenomenon of diversity (Tilman 1997a) and the advantage of diversity, especially in a changing environment where organisms need to be adapted to different conditions. In the case of mycorrhizal fungi, the sampling effect would mean that AMF species exist that are more productive than others (e.g. that have a bigger influence on the growth of all or most plant species within a community and such AMF species could be characterized as being 'all productive'). AMF species indeed differ in their productivity and this is a way in which AMF species composition affects plant community structure. However, all productive AMF species do not exist since the productivity of an AMF species depends on the hosts it is colonizing (van der Heijden *et al.* 1999). This also means that the influence of a certain AMF species on the productivity of a plant community is being altered when plant composition changes.

The sampling may be a valid mechanism to explain effects of diversity on ecosystem functioning when the diversity of on group of organisms is being manipulated. However, the sampling effect is not valid when different trophic groups (such as plants and AMF or insects and plants) interact and when the identity of the members within each interacting trophic group varies. The productivity in each trophic group varies in such case depending on the identity of the interacting organisms.

5. Host specificity effects: Host specificity of plants or fungi could also explain how mycorrhizal fungal diversity leads to enhanced plant diversity (e.g. Sanders 2002). Host specificity has been observed between EMF and several tree species (Molina *et al.* 1992) suggesting that ecto-mycorrhizal diversity contributes to the coexistence of different tree species.

AMF are thought not to be host specific (Law 1988; Fitter 1990) and a recent experiment suggested that there is also no evidence for the evolution of host specificity for one specific plant species, *Prunella vulgaris* (Streitwolf-Engel *et al.* 2001). These observations suggest that AMF specificity would not contribute to increased plant diversity. However, associative complementarity (see above) may still occur. It remains to be tested whether plants are selectively colonized by those AMF species that provide the biggest benefit.

Overall, the contribution of each of these mechanisms, or their interaction, to the observed positive relationship between AMF diversity and plant diversity is unclear. Moreover, the relative importance of these mechanisms depends on the environmental conditions, spatial variability and on plant species composition. Carefully designed experiments are, thus, needed to identify the mechanisms responsible for the observed effects. These include experiments to overcome problems related to the sampling effect (e.g. see Loreau and Hector 2001), the use of hyphal compartment with labelled minerals to tests for functional differences among different AMF species (e.g. Jakobsen *et al.* 1992) and the use of molecular identification tools to assess which fungi colonize which plants (Clapp *et al.* 2002).

16.6 Conclusions

In this chapter, we have shown that symbiotic microorganisms that associate with plants are abundant and widespread. Symbiotic microorganisms affect a wide range of ecosystem processes related with resource acquisition, carbon cycling, productivity and biodiversity. Given the great diversity in form, function, habitat choice and taxonomic identity of various symbiotic relationships, it seems reasonable to assume that such ecosystem properties are affected by the availability of symbiotic microorganisms, but empirical data are remarkably scarce.

AMF can enhance plant diversity but their effect on diversity can also be negative or neutral depending on plant species composition and nutrient availability. We showed, by re-analysing published material, that AMF increased evenness and diversity in a synthetic calcareous grassland community. In addition, the diversity of mycorrhizal communities can affect plant growth, plant species composition and plant diversity. There obviously is a link between below- and above-ground diversity. Future studies, that include other soil microorganisms, need to test the size and significance of this potentially very important and largely overlooked link.

The studies that have been reviewed here, especially the experimental studies, have used grassland communities as their model and are short term and of small scale, like in other studies that investigated relationships between biodiversity and ecosystem functioning. While longer term, larger scale studies are needed in a greater variety of ecosystems, current evidence is at least suggestive that diversity may matter in important and predictable ways.

Future studies should experimentally manipulate mycorrhizal fungi in conjunction with manipulations of other soil community components. Moreover, patterns in the distribution and abundance of mycorrhizal fungal diversity over space and time for the major ecosystem types need to be investigated. The recent development of molecular identification tools can help us with these questions. Knowledge on ecological function of many symbiotic microorganisms in complex natural communities is almost completely lacking and this should be main focus for further research in biodiversity studies.

We would like to thank P. Inchausti, M. Loreau, S. Naeem and W. van der Putten for constructive and helpful comments on this manuscript. A. Hector, N. Johnson and D. Tilman are thanked for providing data and for their comments. This work has been supported by a personal grant from the Dutch Science Foundation (the Vernieuwingimpuls) to M. van der Heijden.

PART V

Extending the scope to other dimensions

CHAPTER 17

Species diversity, functional diversity, and ecosystem functioning

D. U. Hooper, M. Solan, A. Symstad, S. Díaz, M. O. Gessner,
N. Buchmann, V. Degrange, P. Grime, F. Hulot,
F. Mermillod-Blondin, J. Roy, E. Spehn, and L. van Peer

17.1 Introduction

Experiments assessing the effects of biodiversity on ecosystem functioning initially aimed at establishing whether such relationships exist (e.g. Naeem *et al.* 1995; Tilman *et al.* 1996; Jonsson and Malmqvist 2000; Engelhardt and Ritchie 2001). These phenomenological studies were useful for helping to identify patterns and articulate further questions. However, using species richness as a simple measure of biotic diversity, as they did, had no explicit explanatory power: ecosystem level processes are affected by the functional characteristics of organisms involved, rather than by taxonomic identity (Odum 1969; Pugh 1980; Grime 1988). Therefore, functional attributes of species must be considered if a mechanistic understanding of biodiversity effects is sought. In attempting to understand mechanisms, several subsequent experiments have manipulated either the diversity of functional groups (i.e. functional diversity) alone (Hooper and Vitousek 1997, 1998; Symstad and Tilman 2001) or functional group diversity in concert with species diversity (McGrady-Steed *et al.* 1997; Naeem and Li 1997; Tilman *et al.* 1997a; Hector *et al.* 1999; Petchey *et al.* 1999; McGrady-Steed and Morin 2000; Emmerson *et al.* 2001) (see Díaz and Cabido 2001 for a recent review for plants). Here we discuss species and functional-group approaches and compare their suitability for understanding the effects of organismic diversity on ecosystem functioning.

There is a vast literature on functional classification in both terrestrial and aquatic ecosystems (e.g. see reviews on soil organisms: Faber 1991; Brussaard *et al.* 1997; animals in general, guild concept: Simberloff and Dayan 1991; marine sediment organisms: Swift 1993; Snelgrove *et al.* 1997; stream invertebrates: Wallace and Webster 1996; plants: Smith *et al.* 1997; Díaz and Cabido 2001; general: Lavorel and Garnier 2001). Our goal is not to recapitulate the extant literature, but to address three questions of importance to biodiversity–ecosystem functioning research. First, of the various approaches used for functional classification, which are most useful for investigating diversity effects on ecosystem functioning? Second, how are functional and species diversity related in terms of their effects on ecosystem processes? Third, what is the relevance of using a functional versus species diversity approach for understanding the implications of recent experiments to ecosystem management?

17.2 Defining functional groups

Ecologists have used a variety of ways to define functional groups, and such delineation has a long history in ecology (e.g. Raunkiaer 1934). Functional groups have been defined as sets of species showing either similar *responses to* the environment or similar *effects on* major ecosystem processes (Gitay and Noble 1997). In addition, functional groups can be identified as clusters in trait space through multivariate statistics, without *a priori* classifications regarding particular responses to environment or influences on ecosystem processes (i.e. emergent groups, Lavorel *et al.* 1997). The terms 'functional group' and 'functional type' are sometimes used synonymously. We will generally use the term 'functional group' for simplicity. Other related terms include 'guild' and 'ecological groups' (Root 1967; Simberloff and Dayan 1991; Wilson 1999), and there is a close relation of these to various forms of the niche concept (Leibold 1995). Functional diversity refers to the range and value of organismal traits that influence ecosystem properties (Tilman 2001). This can be expressed in a variety of ways, including the number and relative abundance of functional groups (e.g. Tilman *et al.* 1997a; Hooper 1998; Spehn *et al.* 2000b), 'the variety of interactions with ecological processes' (Martinez 1996), or the average difference among species in functionally related traits (Walker *et al.* 1999). In this section, we examine different approaches to delineating functional groups, whether these approaches might be merged and whether functional classifications are hierarchical.

17.2.1 Multiple approaches

A number of approaches for defining functional groups have been used in different ecosystems, at different scales, and for different types of organisms (plants, microorganisms, soil mesofauna, etc). This is not necessarily a problem; it usually reflects current knowledge of organisms and ecosystems and the particular questions being addressed. Indeed, it is unlikely that there will be a single functional classification that is appropriate universally. Instead, what might be called a 'trait toolkit' may be more appropriate, whereby the organisms, their traits, and the scales of diversity (genotype, species, higher taxa, community type) for functional classification (the tools) will be defined in accordance with the job at hand: the processes of interest, the ecosystem type, and the suitable spatial and temporal scale (Fig. 17.1). The number of traits in such a toolkit are not infinite, however, because there are often correlations among traits due to physiological or fitness tradeoffs. For example, drought-tolerant plants may share traits such as position of stomata, cuticle thickness, and photosynthetic pathways even though taxonomically quite different (e.g.

cacti and euphorbs). Exhibiting sets of traits that are collectively associated with adaptation to particular environmental challenges is known as an 'ecological syndrome' or 'primary strategy' (Lavorel *et al.* 1997; Grime 2001). Primary strategies among many different types of organisms yield predictable effects on ecosystem properties (Chapin 1980; Chapin *et al.* 1993; Elser *et al.* 1996; Grime *et al.* 1997b; Reich *et al.* 1997) and may help simplify functional classifications in the trait toolkit. We discuss this approach more in the following sections.

17.2.2 Effect and response groups

Functional classification often has two relatively distinct goals, one of which is to investigate the effects of species on ecosystem properties (functional effect groups) and another which is to investigate the response of species to changes in the environment, such as disturbance, resource availability, or climate (functional response groups) (Landsberg 1999; Walker *et al.* 1999). The distinction between functional effect groups and functional response groups is directly analogous to the distinction between the functional and habitat niche concepts (e.g. Leibold 1995), where the functional niche encompasses the effects that a species has on community and ecosystem dynamics, and the habitat niche encompasses the environmental parameters necessary for a species' survival. Most studies on biodiversity/ecosystem functioning have focussed on functional effect groups, rather than using groupings based on species' responses. We suggest, however, that merging these two perspectives is useful for understanding biodiversity effects on ecosystem properties.

Functional effect groups
Two alternate approaches have been applied for categorizing species into functional effect groups. The first uses *ad hoc* groups based on physiognomic attributes of organisms in the ecosystem studied (Table 17.1), while the second approach looks for general tradeoffs in organism traits as a way of constraining the axes of differentiation for functional classification (e.g. Grime 1979, 2001). These approaches are described below.

Table 17.1 Examples of *ad hoc* functional groups for a variety of organisms and ecosystems. Many of these examples at least partially follow taxonomic lines. This list is meant to be illustrative, not comprehensive. See Díaz and Cabido (2001) and Smith *et al.* (1997) for a more complete listing for plants; See Simberloff and Dayan (1991) for more animal examples. ANPP = Above-ground net primary production

Organism type	Ecosystem	Functional effect groups	Ecosystem properties influenced	References
Plants	Perennial grasslands	C₃ grasses, C₄ grasses, forbs, N-fixers	ANPP	Tilman *et al.* 1997; Hector *et al.* 1999
Plants	Perennial grasslands	C₃ grasses, C₄ grasses, forbs	Soil food webs, microbial dynamics	Wardle *et al.* 1999
Plants	Annual grasslands	Early season annual forbs, late season annual forbs, perennial bunchgrasses, N-fixers	ANPP, soil nutrient pools, ecosystem nutrient retention	Hooper and Vitousek 1997, 1998
Plants	Alaskan tundra	Evergreen shrubs, deciduous shrubs, forbs, sedges, mosses	Decomposition, nutrient cycling, primary productivity	Hobbie *et al.* 1993; Chapin *et al.* 1996
Plants	Costa Rican rainforest	Short and long lifespan monocots and dicots	ANPP	Haggar and Ewel 1997
Herbivorous insects	Various	Leaf-chewers, leaf miners, seed-feeders, phloem-feeders, xylem-feeders, root-feeders, whole-cell-feeders	Plant consumption, secondary production, nutrient cycling	Bezemer and Jones 1998
Ungulates	Savanna	Grazers, browsers	Plant consumption, nutrient cycling	Du Toit and Cumming 1999
Soil organisms	Boreal forests	Fungi, bacteria, saprophytic animals, fungal feeders, protozoa, nematodes, intermediate predators (e.g. mites, nematodes), large, top predators (mites, spiders, coleoptera, ants)	Decomposition	Bengtsson *et al.* 1996
Aquatic consumers (zooplankton and fish)	Freshwater pelagic	Size of prey consumed	Energy flow Response to increased productivity	Hrbácek *et al.* 1961; Brooks and Dodson 1965; Dodson 1974; Hulot *et al.* 2000

Ad hoc *groups*. Many proposed functional classifications have been on an *ad hoc* basis, depending on the ecosystem in question and the major physiognomic forms of the organisms present (Table 17.1). Often the most general grouping in functional effect classifications, either implicitly or explicitly, is by trophic level. Trophic groupings are fundamental to carbon and energy fluxes through ecosystems, and linked to nutrient cycles as well (Naeem, in press). Microbial functional groups based on metabolic capacity and its biogeochemical consequences (heterotrophs, nitrifiers, denitrifiers, nitrogen (N) fixers, etc.) are in some ways similar to trophic groups in that they are based on who consumes what resources. Trophic groupings are not always clear-cut for either micro or macroorganisms. Omnivores are common (Persson *et al.* 1992; Power 1992; Strong 1992; Mittelbach and Osenberg 1993) and even some relatively clear-cut groups include multiple trophic types or levels. For example, nitrifying bacteria include autotrophic, heterotrophic and mixotrophic nitrifiers (Steinmüller and Bock 1976; Degrange *et al.* 1997). Even with these complexities, however, trophic groups are often the most obvious place to start.

Studies explicitly manipulating diversity within and across multiple trophic levels are most common in micro and mesocosms for aquatic (Naeem *et al.* 1994; Degrange *et al.* 1997; McGrady-Steed *et al.* 1997; Naeem and Li 1997; Petchey *et al.* 1999; Hulot *et al.* 2000) and soil ecosystems (e.g. de Ruiter *et al.* 1994; van der Heijden *et al.* 1998; Mikola *et al.*, Chapter 15; Wardle and van der Putten, Chapter 14), although there are some examples from natural and semi-natural systems as well (e.g. Ingham *et al.* 1985; Mulder *et al.* 1999; van der Heijden and Cornelissen, Chapter 16; Raffaelli *et al.*, Chapter 13). However, many biodiversity–ecosystem functioning studies have investigated effects of diversity in only one trophic level, so that a hierarchy of functional effect classification starting with trophic groups is implicit rather than explicit.

A major exception to functional characterizations that either implicitly or explicitly start with trophic categories is that of ecosystem engineers (Jones *et al.* 1994; Lavelle *et al.* 1997). For example, at the water–sediment interface of all aquatic systems, benthic invertebrates living in the sediment regulate a variety of processes, including organic matter degradation, carbon burial, microbial grazing and gardening, bioturbation, and biogenic structure formation (Aller 1983; Krantzberg 1985; Van de Bund *et al.* 1994; Mermillod-Blondin *et al.* 2000). The large heterogeneity of activities necessitates classifying these invertebrates into groups with distinct attributes. For marine bioturbators, organism size, type of biogenic structure produced, and feeding location (sediment surface, within sediment, or both) are primary axes of differentiation and these traits cut across multiple trophic levels (bioturbation groups, François *et al.* 1997) (for feeding groups of invertebrates in rivers, see Cummins 1974; Cummins and Klug 1979).

For some applications, the trophic level of resolution is clearly quite coarse and the crux of functional classification comes with trying to delineate groups within trophic levels. Most approaches to defining functional groups within trophic levels have started with *a priori* designations based on combinations of anatomy, physiology, or behaviour. Plant functional classifications often rely on combinations of physiognomy, phenology, and photosynthetic pathway (e.g. associations with N-fixing bacteria, woodiness, phenology, rooting depth, C_3, C_4 or CAM photosynthetic mechanism and associated tissue quality), whereas functional groupings for animals often reflect guilds based on consumption (Simberloff and Dayan 1991) (Table 17.1). Many animal studies, however, have focused on guilds in relation to forces influencing community composition and trophic structure, rather than effects on ecosystem properties—with the notable exceptions of soil and stream fauna (e.g. de Ruiter *et al.* 1994; Wallace and Webster 1996), pelagic foodwebs (e.g. Carpenter and Kitchell 1993; Schindler *et al.* 1997; Hulot *et al.* 2000), and benthic invertebrates (e.g. Emmerson *et al.* 2001).

Functional effect groups based on complementary resource use (niche differentiation) among species of the same trophic group provide a method to test for effects of functional diversity on ecosystem-level resource use and productivity. If species use different portions of the total resource pool, then greater species diversity should lead to greater utilization of resources and a corresponding increase in productivity (Trenbath 1974; Harper 1977;

Ewel 1986; Vandermeer 1989; Haggar and Ewel 1997; Hooper 1998; Loreau and Hector 2001; Tilman *et al.* 2001). Although complementarity has been applied mostly to plants, it applies equally well to animals (e.g. Simberloff and Dayan 1991; Fox and Brown 1993; Kelt *et al.* 1995).

Competition among plants is presumed to be a common feature of communities given the frequency of shared resources, such as light, space, and nutrients (Tilman 1988). Questions about the degree of overlap (competition) and non-overlap (complementarity) in resource use in a number of studies have raised debate about the effects of plant diversity on ecosystem processes (e.g. Tilman *et al.* 1996; Aarssen 1997; Huston 1997; Wardle 1999; Tilman *et al.* 2001). Distinguishing among complementarity, facilitation, and sampling effects in observed responses to biodiversity requires careful attention to experimental design and analysis (Allison 1999; Loreau and Hector 2001; Hector *et al.*, Chapter 4). Clearly, identifying complementary functional groups should be a priority for a better understanding of how diversity affects ecosystem functioning, particularly primary production, secondary production, and ecosystem-level resource use.

Primary strategies: general tradeoffs in organisms' traits. The search for functional groups that are applicable across ecosystem types focuses on tradeoffs among traits based on evolutionary constraints on the trait space that organisms occupy. The search is based on the simultaneous consideration of multiple individual traits and observations of different species' responses to environmental gradients and effects on ecosystem processes. The traits usually involve key aspects of the organisms' life history, resource use, reproduction, and responses to external factors. Rather than attempting to identify discrete groups, species can then be placed across continuous axes or planes that define evolutionarily realized combinations of interrelated traits. This approach has been applied most often to terrestrial plants (e.g. Chapin *et al.* 1996b; Díaz and Cabido 1997; Grime *et al.* 1997b). For example, plant growth form, leaf turnover, and nutrient status covary with maximum photosynthesis, defence against herbivory, and effects on decomposition and mineralization (Chapin 1980; Coley 1983;

Grime and Campbell 1991; Chapin *et al.* 1993; Reich *et al.* 1997; Grime 2001). However, a similar approach has also been applied to stoichiometry of organismic element ratios and their ecosystem consequences, especially in pelagic systems (Elser *et al.* 1996, 2000). Zooplankton growth rates may define the $C:N:P$ ratios of their cells, with consequences for ecosystem nutrient cycling.

Such correlated suites of traits may help simplify functional designations because one suite of traits may influence several related ecosystem processes in similar ways. In such cases, so-called 'soft traits', those that encapsulate a suite of 'hard traits' (those that actually affect the process), are often comparatively easy to measure and may be useful for designating functional groups (Hodgson *et al.* 1999). For example, increasing leaf toughness or sclerophylly (a soft trait) is often correlated with greater amounts of carbon-based defences, slower decomposition rates and slower rates of N mineralization (hard traits; Coley 1983; Herms and Mattson 1992; Cornelissen *et al.* 1999). Correlation among traits will limit the number of axes across which species are differentiated, and thus the number of functional groups. If these associations among traits can be proven consistent, it would not be necessary to measure all traits to classify taxa into functional groups.

Work on primary organism strategies suggests that the resource environment of a site may select for suites of covarying traits that control entry of species into a given community (Grime *et al.* 1997a). For example, low resource environments select for plants that have low growth rate, high nutrient use efficiency, low litter quality, and high allocation to defence (Chapin 1980). In other words, the resource environment acts as an environmental filter on community composition (Pearson and Rosenberg 1978; Weiher *et al.* 1995; Díaz *et al.* 1998). (These traits may then feed back to further alter resource availability as well; Chapin *et al.* 1986; Hobbie 1992.) One hypothesis in this case is that because traits relating to such a primary strategy will be similar among species, the traits of the dominant species, rather than species richness and complementarity, will exert the strongest control on ecosystem properties (e.g. Solan and Kennedy, in press). However, other forces, such as variability in environmental conditions, selection for trait differentiation due to competition (Bazzaz 1987; Weiher *et al.* 1995), trophic dynamics, or disturbance may counteract such a trend. Empirical studies are needed to address these issues.

Functional response groups

Identification of functional response groups can help understand and predict how communities and ecosystem properties might be affected by environmental change, variability, or disturbance. The task is to define the potential disturbances or environmental fluctuations to which a given system may be subjected, and identify the functional traits relevant to either tolerating or recovering from those conditions (Landsberg 1999; Walker *et al.* 1999). Examples include differential response to extreme climatic events, directional climatic change, grazing, or pathogens; differential recruitment abilities, differential sensitivity to pollutants, or other traits that influence an individual's or population's sensitivity to or recovery from different stresses (Noble and Slatyer 1980; McIntyre *et al.* 1995; Box 1996; Chapin *et al.* 1996a; Buckland *et al.* 1997; Westoby 1998; Díaz *et al.* 1999; Walker *et al.* 1999).

Traits useful for delineating functional response groups may vary independently from those used for delineating functional effect groups. For example, in plants, regeneration traits (e.g. seed size, number of seeds per plant, dispersal mode, pollination mode), which often affect response to disturbance, tend to be only loosely correlated with vegetative characteristics, which often have more direct effects on process rates (Grime 1979; Díaz and Cabido 1997). However, because traits that affect response to disturbance also influence an individual's or population's sensitivity to or recovery from different stresses (e.g. seed size and shape are related to seed persistence in the soil bank: Thompson *et al.* 1994; Funes *et al.* 1999), they may indirectly influence an ecosystem process under consideration.

While functional response groups often have been delineated independently of functional effect classifications, a better integration of these approaches could help understanding of how diversity within functional effect groups influences stability of processes to non-equilibrium conditions or new disturbance regimes (Naeem 1998; Walker *et al.* 1999). For example, as species richness within functional effect groups (trophic groups of producers, bacterivores, herbivores, and predators) was increased in an aquatic microcosm experiment, total respiration became more predictable and temporal variation in the abundance of trophic groups declined (McGrady-Steed *et al.* 1997; McGrady-Steed and Morin 2000; Petchey *et al.*, Chapter 11).

17.2.3 Testing the predictive value of functional traits

An additional, but necessary step for refining knowledge of functional classifications is iteration of the testing process. If initial classifications, do not accurately reflect species' effects on ecosystem properties, re-classification and re-testing help to more closely delineate which particular functional traits are important for which processes. Two approaches that explicitly incorporate this step are the Integrated Screening Program (ISP) for plants (Grime *et al.* 1997b; Díaz and Cabido 1997), and screening for bioturbator functional groups on estuarine mudflats (Swift 1993; Solan 2000) (Fig. 17.2). Grime's ISP starts with individual species and measurements of many functional traits related to life history, physiology, and morphology (Fig. 17.2(a)). Following ordination of these traits, the ISP allocates species into emergent groups based on similar traits, which are then used to predict effects on ecosystem processes or responses to perturbation. Researchers then test those predictions by long-term monitoring in natural systems and experiments in the field or microcosms (e.g. Leps *et al.* 1982; Grime *et al.* 1987; MacGillivray *et al.* 1995). It is an explicitly iterative process: if initial predictions don't hold up, functional classification and testing start again. For example, laboratory screening of plant traits was used to predict the resistance and resilience of five grassland ecosystems subjected to drought, late frost and burning treatments (MacGillivray *et al.* 1995). Results indicated that the same traits were good predictors of both drought and frost resistance, but also that it was necessary to weight predicted ecosystem responses according to the abundance of species in the vegetation (i.e. the mass ratio hypothesis of Grime 1998).

Because of the large amount of effort involved in these studies, it is not practical to apply these techniques whenever ecologists need functional information for research or management purposes. The hope is that intensive studies on multiple species in one system or a small number of systems will provide enough experience to recognize predictive traits in other species and other systems. For example, for bioturbators, the iterative testing process has identified four primary groups into which organisms can be placed without detailed taxonomic information (Fig. 17.2(b)). Classification is based on whether the species primarily acts above the sediment (epifaunal), within the uppermost sediment layer (surficial modifier), throughout the sediment (biodiffusers) or within select areas of the sediment (advective mixing) (Gardner *et al.* 1987; François *et al.* 1997). For both the ISP and the bioturbator example, the functional effects and responses of the organisms cut across taxonomic boundaries.

17.2.4 A simple functional hierarchy

Hierarchical classifications have a couple of advantages. First, they allow researchers to identify different traits of interest for particular groups (i.e. the same set of traits might have different response implications for different life forms). Second, additional levels of detail on traits may be necessary to understand mechanisms of species' effects on processes or responses to environmental changes (Lavorel *et al.* 1997). While there is likely no single correct functional hierarchy, even *ad hoc* hierarchies can be useful. For example, a hierarchical functional classification for plant responses to disturbance (grazing) effectively predicted species' responses to

altered grazing regime (Mcintyre and Lavorel 2001).

We propose a simple hierarchy of nesting response groups into effect groups as a good strategy for understanding effects of functional diversity on ecosystem processes (Fig. 17.1). Studies of biodiversity and ecosystem functioning often have two different, albeit related, goals: investigating diversity effects on process rates and investigating diversity effects on stability of processes. The first takes a short-term, equilibrium view—a necessary simplification for initial understanding of species effects on ecosystem properties. The second takes an explicitly non-equilibrium view, allowing a more complex and realistic perspective for understanding how environmental fluctuations, mediated through changes in species composition, might influence those processes (Chapin *et al.* 2000). Once functional effect groups have been delineated, species within each group can be characterized by their responses to various environmental perturbations, which will determine how species diversity within those functional effect groups influences stability of ecosystem processes (McNaughton 1977; Yachi and Loreau 1999). This approach forms the basis of many discussions of diversity effects on ecosystem process rates and stability (e.g. Naeem 1998; Walker *et al.* 1999; Griffiths *et al.* 2000 to name just a few).

We acknowledge that there is no *a priori* reason for nesting response groups within effect groups. In many cases, response strategies will cut across effect groups, and vice versa. If the primary goal is to understand how global environmental change will affect species' distributions (e.g. Cramer 1997),

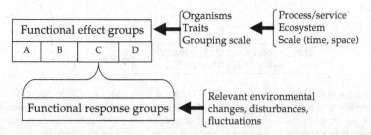

Figure 17.1 A simple hierarchy of functional groups for relating the effects of biodiversity on ecosystem functioning. In general, the process in question, the ecosystem type and the spatio-temporal scales will determine the appropriate organisms, traits, and levels of grouping into functional effect groups. Functional response groups within each effect group are determined by species' responses to relevant environmental stresses.

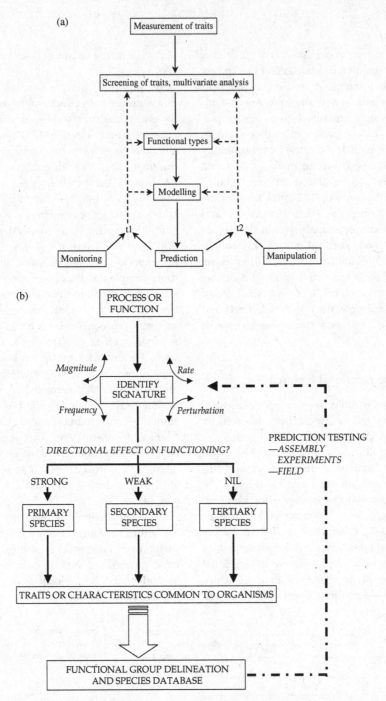

Figure 17.2 Iterative processes of functional group classification. (a) Integrative Screening Process (ISP) for plants. The ISP leads to predictions of functional groups and their effects on ecosystem processes. These predictions are then tested by monitoring and experiments (t1 and t2), whereupon the relevant traits, initial functional groupings, or models of predicted effects or responses may be modified. Redrawn from Grime *et al.* (1997a) (© Cambridge University Press). (b) Definition of functional groups for estuarine macrofaunal invertebrates based upon body size, mobility, and bioturbation reworking mode (Solan 2000). The process starts with a priori predictions as to what traits are likely to modify a given ecosystem process and in which ways they are likely to make these modifications (the 'signatures'). Species are then categorized according to their potential effects. The actual effects on ecosystem processes of species in isolation and in mixture are tested in mesocosm experiments, and predictions are refined.

response groups will most likely be at the top of the hierarchy. On the other hand, if understanding effects on ecosystem properties is the goal, then effect groupings will be foremost.

17.3 Relationships between functional and species diversity

Experiments investigating the effects of diversity on ecosystem properties have manipulated functional composition (presence of certain plant functional effect groups or functional traits), functional richness (number of different plant functional effect groups), and species richness. However, separating the effects of species diversity from those of functional group diversity in experiments manipulating both can prove difficult: in many experiments using randomly assembled communities, the two types of diversity are correlated across much of the experimental space (Tilman *et al.* 1997b; Allison 1999; Schmid *et al.* 2001) (Fig. 17.3). Here, we discuss the importance and difficulties of empirical evaluations of the relative contributions of functional and taxonomic diversity to ecosystem functioning.

17.3.1 Disentangling diversity components in biodiversity–ecosystem functioning relations

Correlation between species and functional diversity leads to a trade-off between two common experimental goals in biodiversity/ecosystem functioning studies: (1) that of examining the broadest range of species richness possible and (2) that of examining the relative effects of functional richness and species richness. One typically cannot have more functional effect groups than species (though see below for counter-examples), and most experiments have limited numbers of species to add within functional groups. For example, if researchers are working with a total pool of four species from each of four functional groups, and examine a range of species richness from 1 to 16 in a logarithmic series, species and functional richness will be strongly correlated (Fig. 17.3(a)). This problem can be reduced by using many species in each functional group and constraining the total range of species richness

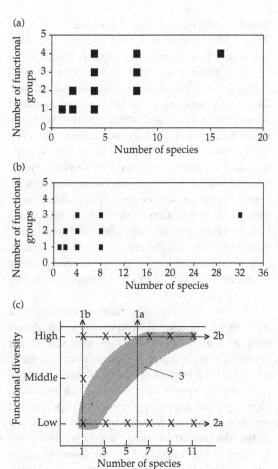

Figure 17.3 The correlation between species richness and functional richness in biodiversity experiments. Graphs show (a) the arrangement of treatments for a hypothetical experiment using four species in each of four functional groups across a logarithmic species richness gradient from one to sixteen species; (b) the arrangement of treatments in the Swiss BioDEPTH site where researchers attempted to minimize the correlation between species and functional group richness (Spehn *et al.* 2000); and (c) a general scheme for how species and functional richness might be related. In (c), along the vertical lines, the number of species is kept constant but the number of functional groups (1a) or the phenotypic or genetic variability of species (1b) is increased. Along the horizontal lines, the number of functional groups is kept constant but the number of species ((2a) narrow-niched species, (2b) broad-niched species) is increased. Most empirical studies lie in the shaded area, 3 (from Schmid *et al.* 2001) (© Princeton University Press).

examined. For example, in an experimental study of grasslands in Switzerland, researchers used a total pool of 48 species in three functional groups (grasses, legumes, and non-leguminous forbs) across a range of species richness from 1 to 32 (Fig. 17.3(b)) (Diemer *et al.* 1997; Spehn *et al.* 2000b). The relatively large number of species compared to the low number of functional groups allowed testing various species compositions within richness levels and increasing species richness without adding more functional effect groups. With the hierarchical structure discussed above, however, adding more species within functional effect groups may add a diversity of functional response types. Alternatively, experiments may be replicated with entirely different sets of species (e.g. MacGrady-Steed *et al.* 1997). Constraining the species pool, however, does not solve all problems. For example, there can still be a high correlation between species and functional diversity at low diversity levels where much of the effect on ecosystem properties often occurs.

However, a direct correlation between species richness and functional diversity may not always hold at low diversity ends of a gradient. For example, morphological or behavioural plasticity could lead to a breadth of functional attributes with relatively few species (Fig. 17.3(c)). Such a relationship does not necessarily solve the experimental problems addressed above, however, because it just raises the important (and largely unanswered) question of how important genetic diversity within species is in affecting ecosystem processes. Differences in behaviour, size, diet, and habitat preferences between sexes and life-history stages (and social status in social arthropods such as termites) also contribute to functional diversity within species. Furthermore, animals may significantly change behaviour patterns in response to external stimuli, behavioural variation, or season/time, which could alter their functional roles. For example, a common polychaete, *Nereis diversicolor*, has several feeding modes that directly affect its bioturbatory capacity. When deposit feeding, it is actively foraging, but when suspension feeding, it is relatively sedentary and has little impact on the sediment profile. Therefore, presence of a single species may equate to the presence of several functional groups, although they may not all occur simultaneously (Solan 2000). If functional groups are to be meaningful in those cases, they must be conditioned on environmental factors that influence functioning.

17.3.2 Is species diversity a surrogate for functional diversity?

While correlation between species and functional group richness often occurs in experiments, the relevance of this correlation to natural communities is subject to debate. With little prior knowledge of a system, and for practical reasons, species diversity may serve as a surrogate for functional diversity. From a bottom-up approach, the concepts of niche differentiation and limiting similarity imply that functional characteristics of coexisting organisms must differ at some level, which means that increasing species diversity should lead to increasing functional diversity, especially if it also broadens the total range of functional traits present (Schmid *et al.* 2001).

Whether species richness is an adequate surrogate for functional diversity in natural systems depends in part on patterns of community assembly. Species richness and functional diversity will tend to correlate if there is a linear increase in niche space 'coverage' as species richness increases (Díaz and Cabido 2001). This situation could happen either if species are assembled at random, as in situations in which disturbance regimes lead to a predominance of stochastic colonization (e.g. Grime 1979; Hobbs and Mooney 1991; Fridley 2001), or if complementarity in species' functional traits is an important component of community assembly (Bazzaz 1987; Weiher and Keddy 1998; Kelt and Brown 1999). On the other hand, Diaz and Cabido (2001) argue, based on the concept of environmental filters, that plant communities are non-random assemblages from the regional species pool. They suggest that climate, disturbance, and biotic interactions impose increasingly fine-grained constraints on the composition of communities (Pearson and Rosenberg 1978; Díaz *et al.* 1998). In this scenario, the same amount of functional variation (or niche space) is just more finely divided as more species are added (Schmid *et al.* 2001), and therefore, functional diversity may not increase with increasing species richness. The crux of the question with regard to diversity effects on ecosystem processes is the degree to which abiotic conditions constrain the functional variation within communities that influences processes within that system. Merging our understanding of ecosystem level controls with our understanding of community dynamics and assembly is an important focus of future study (Thompson *et al.* 2001).

The relationship between diversity of taxa and functional diversity in natural systems will also depend on the level of taxonomic resolution. Adding increasing numbers of genotypes of a given species will likely add some degree of functional diversity (Fig. 17.4). On an average, randomly adding new species within a genus or family would add more functional diversity than adding new genotypes of the same species, and adding new species from different families would likely add even more. Adding species from known different functional groups would give the greatest increase in functional diversity per species added. The relationship of number of genotypes, species, families or functional groups with functional diversity for these different taxa are not likely to be simple lines, but rather broad, potentially overlapping areas. For some processes and species, adding more genotypes of the same species might add relatively more functional diversity than others (Fig. 17.4).

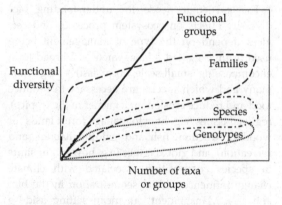

Figure 17.4 Hypothesized relationships between richness of various taxa in an ecosystem and the total amount of trait variation, i.e. functional diversity. On an average, adding more families within a trophic type would add more functional trait variation than adding more species within a genus or family, which in turn would add more variation than adding genotypes within a species. Variability in the general relationship results from genetic variation and plasticity within species, and the variation in traits within families or genera. By definition, the number of functional groups would have the strongest and most linear relationship with total trait variation.

17.3.3 Limitations of species and functional groupings

Species richness has been the most common measure of diversity in biodiversity–ecosystem functioning experiments. While often done for practical reasons, this approach starts from the premise that species' delineations both embody functionally significant information and are distinct. While this approach works in some cases, it clearly will not in many others. Reflecting population and evolutionary processes, species are usually delineated based on genetic or morphological traits. Functional traits are more directly related to ecosystem processes, but organisms with vastly different phylogenies can be very much alike functionally—one of the original reasons for using guilds (Root 1967). Hybridization and variability in mutualistic associations can also confound species designations in diversity/ecosystem functioning experiments. The definition of what is a species also has limitations, both of a theoretical and practical nature, and these have given rise to a multitude of alternate species concepts (e.g. Bisby and Coddington 1995; Hey 2001). In addition, notorious identification problems exist within many groups. Greatest problems are probably encountered with microorganisms, notably bacteria and fungi, whose body forms are simple and whose sexuality and genetics fundamentally differ from those of plants and animals (e.g. Kück 1995; Pace 1997; Staley 2001). In many cases, basing functional classifications on taxonomy is difficult, especially because the number of microbial functions known to be affected by horizontal gene transfer is increasing (e.g. plant pathogenicity supported by the plasmid Ti in *Agrobacterium*; Teyssier-Cuvelle *et al.* 1999). On the other hand, some microbial functional groups in fact define taxonomic families. For example, N-fixers associated with legumes belong to the Rhizobiaceae family, and nitrifiers belong to the Nitrobacteriaceae family (Krieg and Holt 1984). All these reasons suggest that a mechanistic understanding of biodiversity–ecosystem functioning relationships ultimately requires a functional approach to classifying organisms even when species richness may serve as a convenient starting point.

A common criticism of studying the effects of functional group diversity on ecosystem processes is that the underlying rationale is circular. If the functional effect groups in question have been defined by their influence on an ecosystem process, then by definition, adding and removing these groups will alter that process. In most experiments, however, the functional effect groups used are based on (1) traits that are known to influence a process but the generality of the functional grouping has not been tested, or (2) a mixture of taxonomic and physiognomic features ('soft traits') that are more general, and not specifically related to any particular process. As such, these groupings might be more appropriately termed 'candidate functional groups' (cf Vitousek and Hooper 1993). A goal of such experiments should be to verify if the functional effect groups are accurate by testing if they have, both alone and in combination, the hypothesized influences on ecosystem processes (see Section 17.2.3 above). This approach differs from the tautology described above, though the design of the experiment is critical for accurately discriminating functional group versus species richness effects (Allison 1999). In the past, however, many biodiversity–ecosystem functioning experiments have not actually carried out explicit tests of candidate functional effect groups (but see Hooper and Vitousek 1998; Symstad and Tilman 2001).

Reliance on functional classification (effect or response) can have several difficulties as well, however. Functional classifications are often not discrete—many traits vary continuously and cutoffs for inclusion/exclusion for a given group may be arbitrary. Recognizing axes of general functional tradeoffs (Chapin et al. 1993; Grime 2001), and using continuous, quantitative trait axes (Walker et al. 1999) could help remedy this problem. Even more difficult are situations in which it is not possible to discern the summary traits by which to classify species with respect to a given process or response (e.g. inability to predict which types of plants will have strong growth responses to elevated CO_2; Körner 2000). In addition, microcosm or greenhouse experiments may not be adequate to identify functional groups if interactions with other species are important in affecting certain processes (e.g. Chapin et al. 2000; Newton et al. 2001).

Despite these difficulties, functional characterizations offer the best hope of gaining a mechanistic understanding of diversity effects on ecosystem properties. Should we be able to clearly delineate both the functional effect groups and functional response groups for a given system, future management could rely on the functional traits toolbox rather than the black box of random species diversity.

17.4 Implications for management and policy

Although the conservation of biodiversity has been linked with the sustainability of ecosystems (Naeem et al. 1994a, 1995; Tilman et al. 1996), the relevance of these experiments to ecosystem management remains controversial (Schwartz et al. 2000; Wardle et al. 2000b; but see Hector et al. 2001b). Changes in species composition have been shown to influence ecosystem processes and services, often with large economic impacts (Chapin et al. 2000). The primary cause is often gain or loss of single species with key functional traits. The following discussion focuses on management and policy objectives concerning functional groups and ecosystem processes and services.

The particular needs for understanding biodiversity effects on ecosystem processes and services depend on the type of management being used or investigated. Management is a broad term encompassing small-scale, local issues (e.g. how many and which species are necessary to produce food and reduce erosion for a farmer in the tropics), regional issues (e.g. how to manage forest lands for wood production, fish and wildlife habitat, and recreation), and global issues (e.g. how might shifts in species composition associated with climate change influence carbon sequestration in the biosphere?). Management can mean setting aside a parcel of land or water and doing nothing, or it can involve intensive manipulation of both biota and the physical environment. The common thread is that whatever action (or inaction) takes place is done by human choice to achieve some goal or maximize a particular ecosystem service (this includes setting aside wilderness areas or wildlife refuges). At the same time, lack of management can affect biotic diversity outside of the manipulated system (e.g. agricultural runoff affecting water quality in nearby lakes and rivers) (Silver *et al.* 2001), whereupon policy decisions should be made about how to incorporate those effects into more comprehensive management. From the practical standpoint of maintaining ecosystem services in all of these situations, it may be necessary to maintain species and functional diversity for several reasons:

1. To ensure that for key services, important functional groups are present and active;
2. To ensure survival of rare or occasional species, which may resume critical processes following disturbance or gradual changes of environmental conditions (Grime 1998);
3. To maintain a diversity of services in natural or semi-natural systems (e.g. provision of food and fibre, recreation, wildlife habitat, catchment protection and maintenance of water quality, protection against natural hazards).

Managers need to be aware of species composition as well as of richness. Management- or disturbance-related species losses may not be at random, nor are they necessarily equally distributed among different functional groups (Díaz and Cabido 2001). Loss of species diversity to the point that entire functional effect groups disappear will clearly have the greatest influence on ecosystem processes. Such loss of entire functional effect groups is most probable when only one or a few species are responsible for a given process in a given ecosystem (Hooper *et al.* 1995). Loss of functioning in ecosystems can sometimes be restored by changes in environmental conditions if the organismal diversity has remained intact. When viewed from a functional perspective, management practices should aim to ensure that the species present are the ones with traits that will maintain the desired ecosystem properties within acceptable bounds. (Definition of what those desired properties are raises a variety of questions at the intersection of ecology and societal values, but we will not address those issues here: Rapport 1995; Wicklum and Davies 1995).

The key question for all of these issues is what level of functional diversity is needed to sustain the ecosystem services in question in response to loss of biodiversity resulting from a variety of global

changes (Sala *et al.* 2000). This issue has two components, both of which require functional group classifications: (1) identification of which species and traits have a large influence on processes under current conditions, and (2) the delineation of functional response groups to improve our ability to predict which organisms might be lost from ecosystems in response to given environmental perturbations.

The pattern of response of ecosystem services to altered diversity will likely depend on patterns of loss of diversity. For example, gradual losses of species as abiotic conditions begin to exceed tolerance limits (e.g. with climatic change) could result in random losses of functional effect groups if functional effect and response groups are independent from one another. In such a situation, average patterns of process response to changes in diversity could be similar to those observed in randomly assembled communities (Tilman *et al.* 1996, 1997a; Hector *et al.* 1999; but see also Wardle 1999). At the other extreme, as in situations involving land-use transformation, gross changes in abiotic conditions and loss of a majority of the functionally important biota may have a greater impact on ecosystem processes and services than the decline of species richness *per se*. Between the two extremes of complete dependence on abiotic conditions and complete dependence on species or functional richness, is probably where many real world situations will fall. For example, landscape fragmentation involves both gross transformation in some areas and subsequent more gradual species loss in remaining fragments (under island biogeographic models), so both abiotic and diversity drivers could apply depending on the area involved.

In other cases, species losses may not be random with respect to species effects on ecosystem processes because certain traits related to response to the environment also affect ecosystem processes (Díaz and Cabido 2001). For example, there has been debate about whether plant diversity or traits of certain species are responsible for decreased resistance of grassland production to drought in the experiment by Tilman and Downing (1994) because the gradient in species richness was caused by N fertilization (Givnish 1994; Huston 1997). The mechanistic debate may be moot from the practical perspective of managing N within landscapes, however, if changes in both diversity and traits of the dominant species are consistently correlated across gradients of N deposition (Berendse *et al.* 1993; Tilman 1996). Thus, effective ecosystem management requires an integrated understanding of the relative effects of individual species traits, species or functional richness, and abiotic drivers of ecosystem processes.

17.5 Conclusion

A hierarchy of functional response groups nested within functional effect groups is one way to approach questions of how changes in biotic diversity might affect ecosystem properties, both on short time scales and in response to changing environmental conditions. Progress in three key areas will substantially further efforts to gain a rigorous understanding of how functional attributes of species, and their interactions, influence the response of ecosystem properties to changing biodiversity:

1. Synthesis of the *ad hoc* and primary strategies approaches for defining functional groups, in concert with development of methods for quantitatively measuring functional diversity (e.g. Walker *et al.* 1999);
2. Better understanding of which functional response and effect traits are correlated versus independent, particularly with respect to the predominant forces of global change; and
3. Better understanding of how patterns of community assembly influence relationships between species and functional diversity in natural communities, and how this might differ in different environments.

Knowledge of the effects of species and functional diversity on ecosystem services, particularly in the context of abiotic drivers, individual species effects, and global change, will be critical where management priorities seek to manipulate species composition directly. Intensive management often relies on the functional characteristics of one or a few species and substitution of human inputs for biotic processes. Clearly this reliance decreases the planned diversity of these systems, but the unplanned (i.e. associated) diversity may also decline (Ewel 1991, 1999; Vandermeer *et al.*, Chapter 19). However, the insurance hypothesis (Naeem 1998; Yachi and Loreau 1999) and the precautionary principle emphasize that land managers and policy makers also must be prepared for unpredictable events and a changing world. Faced with the unpredictable, preserving species or taxonomic diversity (e.g. Clarke and Warwick 1998; Warwick and Clarke 1998; von Euler and Svensson 2001) within functional effect and response groups may better allow long-term, internal dynamics and evolution of managed systems as they face new environmental conditions.

A critical question for the future is how to balance patterns of human use and biotic diversity at the landscape scale to maintain (a) local diversity within sites, (b) regional diversity among sites, (c) ecosystem services that depend on small-scale functions (e.g. crop productivity in a field), and (d) ecosystem services that depend on interactions among different landscape components (e.g. nutrient transformations in riparian zones). Answering this question will require a rigorous synthesis across all scales of ecological organization, from physiological to landscape levels.

The authors would like to thank the organizers of the Paris conference for their efforts to bring together and encourage discussion among researchers with diverse views on biodiversity and ecosystem functioning. We also thank the support of the European Science Foundation, the National Science Foundation (USA) and the National Centre for Scientific Research (France) for making our participation possible. Finally, we thank the editors for their continuous efforts to make this book a strong, synthetic contribution to the field.

CHAPTER 18

Slippin' and slidin' between the scales: the scaling components of biodiversity–ecosystem functioning relations

J. Bengtsson, K. Engelhardt, P. Giller, S. Hobbie, D. Lawrence, J. Levine, M. Vilà, and V. Wolters

18.1 Introduction

Whilst the problems of scale are well-known to most ecologists, many subdisciplines of ecology are still struggling with its central concepts—how the dimensions of space and time, and levels of organization, influence ecological patterns and processes (e.g. Peterson and Parker 1998; O'Neill and King 1998; Petersen and Hastings 2001; Schneider 2001). Individual ecologists may, at their own peril, ignore the problem and focus on one scale and one level of organization only. However, the synthesis of many individual studies conducted at different scales requires a better understanding of the influence of scale on ecological processes. This is particularly evident in the study of biodiversity and ecosystem functioning. Although studying at least two levels—community and ecosystem—most research has been performed at small spatial and short temporal scales (Loreau et al. 2001; Naeem 2001), and it is not clear if such studies can be used to inform the public and policymakers about the large-scale consequences of biodiversity loss.

An example of the small spatio–temporal scales often used in biodiversity–ecosystem functioning studies are the grassland experimental plots, which have formed the core of the recent examinations of such relationships. These plots usually have a spatial extent of $<100 \, \text{m}^2$ and the experiments span a period of a few to ten years (<1 to 10

generations depending on the species) (Hector et al. 1999; Tilman et al. 1996, 1997a). Similarly, microcosm experiments are by definition small in space in comparison to the systems they are modelled after, such as using Ehrlenmeyer flasks as model fresh-water ecosystems (McGrady-Steed et al. 1997; Naeem and Li 1997). However, microcosms can encompass a large number of generations for small organisms such as prokaryotes, protists, or other members of phytoplankton and zooplankton communities, and can have a large spatial scale relative to the organisms concerned (see Petchey et al., Chapter 11).

A second major issue regarding scale and biodiversity–ecosystem functioning research is that most experiments carried out so far have been more or less closed to their surroundings. This precludes processes varying at larger spatial scales from influencing the results, leaving only local interactions to determine community composition and diversity. In systems where disturbances and landscape configuration strongly influence diversity and community structure, such small-scale experiments may only reveal part of the story.

The term 'scale' usually refers to dimensions of observed entities and phenomena in space and time (i.e. extent, grain, size and resolution; O'Neill and King 1998; Schneider 2001). Scale does not describe the level of organization of a system (populations, communities, and ecosystem), although scale is sometimes referred to in this context. Populations or ecosystems have no general scale, although specific systems may have dynamics on particular scales in relation to organism size and generation time. Naeem (2001) uses the ambiguous term 'biotic scale' for genetic, population and ecosystem properties, admitting that these lack standard units for measurement and that these operational units vary among studies. When the scale of investigation changes properties of communities and ecosystems do not just change in any coherent fashion. That is, if one investigates a large-scale phenomenon using a small-scale experiment, or conducts an experiment for two generations to study a phenomenon that occurs on a scale of 10 generations, there is seldom any justification in the assumption that one just has to multiply up or down to make predictions about the phenomenon from experimental results. Rather, the extrinsic and intrinsic forces driving the dynamics of populations or ecosystems may change (O'Neill and King 1998). A well-known example from population dynamics is metapopulation dynamics (e.g. Hanski 1999). Resource competition or predation may determine local-dynamics, but at the regional scale patch relationships, landscape configuration and dispersal are more important.[1]

Not only do ecologists struggle with space–time scaling problems in their attempts to understand relationships between communities and ecosystems, but they must also take into account the various degrees of artificiality imposed on study systems by their experiments (Naeem 2001; Fig. 18.1). It is well known, for example, that enclosures and exclosures impose artificial boundaries on the study systems. This creates edge effects on dynamics that may seriously confound the interpretation of results

from such studies (e.g. Peterson and Hastings 2001). Theoretical approaches, model systems like microcosms, experiments on plots in the field, and natural ecosystems have their own space–time domains. How to extrapolate from, for example, theory to protist microcosms to freshwater pond or temporary pool ecosystems to larger major ecosystems, such as lakes or forests, is not clear at all. In addition, within each space–time–artificiality domain, ecologists can study populations, communities or ecosystems, or combinations of these levels (Fig. 18.1). This adds further complication, because studies at one level of organization may not be directly relevant at other levels, even if performed at the same scale.

Given this multiplicity of scales and levels of organization, there are some central questions that must be tackled (e.g. Levin 1992, 1999; Anderson 1995; Peterson and Parker 1998; Loreau 2000a; Petersen and Hastings 2001; Gardner *et al.* 2001) if we want to improve our understanding of the nature and functioning of ecosystems:

- How do we know that what we learn at one scale and level is relevant at other scales?
- How do we (and can we) extrapolate to other scales?
- Can we generalize diversity effects on ecosystem functioning across different ecosystems and scales?
- What are the perils when ignoring scaling issues?

An additional question, especially important as ecologists are increasingly asked to provide information for policy-makers, is:

- Do the studies we conduct at smaller scales inform us about the consequences of biodiversity loss at the landscape and regional scales that policy-makers are most interested in?

In this chapter, we discuss these scaling questions in the context of studies on biodiversity and ecosystem functioning. We place special emphasis on the relationship between studies of local within-site dynamics and regional (landscape) dynamics, and consider the extent to which we can scale up from local plots to landscapes, from the laboratory to the field, and from artificial to natural systems.

---

[1] In this chapter, our use of the terms landscape and region are context-dependent. When discussing diversity and dynamics of populations and communities, we use 'regional' as an opposite to 'local', e.g. regional metapopulation dynamics versus local within-patch dynamics. However, when referring to spatial scales, we use landscape as an intermediate scale, larger than local but smaller than a geographic region, encompassing several habitats, and to a large degree defined by human perception. It is only from a human perspective that the landscape scale makes good sense, and hence model systems with model landscapes (see below) may not necessarily be relevant for our management of ecosystems.

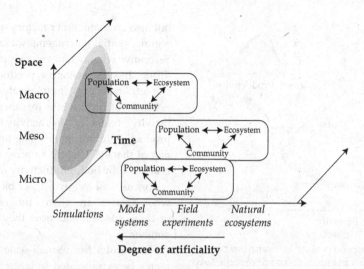

Figure 18.1 Ecological studies can be made at many combinations of scales in space and time, degrees of artificiality, and on several levels of organization. Studies made in one space–time–artificiality domain are not necessarily relevant in other domains and other levels. Experimental studies on diversity and ecosystem functioning have mainly been made in a small subset of the possible space–time–artificiality combinations (mainly model systems and field experiments as indicated; e.g. McGrady-Steed *et al.* 1997; Naeem and Li 1997; Laakso and Setälä 1999b; Tilman *et al.* 1996, 1997; Hector *et al.* 1999), while simulations seem most relevant in a different domain of long time scale but undefined spatial scale (e.g. Norberg *et al.* 2001). In all cases, diversity–ecosystem functioning studies have studied the linkage between population/community and ecosystem levels. (Modified after Naeem 2001)

18.2 Relating local plot experiments to regional patterns

One of the sources of controversy in the biodiversity debate has been the conflict between the results of small-scale experiments and comparative multi-site studies (Loreau *et al.* 2001). Take for example the relationship between productivity and diversity. This relationship is often described in the literature by a hump-shaped curve, with peak diversity at some intermediate level of productivity (Fig. 18.2, small graph). A large number of mechanisms have been posited to explain why diversity depends on productivity, but no single explanation seems to have unequivocal support (Rosenzweig and Abramsky 1993; Huston 1994; Abrams 1995a).

The first approach taken in most examinations of this relationship compare diversity between sites with different primary productivity or other variables, e.g. nitrogen availability, that indirectly indicates productivity. The second approach is that of the more recent experiments on the effects of diversity on ecosystem functioning (Naeem

et al. 1993; Tilman *et al.* 1996, 1997a; Hector *et al.* 1999, Chapter 4) that have turned the diversity–productivity question on its head (Loreau *et al.* 2001; Fig. 18.2). Here diversity is suggested to be one of the factors that productivity depends on. Experiments examining this hypothesis manipulated diversity at single or several sites on which there was no perceived variation in other factors affecting productivity, such as water or nitrogen availability.

This second approach, in which productivity is the response or dependent variable rather than the independent variable as in the first approach, examines whether variation in diversity affects productivity within sites, keeping other factors constant. This corresponds to experimentally varying diversity at a particular potential productivity, but should not be confused with between-site comparisons. One way of reconciling the results from the two approaches is to superimpose the results of the recent experiments on the area defined by the hump-shaped curve (Fig. 18.2, large graph) (Lawton 2000; Loreau *et al.* 2001). The experiments could be interpreted as examining whether a local decrease

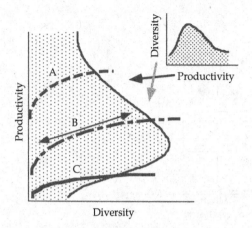

Figure 18.2 The two approaches to the relationship between diversity and plant biomass productivity examine different questions. In one approach, the relationship between productivity and local diversity is usually described by a hump-shaped boundary line, with a large number of data points in the area underneath this line (small graph). This approach mainly attempts to explain large-scale (between site) variation in diversity by factors such as productivity, climate or disturbance regime. Recent experiments on biodiversity and ecosystem function turn this question around and examine if variation in diversity at single sites affects productivity, keeping other factors constant. If the axes are switched, these experiments can be viewed as examining the trajectory of productivity response as diversity is experimentally decreased (or increased) at a single site (large graph; A, B, and C indicates three sites with different potential productivity, the arrow indicates how species richness can be varied). (Modified after Lawton 2000 and Loreau et al. 2001)

in diversity reduces the ability of a community to realize the potential productivity of a site. Such decreases in local diversity may arise from either local or regional processes, such as changes in management and land use, or habitat fragmentation.

It should have been clear from the start that these two approaches are different for two reasons. Firstly, the humped curve of productivity versus diversity is actually a boundary with a large amount of data points filling the space underneath, rather than a single line (Fig. 18.2, small graph).[2] At any point, species can be removed, species composition manipulated, or nutrients added or removed. The ensuing effects can then be examined as a trajectory in the graph. Consequently, the first approach, in which diversity is the response variable, attempts to explain larger-scale variation in diversity, which may depend on productivity (resource availability)

[2] We owe this point to B. Schmid.

but also on many other factors such as disturbance regime, spatial heterogeneity, climate or herbivory. Secondly, what is meant by 'productivity' often differs. In some instances, productivity refers to the site fertility, energy inputs, water regimes, or other extrinsic factors, while in other instances 'productivity' refers to the activity, such as biomass production or nutrient cycling rates, of the community at the site. Clearly an ecosystem's functioning is affected by both productivity in terms of extrinsic factors as well as intrinsic, or biotic factors such as what species are there, the relative abundance of these species, and how they interact with one another.

Can results from small-scale local plot experiments be extrapolated to larger scales and inform us about the consequences of biodiversity loss at the landscape or regional scales? Under special circumstances, the answer can be 'Yes!'. When the drivers[3] of productivity and diversity are the same at the different scales, small-scale experimental studies will also shed light on patterns at larger scales. However, if the drivers are different, then the extrapolation is not valid, and studies addressing the landscape or regional scale are required for answering questions at this scale (O'Neill and King 1998).

From the field experiments of Hector et al. (1999) and Tilman et al. (1996, 1997a), three mechanisms emerged as potential candidates to explain the patterns between diversity and productivity these studies observed—complementarity, positive interactions, and sampling effects. All act primarily at small spatial scales. The former two are local deterministic explanations for both increased diversity and increased productivity based on niche partitioning theory. The three mechanisms are based on the assumption that species interactions and the functional traits of species drive the relationship between diversity and ecosystem functioning. However, on larger scales and when comparing the same local habitat type across space, variation in productivity and other ecosystem processes is mainly caused by variation in resource availability

[3] In the following, we use the term 'driver' as a convenient abbreviation for 'factors or mechanisms explaining variation in community or ecosystem characteristics such as diversity or productivity'.

and abiotic factors (Huston 1994; Anderson 1995). At the same time, factors other than niche partitioning drive diversity at larger scales, e.g. dispersal and propagule production, disturbances, and spatial patterns in environmental factors can all regulate biodiversity. This suggests that many small-scale experiments may be of limited relevance for understanding biodiversity–ecosystem functioning relations on larger scales, unless they incorporate spatial and temporal heterogeneity, disturbances, or dispersal limitation of species in the experimental design, and possibly also other extrinsic drivers of biodiversity.

At local sites, positive effects of diversity on ecosystem processes such as productivity and nutrient mineralization are most likely to be found when the process as well as the maintenance of diversity are linked to resource utilization and availability, and mechanisms such as complementarity, positive interactions, and competition operate. However, not all local ecosystems are dominated by these factors and biotic or interaction-based mechanisms. In early successional systems that are strongly affected by frequent disturbances, the relationship between diversity and ecosystem functioning may be very different. Because resource limitation is not strong under such conditions and dispersal has a large stochastic component, local productivity patterns may be driven by the traits of colonizers and dominant competitors in earlier stages of succession, yielding little or no effect of diversity on ecosystem functioning. Engelhart and Ritchie (2001) suggested that disturbances in wetlands might allow less competitive species with strong effects on ecosystem processes to coexist with competitively superior species. Because competitive dominants with lower productivity form largely monospecific stands in the absence of disturbances, plant diversity needs to be increased to enhance ecosystem functioning. In this case, the mechanism is related to the interaction between disturbance regime and competition, rather than niche partitioning. Loreau and Mouquet (1999) pointed out that when diversity is maintained by immigration, productivity should on an average stay constant or decrease with increases in diversity.

Regional diversity losses could, however, decrease local productivity in systems where productivity depends on recruitment of appropriate dominants from the regional species pool after disturbances. If populations of appropriate dominants for local sites become too rare or disappear regionally, the potential productivity of local sites may not be realized because dominants, assumed to be most efficient in production, colonize too slowly or not at all. The trade-off between competitive and colonization ability (e.g. Levins and Culver 1971; Hanski and Ranta 1983; Nee and May 1992; Yu and Wilson 2001), suggests that this may result in greater losses of competitive species with strong effects on ecosystem processes (e.g. Tilman *et al.* 1994). Thus, plant communities in fragmented landscapes may increasingly consist of species with traits enhancing dispersal and rapid colonization, rather than species with traits leading to high competitive ability, such as the ability to deplete soil resources that may enhance nutrient retention (assuming that competitive abilities and dispersal represent trade-offs in organisms). This change in biodiversity–ecosystem functioning is just one consequence of larger and larger areas becoming part of 'a weedy world'.

18.3 The degree of similarity across ecosystem types in diversity–ecosystem functioning relationships

If we move from local plots to landscapes, our study area may include more and more ecosystem types (grasslands, forests, lakes, river banks, etc). This leads to a second controversy in diversity–ecosystem functioning studies, which has been based on the fact that results from one ecosystem do not necessarily inform us about other systems. This is perhaps a trivial finding but nonetheless one that has often been overlooked when ecologists have entered the area of policymaking.

Most empirical studies on diversity-functioning relationships have been conducted in mesic grasslands. Is there any reason to expect that other ecosystems would behave differently? Studies in other systems both support and contradict the mesic grassland studies. The majority of experiments do indeed show some relationship between species diversity and functioning, usually at the low end of the diversity gradient (Schwartz *et al.* 2000), but there are some notable exceptions. Even in grasslands, single functional groups can outperform functional group mixtures, and Troumbis *et al.* (2000) observed an inverse sampling effect, in which superior competitors do not have the greatest effect on ecosystem processes (Loreau 2000a). Engelhart and Ritchie (2001) also observed such an effect in wetlands.

Studies of soil systems have yielded little support for effects of species diversity in soils on decomposition and production, although several experiments clearly show that functional diversity of soil fauna increases nitrogen mineralization and plant growth (e.g. Laakso and Setälä 1999b; Hooper *et al.* 2000; Mikkola *et al.*, Chapter 15; Wardle and van der Putten, Chapter 14). Also, soil process rates seem to be more affected by the properties of individual plant species, e.g. litter chemistry, than by plant diversity (Wardle *et al.* 1997a; Wardle and van der Putten, Chapter 14).

It may be too early to conclude that there is any consistent variation in the shape and existence of diversity–ecosystem functioning relations among major ecosystems or ecosystem processes. Before we, as ecologists, are in a position to be able to start making scientifically robust predictions on these relationships, there is an enormous amount of research necessary that is difficult in both its extent and complexity. Figure 18.3 outlines some basic requirements (see also e.g. Bengtsson 1998; Schläpfer and Schmid 1999).

Starting from the (hitherto) paradigm system 'mesic grassland plant diversity and productivity', there are a number of avenues for further studies. We can examine the effects of primary producer diversity (or microbes in decomposition systems) on production in a variety of ecosystems, such as forests, soils, lakes, streams, etc. We can also begin

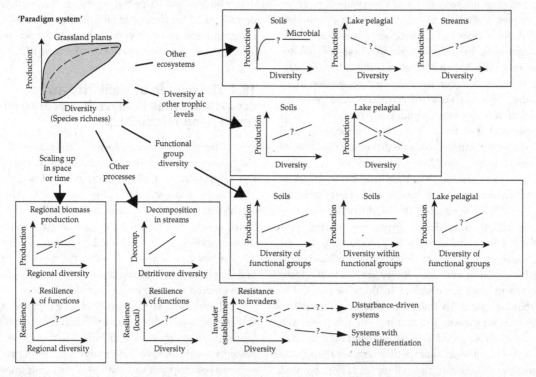

Figure 18.3 Studies on biodiversity and ecosystem functioning have been made on a small number of ecosystems, ecological levels, and scales. The major 'paradigm study system' has been plant biomass production and species diversity in mesic grasslands (e.g. Tilman *et al.* 1996, 1997; Hector *et al.* 1999). These studies should be complemented with studies (a) in other ecosystems, (b) manipulating diversity at other trophic levels, (c) manipulating the diversity of functional groups, (d) examining other ecosystem processes, and (e) at larger scales than local plots or sites. The panels in the figure give some examples of such complementary studies that are needed. Hypothetical relationships between diversity and ecosystem functioning are indicated (based on studies referred to in the text). See also text for explanation.

to study systems where higher trophic levels may affect diversity (Paine 1966; Connell 1975), production (e.g. Setälä and Huhta 1991; Laakso and Setälä 1999b; Wedin 1995) and the shape of the relation between primary producer diversity and production (Mulder *et al.* 2000). It would hence be highly appropriate for future studies to incorporate diversity at higher trophic levels (see Raffaelli *et al.*, Chapter 13). How we treat the diversity of functional groups (Hooper and Vitousek 1997; Tilman *et al.* 1997; Hector *et al.* 1999; Laakso and Setälä 1999b) in a multitrophic system is particularly challenging.

The world does not consist of primary production alone. Other ecosystem processes of interest include other rate processes such as decomposition, nutrient mineralization and retention, and secondary production. In soils, decomposition and nutrient mineralization were positively related to the diversity of soil animal functional groups (Laakso and Setälä 1999b), while in streams decomposition was positively affected by the number of species (Jonsson and Malmqvist 2000), albeit at the low end of the diversity gradient.

We might also consider the effect of diversity on the stability of the systems, the variability (or conversely reliability) of ecosystem processes (see Loreau *et al.*, Chapter 7 and Hughes *et al.*, Chapter 8), the resilience of processes or of community composition, or invasion resistance (e.g. McGrady-Steed *et al.* 1997; Naeem 1998). Stability and resilience in most local ecosystems are likely to depend on an interplay between local and regional processes, with diversity as a key component (Holling *et al.* 1995; Petersen *et al.* 1998; Bengtsson *et al.*, in press; Levine *et al.*, Chapter 10). Finally, local relationships may or may not scale up to landscape or regional relations, so each system and process ought to be studied at several scales.

These short paragraphs have sketched a multi-dimensional problem with almost infinite possibilities for combinations of study objects. In spite of this, comparable studies of diversity-functioning relations in a number of different major ecosystems should be of high priority, with due considerations of the limitations that face almost any such study. The study of soil fauna effects on plant growth by Laakso and Setälä (1999b) is exemplary for

addressing several of the dimensions at the same time—species and functional group diversity, several trophic levels and several processes. As we scale up in space and time, the issue of replication becomes a difficult one, and statisticians may well need to come to grips with issues of what constitutes statistically acceptable designs in the face of pragmatic and realistic experimental constraints in a variable world.

In the absence of data and theory for many ecosystems and processes, what can ecologists say at this time about the expected relations between diversity and ecosystem functioning in different systems? Differences among studies, sites or ecosystems in the effects of diversity on ecosystem functioning are likely to be related to the major factors driving diversity, species abundance and process rates in different systems (Table 18.1). If resources are limiting, exploitative rather than interference competition dominates, and if the potential for partitioning of resources among species exists, then niche-based complementarity is a reasonable mechanism explaining a positive relationship between diversity and the rate of a process related to that resource (Loreau 2000a; Loreau and Hector 2001). Positive interactions among species that increase resource utilization or the amount of available resources would lead to similar positive relations.

However, in other circumstances these mechanisms may not be the driving factors, in which case the effects of species richness on ecosystem functioning may be relatively minor. This is nicely illustrated by contrasting the species diversity-invasion resistance results of Levine (2000a) for plant assemblages along rivers and McGrady-Steed *et al.* (1997) for protist microcosms. In McGrady-Steed's relatively homogenous and closed microcosms, where diversity was the only factor varying among treatments, diversity was found to enhance invasion resistance. Levine (2000a) found a similar effect of diversity locally, but in contrast, at the 8 km scale of his riparian system, factors other than diversity such as seed supply drove invasion success. This is not to say that diversity is inconsequential, but at community-wide scales in heterogeneous systems, the effects of diversity may be overwhelmed by other factors (see Levine *et al.*,

Table 18.1 Possible relations between diversity and ecosystem functioning (in general) in ecosystems with different driving forces (mechanisms) for diversity and process rates. Systems fulfilling conditions under A are expected to show a positive relation between diversity and ecosystem functioning locally, whereas systems under B are not, based on present knowledge. Functions under C and D are related to stability rather than process rates, C refers to local diversity, D to regional diversity

Systems	Expected effect of diversity on function	Mechanisms
A		
Systems with exploitative competition for limiting resources that can be partitioned	Positive	Complementarity, Niche partitioning
Systems with positive interactions between species (mycorrhiza, N-fixation)	Positive	Increased resource availability or amount of resources with diversity
B		
Disturbance-driven systems	No?	Growth rates and colonization rates of individual species drive functions (initially)
Top-down effects dominating	Positive, no or negative (?)	Consumers preclude resource partitioning, resource limitation may not occur
Interference competition dominates	No or negative?	Traits of dominants drive function
Processes not resource-limited, or resources cannot be partitioned (examples: soil fertility, maintenance of atmospheric composition, etc)	No? unclear, or meaningless	Various
C		
Invasion resistance in systems with resource limitation and niche partitioning	Positive	No 'empty niches' available for colonizers
Invasion resistance in disturbance-driven systems	Negative	Disturbances enhance diversity and establishment of all species
D		
Resilience of functions: larger spatial and temporal scales, mosaic landscapes, disturbances, species substitutable, etc.	Positive (?)	Portfolio effect (spatial and temporal), spatial resilience, dispersal among patches possible, several species in functional groups provides redundancy

Chapter 10). More generally, we find it likely that in early successional plant systems establishment through dispersal from nearby patches and the traits individual species of colonizers, e.g. growth rates, will be more important than complementarity in driving ecosystem processes following the disturbance. How organism size and generation time relate to the diversity–ecosystem functioning relationship must also affect the relative importance of these mechanisms. The generation time in relation to disturbance frequency and organism size in relation to resource heterogeneity will produce variable responses even within similar ecosystems.

Some ecosystem processes are not resource-based and there may be no potential for resource partitioning to influence them. Examples may be soil fertility, maintenance of atmospheric composition, erosion control, and possibly resilience and resistance to disturbances. In the case of resilience, diversity may still be important because of other mechanisms (the insurance hypothesis, spatial resilience; see below) and at larger scales. Finally, positive effects of diversity on rates of ecosystem processes may be less common in systems where interference competition is strong and consequently only a few species are dominant. In this case, niche partitioning is not an important factor structuring the community.

Future empirical and theoretical studies should seek a more unified framework in which to place the different types of diversity–ecosystem functioning results that may emerge. This would aid the process of generalizing from the present case by case nature of the subject. Figure 18.3 and Table 18.1 is a first attempt to develop such a framework, and it would be informative to be able to compare results from a number of studies of diversity-functioning relations in systems which differ according to the criteria in Table 18.1.

18.4 Regional diversity maintaining resilience and local ecosystem functioning

Environmental variation and disturbances are ubiquitous in ecosystems. Much of classical community ecology was developed from a perspective of local species interactions and succession at single sites. Disturbances, if they occurred, were viewed as small-scaled, and disturbed areas were assumed to be recolonized from within the local system. The surrounding landscape and its diversity was regarded to be of minor importance. However, theories of metapopulations and spatial dynamics, and the still undigested perspective from landscape ecology, are beginning to change the emphasis from the local to larger scales in many communities (some recent examples are Hanski 1999b; Lawton 2000; Nyström et al. 2000; Hubbell 2001; Bengtsson et al., in press). Yet, studies of biodiversity and ecosystem functioning have been almost entirely performed within a fairly narrow scale paradigm emanating from classical (local) community ecology. If we want to answer questions about the consequences of biodiversity loss on larger scales than plots or microcosms, we need to incorporate insights from other perspectives emphasizing regional and landscape scales.

A growing number of studies across many different ecosystems suggest that local and regional diversity are related to each other, and that the surrounding landscape can have a profound impact on local communities (e.g. Srivastava 1999; Lawton 2000; Loreau 2000a,b; Bengtsson et al., in press). A recent review by Lawton (2000; but see Loreau 2000b) suggests that local communities may be arranged along a continuum from those that are proportional samples from the regional species pool to those in which species interactions sets limits to species richness. In the former systems (termed 'type I systems'), processes driving regional diversity should be particularly important for local diversity. For example, the disappearance of species in intensively managed landscapes as a consequence of the ongoing landscape fragmentation and habitat loss will affect the diversity and abundance of those species that remain (Bascompte and Rodriguez 2001). This could have effects on important ecosystem services such as biological control. In fact, in agroecosystems Thies and Tscharntke (1999) and Östman et al. (2001) showed substantial effects of landscape heterogeneity on biological control by parasitoids and generalist predators, respectively. In both cases, biological control was more efficient in small-scaled heterogeneous landscapes. A similar case might also be made concerning pollinator declines because of fragmentation (e.g. Buchman and Nabhan 1996; Thomson 2001).

In ecosystems where disturbances are frequent and regional species composition is important for local communities, ecosystem reorganization after disturbances will be affected by both local conditions and the landscape composition in the surroundings of disturbed patches (Holling et al. 1995; Bengtsson et al., in press). Important local factors affecting ecosystem reorganization after disturbances include seed banks, structures allowing recolonization (often termed 'biological legacies'; Turner et al. 1998; Franklin and MacMahoon 2000) and species interactions during succession. The surrounding landscape, on the other hand, contains the areas that serve as source patches for propagules, the species that transport propagules, and species that interact and interfere with the dispersal, dynamics and succession in local patches (Bengtsson et al., in press). This was termed 'spatial resilience' by Nyström and Folke (2001), and the processes are likely to be important no matter the shape and existence of a relation between diversity and rate processes. The composition of the landscape will affect the recruitment of the dominant species locally driving ecosystem processes during earlier stages of succession, and the recruitment of appropriate species to fill 'local niche space' and increase local resource utilization through increased diversity. Landscape composition will also influence the modification of the diversity–ecosystem functioning relations through redistribution of water, soil particles and seeds, and interactions between vegetation and higher trophic levels. Species at higher trophic levels are more likely to have large home ranges and their dynamics depend on landscape composition at larger scales than local patches (e.g. Ritchie and Olff 1999).

One larger-scale perspective on the relation between diversity and ecosystem functioning that might be useful to incorporate in future studies comes from island biogeography and metapopulation dynamics. For example, species-area relations imply that the long-term maintenance of local diversity at a given level may require much higher diversity at regional scales (Tilman 1999b). The mechanism is the classical one—local extinctions must be balanced by new colonization events from the species pool for diversity to be maintained (MacArthur and Wilson 1967). However, in the original island biogeography theory, the species pool on the mainland is constant and not affected by the islands in question (but see Schoener 1976). In mainland systems, however, the species pool is continually produced by the component local communities sending dispersing propagules to other patches. Changes in local community composition, in particular directional ones such as abundance déclines and extinctions, will sooner or later feed back to produce the new regional species pool. Hence, if the temporal scale is extended, regional diversity and species composition is a dynamical variable in real ecosystems (Hubbell 2001). Systems with this kind of dynamics were modelled by Wilson (1992), but have received relatively little attention from theoreticians and empirical ecologists.

Another perspective is the body of theory developed by Holling and co-workers about multiple stability domains in ecosystem dynamics (e.g. Holling et al. 1995; Petersen et al. 1998; Gunderson 2000; Scheffer et al. 2001). This theory differs somewhat from much of textbook theoretical community ecology in its emphasis on multiple stable states in ecosystems, and how transitions between the states are produced. For Holling and co-workers, ecological resilience is the magnitude of disturbance that an ecosystem in one stability domain can absorb

(resist) before it moves into another stability domain (Gunderson 2000).[4] Biodiversity is assumed to play a key role in this respect (Holling et al. 1995; Walker et al. 1999), because losses of biodiversity are regarded as making systems more vulnerable to disturbances that move them into new stability domains (Scheffer et al. 2001).

The consequences of this theory are related to the ecological redundancy and its role in insuring ecosystem functioning (e.g. Naeem 1998; Walker et al. 1999), but extended from local to larger scales. For an ecosystem to reorganize, i.e. remain within the same stability domain, after a disturbance, a diversity of species and structures within the disturbed area and in the surrounding landscape is needed. Ecosystem functions, being performed by species, are maintained in large-scale mosaic systems if a number of species exist that are to a large degree substitutable (Walker et al. 1999). Such substitutable species are regarded as being able to coexist in the landscape primarily because they have dynamics at different scales in space and time (Petersen et al. 1998; see also Ritchie and Olff 1999). In this way, they contribute to the resilience of the system even though they may appear to play no functional role at the moment. Ecosystem resilience is maintained in the long term by preserving biodiversity on the landscape and regional scales (Holling et al. 1995; Petersen et al. 1998; Bengtsson et al., in press).

Another component of the Holling theory is the adjustment of ecosystems and the functions performed in them to environmental changes (termed ecosystem adaptation by the proponents of the theory). When regional diversity is high, there will be not only many substitutable species, but also many species with different traits and requirements. For example, if spatial heterogeneity is maintained in managed landscapes, species with different niches can coexist more easily. When landscapes are subjected to environmental changes, such as global warming or decreased rainfall, a high regional diversity would allow ecosystems to adjust to these changes more easily by recruiting the appropriate species from the species pool, thus maintaining

---

[4] Note that Holling's 'ecological resilience' is termed 'robustness' in Loreau et al., Chapter 7.

ecosystem functioning (Norberg *et al.* 2001; Chapter 7). If this is true, the niche-based arguments advanced for positive local diversity-functioning relations can, with appropriate modifications, also be advanced for a positive relation between diversity and ecosystem resilience at the landscape scale.

In reality, little is actually known about the effects of large-scale diversity losses on ecosystem processes and services. It has been hypothesized that observed declines in pollinator and natural enemy diversity in intensively managed agricultural landscapes could result in less efficient delivery of the ecosystem services crop pollination and biological control of pests (Buchman and Nabhan 1996; Daily 1997; Björklund *et al.* 1999). However, relevant data collected at the relevant scales are largely lacking. Since designing and performing classical experiments at the landscape and regional scales is difficult, to say the least, other means must be used to obtain knowledge on large-scale effects of biodiversity loss on ecosystems.

One way of dealing with this problem is to construct model ecosystems to examine how landscape structure may affect species distributions and diversity (Gonzalez *et al.* 1998). The microcosm approaches of McGrady-Steed *et al.* (1997) and Naeem *et al.* (2000a) could also be extended by using microcosms connected to each other in various ways to mimic variation in landscape configuration (cf e.g. Holyoak 2000).

Apart from such model ecosystem studies, the closest we have to large-scale field experiments is probably the comparison of lands under different management practices, natural landscape differences, natural gradients or areas with different restoration methods. For example, Östman *et al.* (2001) studied the efficacy of biological control and patterns of diversity in paired organic (pesticide-free) and conventional farms along a gradient of landscape heterogeneity. In this case, the ecosystem service of biological control was enhanced in landscapes that are more heterogeneous and on organic farms. However, there was no obvious connection to the diversity of natural enemies, which was higher in heterogeneous landscapes but surprisingly, lower for carabids on organic farms.

To summarize, processes affecting regional diversity patterns will often interact with local processes to produce local relationships between diversity and rates of ecosystem processes. Even if no clear diversity-functioning relation (positive or negative) can be found locally, regional diversity can still be an important determinant of ecosystem processes at local sites. This may be self-evident for many researchers studying ecosystems and population dynamics at larger scales, but knowledge of large-scale processes still needs to be actively incorporated into studies of diversity and ecosystem functioning.

18.5 The way forward

It is unlikely that scaling problems will be easily solved. However, it should be possible for ecologists to provide better guidelines for scaling issues than they do at present. We still lack a comprehensive treatise on scales in ecology (Bengtsson 2000 reviewing Peterson and Parker 1998) that could help us. What can be done in the meantime?

Firstly, more studies should not only repeat the same approach at different sites (cf Hector *et al.* 1999) but also at different spatial scales, for example conducting the same diversity treatments in nested experiments. Studies scaling up from local plots to incorporate interactions between local, landscape, and regional diversity, and the ensuing effects on ecosystem functioning, are critically needed.

A drawback of most field studies is their short duration relative to the dynamics of organisms and processes affecting community composition. Some microcosm studies have attempted to ameliorate this situation (McGrady-Steed *et al.* 1997; Naeem and Li 1997; Naeem *et al.* 2000; see Petchey *et al.*, Chapter 11). However, these systems still have to be scaled up to larger spatio–temporal scales, which is by no means easy. Moreover, there is a general, but perhaps unfounded, scepticism to microcosm studies among many ecologists. This means that microcosm studies at best can be used to complement other approaches. Long-term studies that simultaneously examine how community structure and processes change through time will be informative in this respect.

Clever and innovative use of natural experiments and gradients, variation in management practices, and restoration projects to examine hypotheses at landscape and regional scales should increase. For example, analysis of diversity–ecosystem functioning relationships in woody ecosystems could be conducted by well-planned observational studies (Troumbis and Memtsas 2000). Effects of regional variation in diversity on other important ecosystem services than biomass production should be of high priority, e.g. nutrient retention, pollination and biological control. More fundamentally, we need a better understanding of how regional diversity influences local diversity in different ecosystems.

At the same time, studies on larger scales necessarily have drawbacks. They need to be complemented by in-depth studies of mechanisms, and these will have to be performed at smaller scales. The degree of replication and control over treatments in large-scale studies will be lower than in most experimental studies. Requirements on statistics and methodology may have to be relaxed (but not abandoned) since control of all confounding factors is impossible. In some cases designs such as BACI (before-after-control-impact) as used in whole-lake manipulations (e.g. Carpenter and Kitchell 1993) will be useful. Because spatial and temporal scales are related, large-scale studies will usually take more time. Hence, reliable results will take some time to obtain, which must be acknowledged. On the other hand, the issues are so pressing that it is impossible to wait until the last sceptic is convinced, before results are used for management and policy.

Theories and model systems that explicitly incorporate several scales are also needed. In order to be helpful for empirical studies, new theory should be developed that incorporates local–regional interactions and generates hypotheses for larger scales than local plots. Then we may ultimately be able to answer whether diversity really is important for the resilience and maintenance of ecosystem services at larger scales in space and time.

18.6 Conclusions

Scaling effects of biodiversity on ecosystem functioning presents problems that need to be solved by both empirical and theoretical studies performed at several scales simultaneously. Species loss at a regional scale may result in each local site not receiving the set of species necessary to realize the potential rate of ecosystem processes such as productivity. We suggest that at local sites, the main effects of diversity are on rates of ecosystem processes, but at a regional scale diversity will mainly influence resilience and stability-related functions. Diversity effects on functioning may or may not be similar across ecosystem types, depending on the process and nature of the system. When systems are driven by disturbances, top-down regulation or interference competition, positive diversity–ecosystem functioning relations may be less common. In contrast, systems with exploitative competition and resource partitioning are most likely to show positive effects of diversity on ecosystem functioning. Studies in different ecosystems examining diversity–ecosystem functioning relations at local, landscape, and regional scales are needed. A better theoretical understanding of the relationships between regional and local diversity will be helpful when interpreting such results. In the absence of true large-scale experiments, we advocate more clever and innovative use of natural experiments, diversity gradients, and managed systems.

This is the report from the discussion group on 'Comparing experimental results across ecosystems and spatial scales' at the Paris workshop. We thank all those unnamed contributors who have provided us with many of the ideas that found their way into this chapter. C. Koerner and the reviewers gave many comments on the manuscript. Christian also communicated the idea of B. Schmid to us.

CHAPTER 19

Effect of biodiversity on ecosystem functioning in managed ecosystems

J. Vandermeer, D. Lawrence, A. Symstad, and S. Hobbie

19.1 Biodiversity function in managed ecosystems

Depending on definition, as much as 90% of the terrestrial surface of the earth is estimated to be maintained in some sort of managed state, usually forestry or agriculture (Western and Pearl 1989). This very abundance should be sufficient to place managed ecosystems at the centre of the analysis of biodiversity and ecosystem function, and indeed a large literature has emerged on this topic in agricultural and forestry ecosystems (Pimentel *et al.* 1992; Paoletti *et al.* 1993; Collins and Qualset 1999). In this chapter, we first cast the major questions about biodiversity and ecosystem functioning in the context of managed systems, introducing the ideas of planned versus associated biodiversity and the intensification gradient, concepts that are unique to managed systems. We then discuss the consequences of intensification on biodiversity. While most of our examples come from agroecosystems and our analysis probably applies most clearly to this particular subset of managed systems, it is our expectation that many of the same concepts will likely apply to non-agricultural managed systems.

Before beginning, it is worth noting that ecosystems are managed for a wide variety of reasons, from maximizing short-term profit from a certain crop to preserving biodiversity. In addition, a given piece of land or water is often managed simultaneously for many goals, including such complicated matters as recreation and sustaining a certain way of human life. Our focus is on agroecosystems, but we use the more general term 'managed systems' with the hope that the concepts and topics discussed

here are applicable to and stimulate thought about other types of managed ecosystems.

It is also worth noting that much of what stimulates concern with 'biodiversity and ecosystem function' in general is the recent dramatic decline in the earth's biodiversity. Yet the bias that exists in much of this research is that in significant areas of the Earth we have substituted species-rich, coevolved communities with depauparate, highly managed systems, and that if we could discover that biodiversity had some utilitarian value, this would provide an argument for stemming the tide of such transformations. Whatever the merits of this view, it is also the case that the same concerns really ought to be focused on the managed systems themselves. Indeed, it would be quite myopic of us to concentrate only on the loss of charismatic megafauna and transformations of pristine ecosystems if, as seems to be the case, most of the biodiversity loss in the world today is occurring within managed systems largely due to the transformation of those managed systems from one form to another. For this reason, the role of biodiversity in the functioning of managed ecosystems should be, we believe, the most critical issue for those concerned with biodiversity loss.

19.2 Biodiversity and ecosystem functioning—some initial considerations

It is well known that some managed systems are very diverse while others are remarkably monotonous, the former usually associated with some form of traditional or 'alternative' production schemes, the latter with modern technified agriculture (or forestry or other managed ecosystems). It is not difficult to form an initial opinion based only on this simple observation. On the one hand since traditional forms are diverse and tradition is wise, it is claimed, there must be some reason that managers plan for high diversity, or in somewhat imprecise language, there must be some function associated with the biodiversity. On the other hand, since the modernization of production always improves things, it is claimed, most of the biodiversity that has been excluded from more modern systems was unimportant. These opposing points of view form the basis of two of the major hypotheses associated with biodiversity function in general; (1) that biodiversity does have importance with regard to ecosystem function (the traditional agriculture position) or (2) biodiversity includes so much redundancy that it effectively has little or no importance with respect to function (the modern technified agriculture position). These hypotheses are standard and have repeatedly appeared in the literature on biodiversity and ecosystem functioning (e.g. Walker 1992; Lawton and Brown 1993; Gitay *et al*. 1996; Naeem 1998).

It is also worth noting that 'managers' may think of function quite differently than academic analysts. There are at least three ways in which the idea of function enters into the biodiversity and ecosystem function debate. (1) Increased (or decreased) biodiversity may cause changes in some ecosystem characteristics (e.g. productivity, nutrient cycling) and those characteristics may be referred to as 'ecosystem function'. (2) Increased or decreased biodiversity may affect some aspects of ecosystems, and the role of biodiversity as a driver may enable a particular conceptualization, namely that of the 'ecosystem function of biodiversity'. In other words, in (1) we have the effect of biodiversity on ecosystem function and in (2) we have the function of biodiversity. In yet another formulation, the ecosystem is seen as having a larger 'function' (e.g. carbon sequestration) and presumably biodiversity could affect those larger functions. For the most part these distinctions are not especially useful and we feel, arise from the word function itself, which has been wisely avoided by the editors of this

volume. On the other hand, for this particular chapter, the word function has a clear meaning (the third one) and thus we retain it.

However a more subtle issue is involved here. There is a hint of normative in the question to start with—normative in the sense that the question 'What is the function of biodiversity?' contains the subtext 'of what use is biodiversity?' Many ecologists would deny that such a subtext exists, while many others would claim it has little significance. Some, however, acknowledge that at a minimum it is a subtext that automatically emerges in the public's mind, and some would be bold enough to acknowledge that funding and fame derive from the subtext rather than the stated scientific question.

How to disentangle the practical subtext from the theoretical scientific question is not always obvious when dealing with pristine tropical rainforests or untouched savannahs of the Sahel. Reference to carbon sequestration, maintenance of a desired local humidity, or provisioning of high quality water sources are clear ecosystem services that may be provided by unmanaged systems, and questions of biodiversity function are sensible. However, other, perhaps more common, questions, such as how biodiversity affects local nutrient cycling or productivity, do not always make normative sense when speaking of unmanaged systems (rapid or slow nutrient cycling or high versus low productivity do not in and of themselves have normative content). However, when dealing with managed ecosystems the subtext emerges as the *raison d'etre*. One may question the wisdom of focusing on productivity, for example, in an unmanaged ecosystem, but raising or stabilizing productivity in a managed ecosystem is a clear human goal and one need not pretend the normative question is being ignored.

As one of the more spectacular examples of high biodiversity in a managed system, consider the case of home gardens. Especially in the tropics, it is well known that home gardens are frequently highly diverse. Indonesian home gardens frequently contain over 30 species of planted crops (Soemarwoto *et al*. 1985; Michon *et al*. 1986) and Mexican 'huertas' are similarly diverse (Gleissman 1993). It would be foolish to suggest that this ubiquitous pattern exists without some sort of function. Sometimes the function is obvious, sometimes a simple question to the farmer shows the function, and sometimes the function is obscure. Nevertheless, despite an apparent lack of systematic ecological study of home gardens, the fact that they are ubiquitously planted with high diversity in the tropics cannot be ignored. Frequently these high diversity gardens seem to be designed with multiple functions in mind, rather than simply higher production (e.g. food, fuel, fibre, medicines, and timber), a crucial factor to take into consideration when evaluating biodiversity in any ecosystem. Furthermore, risk avoidance is a well-known motive for maintaining biodiversity in these traditional gardens, perhaps reflecting ecologists' concerns about resistance and resilience in unmanaged systems. This suggests a third hypothesis regarding the function of bio-diversity, namely risk reduction, or what has sometimes been called the insurance hypothesis (Folke *et al.* 1996; McGrady-Steed *et al.* 1997; Naeem and Li 1997; Naeem 1998; Yachi and Loreau 1999).

These initial considerations suggest that three basic hypotheses about biodiversity function in managed ecosystems may contend for dominance: (1) many components of biodiversity are critical/essential for the functioning of the ecosystem, (2) most components are redundant, and only one species (or type) need be present in each of a small number of functional groups, and (3) varied types are needed in all functional categories in anticipation of maintaining function in the face of future environmental changes—the function hypothesis, the redundancy hypothesis and the insurance hypothesis.

19.2.1 Planned versus associated biodiversity

In the case of managed systems there is another issue that emerges which is not relevant to unmanaged systems (but perhaps ultimately links managed and unmanaged systems in a natural way). In a managed system there are a variety of biological components that are chosen by the manager. These may be the crops chosen to be planted by the farmer, the volunteer medicinal plants that are not planted but nevertheless tended by the farmer, the tree species chosen to be planted by the forester, the trees chosen to be harvested by the logger, the fish species chosen to be harvested, the aquatic species chosen for aquaculture, and so forth. Much litera-ture has been devoted to the question of the function of multiple species managed systems (Kass 1978; Willey 1979), usually aimed at asking whether they yield better than associated monocultures but occasionally asking whether they are more sustainable or provide other more subtle services. This bio-diversity is the 'planned biodiversity' (Lawrence 1996; Swift *et al.* 1996). Focusing on the planned biodiversity in many ways misses a great deal that is important in the discourse of biodiversity. For example, a logger logs about 20 species of trees from a neotropical forest that contains more than 400 species. What is important about the system, the fact that 20 species are logged or the fact that the eco-system contains 400 tree species? To take another example, cotton and wheat are intercropped in China, leading to the control of the cotton aphid by a coccinelid beetle that builds up its population density in the associated wheat. Concentrating on only the fact that there are two crop species misses an important fact about the biodiversity, namely the existence of the coccinellid in the system.

The 'associated biodiversity' can be substantial. Ecosystem function results from both planned and associated biodiversity, but the associated bio-diversity is at least partially determined by the planned biodiversity. Recently, a large literature has been generated on the subject of associated bio-diversity in agroecosystems (Perfecto and Snelling 1995; Perfecto *et al.* 1996; Iverson and Prasad 1998; Linusson *et al.* 1998; Ryan *et al.* 1998; Bjorklund *et al.* 1999; Smedling and Joenje 1999; Holland and Fahrig 2000; MacDonald *et al.* 2000), mainly con-centrating on the relationship between agricultural intensification (see below) and associated bio-diversity. Thus in attempting to answer the ques-tion 'what is the function of biodiversity in managed ecosystems?', we must deal with both concepts of biodiversity, the planned as well as the associated.

19.2.2 The intensification gradient

The framework for thinking about both planned and associated biodiversity has been formulated in terms of what is often termed 'agricultural intensification' (Perfecto *et al.* 1996; Swift *et al.* 1996). The notion of agricultural intensification, originally used in the context of extensive agriculture is here intended to provide a qualitative conceptual framework for thinking about changes in multi-species agriculture. Farmers modify farming practices in response to a wide range of physical, demographic, economic, social and technological pressures. These modifications may be expressed in relation to agricultural intensification, the most fundamental feature of which is increasing the duration for which the same piece of land is used for growing crops. Associated aspects of intensification include increased use of resources, the switch from internal to external regulation through the use of purchased inputs, changes in labour use and management practice, and increased linkage with market economies. Another, almost universal, feature of agricultural intensification is that of increasing specialization in the production process resulting in reduction in the number of crop and/or livestock species utilized and an apparent reduction in other unplanned components of biodiversity (Lawrence 1996; Perfecto *et al.* 1996). With respect to planned biodiversity, a simple count of the types of plants purposefully tended in each of these systems is almost surely a monotonically decreasing function of intensification, although the exact shape of the curve is rarely known. This pattern is not a physical necessity, but it has been true, on an average, over the past 500 years of agricultural development.

There is now an enormous literature on the practical significance of this trend in the loss of planned biodiversity with intensification, and we do not lack hypotheses about the agroecosystem function of the planned biodiversity (e.g. Altieri 1990; Spain *et al.* 1991; Waage 1991; Materon *et al.* 1995; Netting and Stone 1996; Swift *et al.* 1996). However, this focus is perhaps a bit myopic. The crops planted by the farmer represent only a fraction of the overall biodiversity in the agroecosystem. Both planned and associated may be involved in ecosystem function, but the effect of associated biodiversity, which is the indirect effect of the planned biodiversity, may be largely invisible, at least initially, to the farmer and we have little theory that might predict it.

If we now wish to explore what pattern of associated biodiversity exists *vis-a-vis* the intensification gradient, we encounter a dramatic absence of relevant data. It has not been popular to study associated biodiversity *per se* in agroecosystems (Vandermeer and Perfecto 1997), and with a handful of exceptions (Pimentel *et al.* 1992; Lawrence 1996; Perfecto *et al.* 1996) we know very little. We can, nevertheless, speculate with some simple biological intuition. The pattern is likely to take on one of four distinct forms (Fig. 19.1). Forms I or III is what many conservationists suggest is true. As soon as the process of intensification begins, there is a dramatic loss of associated biodiversity, perhaps levelling off at some intermediate intensification level, as in curve III, perhaps not, as in curve I. The other extreme is that as the traditional system is intensified, not much happens to the biodiversity until some threshold level of intensification is attained, at which point a dramatic decline occurs, perhaps occurring at some intermediate level of intensification as in curve IV, or at some higher level, as in curve II. As to which of these curves represents reality, we have no idea.

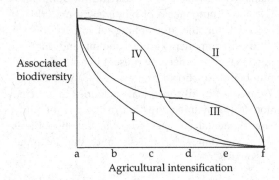

Figure 19.1 Possible scenarios for the relationship between degree of intensification and total biodiversity. The intensification gradient moves from a = an unmanaged system (forest, grassland) b = Casual management (e.g. shifting cultivation, nomadic pastoralism, home garden), c = Low intensity management (e.g. traditional compound farm, rotational fallow, traditional agroforestry), d = Middle intensity management (e.g. horticulture, pasture, mixed farming, traditional cash cropping), e = high management (e.g. crop rotation, multicropping, alley cropping, intercropping), f = Modernism (plantations and orchards, intensive cereal and vegetable production). Pattern I is where any extent of human intervention drastically reduces biodiversity with added effects being relatively minor and progressive. Pattern II is where intensification affects species diversity only at very high levels. Patterns III and IV are intermediate relationships.

Table 19.1 Number of species (and individuals) of beetles, ants, and non-formicid hymenopterans in the canopy of shade trees and coffee plants in coffee farms, as determined by insecticidal fogging (from Perfecto *et al.* 1997). Each number refers to species and individuals (in parentheses) in an individual tree

Species of tree	Type of farm	Beetles	Ants	Hymenoptera
Canopy trees				
Erythrina poeppigiana	Traditional	126 (401)	30 (333)	103
E. fusca	Traditional	110 (393)	27 (1105)	61
Annona chirimoya	Traditional	NA	10 (179)	63
E. poeppigiana	Moderately shaded	48 (107)	5 (64)	46
None	Unshaded (modern)	0	0	0
Coffee bushes				
C. arabica	Traditional	39 (76)	14 (135)	34
C. arabica	Moderately shaded	29 (82)	9 (128)	31
C. arabica	Unshaded (modern)	29 (92)	8 (47)	30

The virtual absence of interest on the part of conservationists in the question of associated biodiversity in agroecosystems has resulted in a virtual absence of research activity aimed at this most common of all ecosystems (Vandermeer and Perfecto 1997). On the other hand, the few studies that have been done are highly suggestive.

For example, Perfecto *et al.* (1997) examined general arthropod biodiversity in the coffee agroecosystem, comparing the traditional coffee production system with more modern or intensified ones. A high level of planned biodiversity has characterized traditional coffee production in Central America. A traditional farm typically has a large mixture of shade trees and fruit trees, coupled with plantains and other understory crops, in addition to the coffee bushes. When a shade tree is harvested for lumber, the light gap so created is sometimes used as a site for growing basic grains while the new shade tree is growing to fill the canopy again, similar to gap phase dynamics in a natural forest. These traditional production systems thus have various features that are reminiscent of a natural forest.

At the other extreme, a highly intensive form of production has been recently promoted, based on modern varieties that respond favourably to light. This form of production has very low planned biodiversity, is heavily dependent on chemical control of weeds, requires high application rates of nitrate, and calls for a great deal of labour for maintenance. Between these two extremes a variety of production forms can be found.

In coffee farms ranging from the traditional to the modern, arthropod biodiversity was sampled, using insecticidal fogging. The results were surprising (Table 19.1). In addition to the data displayed in Table 19.1, the traditional system harbours over 25 species of ground foraging ants (data not shown in Table 19.1). In contrast, the modern system has just eight species, two of which dominate over 95% of the ground area in any given plantation (Perfecto and Vandermeer 1994; Perfecto and Snelling 1995; Perfecto *et al.* 1996). The pattern seems clear. There is a tremendous loss of associated biodiversity as this system is intensified.

Another example comes from an analysis of tree diversity in the managed forests associated with shifting cultivation in West Kalimantan, Indonesia (Lawrence *et al.* 1995; Lawrence 1996). Traditional fallows, fruit gardens and rubber gardens all originate following shifting cultivation of upland rice. They represent a gradient of management intensity. Traditional fallows are relatively unmanaged, fruit gardens are more intensively managed, and rubber gardens are most intensively managed (Table 19.2). In a pattern similar to that observed for arthropods in coffee farms, the number of unplanned, associated tree species is highest in traditional fallows. Thus, intensification leads to a shrinking pool of associated tree species and increasing rarity for those that persist.

Table 19.2 Planned and associated tree diversity along an intensification gradient of fallow management in West Kalimantan, Indonesia. Average values ± standard errors for the number of morphotypes per 1000 m²; total number of stems of planned or associated species in parentheses. Based on trees >10 cm diameter at breast height in 10−11 parcels per forest type (Lawrence, unpublishded data)

Managed forest type	Management regime	Planned tree diversity[a]	Associated tree diversity[b]
Traditional fallow	Some selective weeding, sporadic harvesting	7 ± 1 (28 ± 6)	16 ± 4 (25 ± 5)
Fruit garden	Planting, selective weeding and periodic harvesting	13 ± 1 (29 ± 1)	8 ± 1 (12 ± 2)
Rubber garden	Planting, focused on one commercially valuable species, repeated weeding, and regular harvesting	5 ± 1 (32 ± 4)	1 ± 0 (1 ± 1)

[a] The most conservative definition of planned tree diversity has been applied: any useful species is deemed as planned, despite the fact that some of these are in fact unplanted and untended volunteers, especially in traditional fallows.

[b] Associated tree diversity is defined as those trees for which no use is attributed, usually but not always distinguished by lack of a common name. Unnamed specimens were matched within a plot to provide an estimate of the number of unique morphotypes per plot. It is a plot-level, not a landscape-wide index.

19.3 Consequences of intensification of biodiversity

19.3.1 The competitive production principle

Theoretical ecology has long promoted the idea that if interspecific competition is greater than intraspecific competition two species would not be able to coexist indefinitely, the so-called competitive exclusion principle (Hardin 1960). In fact it has been known for some time (Levins 1979; Armstrong and McGhee 1980) that this simple idea evaporates when one relaxes the deterministic linear assumptions upon which it is based. Nevertheless, it has remained a popular idea, perhaps as important in community ecology as is the Hardy Weinberg law in population genetics (Palmer 1994).

With respect to ecosystem function, the competitive exclusion principle implies that resources will be used more efficiently when more species coexist. A corollary of this hypothesis is that production will be higher when there are more species present. The hypothesis stems from the idea that species A may be competitively superior to species B such that the latter is excluded from the environment, but species A occupies only a fraction (perhaps a large fraction, but nevertheless only a fraction) of the niche space occupied by both species together. Thus, suppose there is R_i resource available in the environment for species C_i, where

the resource utilization efficiency of all species is equal and in monoculture C_i consumes all available R_i. Suppose C_1 eats some fraction of R_2, the alternative resource (call that fraction p_{12}). If p_{12} is sufficiently large, C_2 will not be able to survive in perpetuity (the basic idea of competitive exclusion). With C_1 the only species remaining, the resources are consumed incompletely (only p_{12} of R_2 is consumed), and consequently the production of the consumer trophic level is less than if both C_1 and C_2 were present. Thus the dynamics of competition caused extinction of one of the species, leaving a single species that cannot take full advantage of all resources available. The effect of biodiversity is thus increased production. That increase in production is achieved through so-called 'complementarity', which refers to the fact that p_{ij} is usually less than 1 when i is not equal to j.

This simple idea leads to surprisingly complex questions about organisms in the real world, many of which are summarized in this volume. The central idea is clear. But applying it to the natural world leads to remarkably complicated experimental contingencies (e.g. see the debate concerning complementarity versus the selection/sampling effect as causes of the positive relationship between plant diversity and productivity in recent experiments: Aarssen 1997; Huston 1997; Wardle 1999; Hector *et al.* 2000b; Huston *et al.* 2000). While the normative question referred to above (e.g. it is not clear in the first place why increased production is either good or bad in a non-managed system) is not an issue in managed systems, the experimental contingencies are similar in managed or unmanaged systems. The unmanaged system needs to be analysed from the short term versus long term point of view, where the latter implies the equilibrium state. In managed systems, long term versus short is normally thought of in more practical terms as production versus sustainability. It is thus appropriate to examine the question of competitive production from both the short term and long term point of view.

In the short term, the agronomic literature on intercropping is extensive and usually based on the simple idea of land equivalent ratio (LER) (or equivalently the relative yield total, RYT), where RYT = $(P_1/M_1) + (P_2/M_2)$, where P_i is the production of species i in polyculture and M_i is the production of species i in monoculture. A RYT greater than unity indicates overyielding. However, the conditions under which such a test must be made are stringent and, unfortunately, rarely met. The question is usually posed in a very simple way—does the combination of crop A and B overyield (i.e. is LER > 1.0?). But to ask that question in its most general form requires certain experimental conditions. For examples, if we wish to ask, does the combination of maize and beans overyield in environment X, without further stipulations, we have a rather difficult experimental task. There are two principle issues, monoculture yields and intercrop planting design.

First, the monocultural comparison for computation of LER (or RYT) must be optimal, which usually requires a whole set of experiments in addition to the intercropping trials. That is, of the range of densities possible for each of the two crops, we need to know which density gives the maximum yield, and we need to use that maximum yield in the computation of the LER. Since the monoculture yields are the denominator in the equation for LER (or RYT), it is not difficult to obtain very large LERs by using suboptimal monoculture densities. This problem is illustrated in Fig. 19.2. When plotted on a graph of yield of species 1 versus yield of species 2, if the yield of both species falls above the line connecting the unencumbered yield of each of the

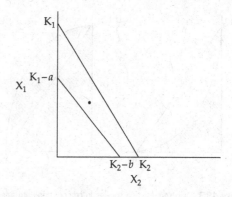

Figure 19.2 Two species in competition, X_1 is the density of the first, X_2 the density of the second.

species (the carrying capacities), overyielding will occur (for the same reason that we expect dynamic coexistence from this condition). In the example in Fig. 19.2, the actual point representing the experimental results falls below the line connecting the two unencumbered yields and thus represents an intercrop that does not overyield. However, if the estimates of the monoculture are $K_1 - a$ and $K_2 - b$ (i.e. they are subestimates of the optimal monoculture yields), this particular intercrop would be judged incorrectly as overyielding.

Second, there are a very large (effectively infinite) number of planting designs possible for the intercrop. Which planting design needs to be used in the computation of LER is not at all obvious. There are two possible solutions to the problem of intercrop planting design, only one of which actually approaches the original question of overyielding. First, one may stipulate a particular planting design. This has usually been done in one of three ways—substitutive design, additive design, or actual design used by a farmer. In all three cases, while there is a clear base on which experiments can be designed, the question being answered is not the original one (Does an intercrop of species A and B overyield) but rather, 'does this particular way of planting species A and B together overyield?' It is certainly an interesting and potentially important question, but not the question originally posed.

The alternative conceptual framework which indeed does approach the original question is through the use of the yield set and adaptive

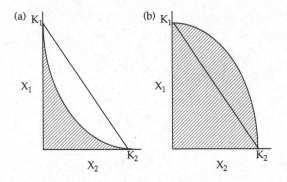

Figure 19.3 Yield sets for two species in competition with one another. Shaded areas represent all possible outcomes from all possible intercropping planting designs.

function (Vandermeer 1989). Here we begin with the simple Lotka-Volterra competition formulation, although in the short term the complications of added stochasticity (Levins 1979) or non-linearities (Armstrong and McGhee 1973; Levins 1979) do not pose the same problem they do when considering the long term. As before we are concerned with whether the intercropping yields fall above the line connecting the two unencumbered yields (and here we presume the unencumbered yields are maximal). But here we are concerned with the set of all possible intercrop designs, which means the set of all possible points on the graph of X_1 versus X_2. The two qualitatively distinct possibilities are illustrated in Fig. 19.3, where the yield sets are the set of points indicated by the shaded regions. In Fig. 19.3(a) is a concave set in which there are no intercropping planting designs that will generate overyielding. In Fig. 19.3(b) is a convex set in which there are many intercropping designs that will generate overyielding, although not all designs will. The practical problem is one of determining the optimal planting design, which depends on the criteria one chooses to use in the optimizing procedure. The details are beyond the scope of this chapter and can be found elsewhere (Vandermeer 1989).

The literature on intercropping is vast and almost always is concerned with comparing a particular intercrop design (substitutive, additive, or actual design from nature) with associated monocultures. Reviews of this literature are not difficult to find (e.g. Kass 1978; Willey 1979; Vandermeer 1989;

Liebman 1995). Unfortunately, it is not always the case that the monoculture comparisons have been properly construed. Thus many of the reported very high LERs are suspect if the experimenters did not assure that the monocultural comparisons were optimal (i.e. from monocultures that give the highest yield). On the other hand, the reverse is also true. Reported low LERs could be due to a particular sub-optimal planting design (Vandermeer 1998).

While the vast majority of studies in the intercropping literature are concerned with only two species, extending the above theoretical framework to the multiple species situation is a trivial exercise. The yield sets of Fig. 19.3 become multidimensional, but the basic arguments are the same. Because of the multidimensional nature of the problem, the criterion for comparison is not simply the best monoculture. True polycultural overyielding implies that there exists an N-species combination that yields more than the best yielding of any $N - i$ (for $i = 0$, $N - 1$) combination of species.

One pair of review articles retains a sort of benchmark significance. Trenbath reviewed a series of mixture experiments with grasses grown in pots, restricting his review to diallel mixtures. From these experiments he concluded that overyielding was at best modest, and possibly could even be explained by some subtle statistical artefacts (Trenbath 1974). That is, there was little evidence for overyielding (or complementarity or competitive production advantage). It would appear that 'a grass is a grass', that there are basic ecological properties associated with being a grass that cause all the grasses in these studies to be effectively in niches so similar that competitive production or complementarity could not occur. Such an idea, competitive neutrality, has been revisited more recently based on fundamental physiological principles (Grime 1979) or associated with particular ecosystems, such as tropical rainforests (Hubbell and Foster 1986; Hubbell et al. 1999; Hubbell 2001). On the other hand, when reviewing the literature on grass legume combinations, a distinct overyielding was the norm, with an average LER greater than 1.0 (Trenbath 1976). The presumption is that the legume either supplied more nitrogen into the system, or there was a niche division with the legume able to tap a source of nitrogen unavailable to the grass (Snaydon and Harris 1979). The experiments generally were not designed to distinguish between these two possibilities, although they are potentially important distinctions as discussed below.

19.3.2 The facilitative effect

There is another way of viewing competitive production. There are many occasions in which one species alters the environment such that it benefits another species. That is, one species 'facilitates' another species. In this case the yield set takes on a distinct qualitative form (see Fig. 19.4), in which some points in the set are larger than the unencumbered yield of one or more of the species. Mutual facilitation (Fig. 19.4(b)) is mutualism. Note that the potential for facilitation actually functioning in a particular situation does not necessarily mean that it in fact does function. For example, in Fig. 19.4(a) the small solid square illustrates an intercropping combination for which X_2 is in fact a facilitator of X_1. But some other intercropping design could produce any other point in the shaded region of the figure (the yield set). Thus, this particular crop combination may or may not actually overyield, and its overyielding may or may not be due to facilitation. For example in Fig. 19.3(b) point 1 illustrates a planting design in which X_2 facilitates X_1, point 2 is from a planting design in which both species are facilitated, point 3 from a planting design which generates overyielding but not due to facilitation, and point 4 from a design that does not

overyield. Note that an LER > 2.0 clearly signifies that some sort of facilitation is going on, but an LER < 2.0 does not necessarily signify that there is no facilitation.

Combining grasses with legumes is frequently thought to be an example of facilitation, with legumes providing extra nitrogen to the system that the grasses then can utilize. However, simple overyielding does not necessarily mean that facilitation actually occurs. Overyielding could result simply from the fact that different nitrogen pools are being utilized, a simple case of competitive production. To demonstrate an actual transfer of nitrogen from the legume to the grass is not a simple task (Vandermeer 1989).

The practice of agroforestry is frequently predicated on the idea that there will be some sort of facilitation involved. Indeed the intellectual organizational base for agroforestry seems to be converging on what it has been for some time in intercropping systems, the balance between competition and facilitation (Ong 1994; Sanchez 1995). Initial enthusiasm for agroforestry seems to have assumed that trees must be bringing something positive to crops or pastures, otherwise farmers would not use them. However, the competitive production principle allows for a net negative effect of trees on the crops yet still provides a rationale for the joint production of trees and crops or pasture, as explained above (see Fig. 19.2(b)). For example, Sanchez surveyed 11 studies in which appropriate treatments were included and found that in only three of the 11 was there evidence that

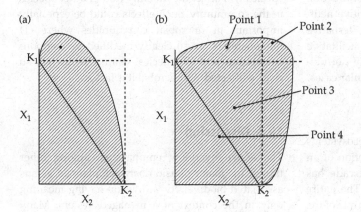

Figure 19.4 Yield sets illustrating the potential for facilitation in a two species system. (a) X_2 potentially facilitates X_1. (b) potentially mutual facilitation (mutualism).

facilitation is greater than competition, suggesting that the combination of trees and crops does not generally provide the production advantage as is sometimes thought. However, Sanchez's analysis was only for the particular planting designs investigated in these 11 cases and can only be thought of as suggestive (Vandermeer 1998). Two crops may have great overyielding potential, but that potential may not be realized because the design is wrong (i.e. there are many points on the yield set that will not overyield even if the potential for doing so is real—Figs 19.3 and 19.4).

In summary, if we are to look for an advantage of polycultures over monocultures in the sense of complementarity (the competitive production principle, or the facilitative effect), the literature in managed systems does not provide much evidence, beyond the effect of legumes growing with non-legumes. The general conclusion seems to be that, although theoretically it seems reasonable to suggest that two ecologically different crops could fit into an environment more efficiently than an equal biomass of the same crop, accumulated evidence offers little support for that idea. It would seem to be equally logical, especially in light of recent theory in plant ecology which suggests that many species may be effectively ecological equivalents (Chesson and Warner 1981; Ågren and Fagerström 1984; Shmida and Ellner 1984; Silvertown and Law 1987; Hubbell 2001), to suggest that different species of annual crops are more or less interchangeable ecologically, and increased production with increased biodiversity not only is not realized empirically, perhaps it should never have been expected in the first place. Maybe the competitive production principle will turn out to be the competitive neutrality principle. Nevertheless, the alternative mechanism for intercrop advantage, the facilitative effect (Vandermeer 1989), while certainly not ubiquitous, seems well-established in particular cases.

19.3.3 Indirect interactions

As recently noted (Polis *et al.* 2000), agroecosystems represent an ecosystem in which the notion of an indirect effect, in the form of a trophic cascade, has been well known for quite some time. The entire endeavour of biological control is an example of the application of an indirect effect. The introduced predator or parasite controls the herbivore (pest) and thus has a positive effect on the plant (crop). The many excellent reviews of biological control (Murdoch *et al.* 1985; Waage and Greathead 1988; Murdoch and Briggs 1996) are thus examples of the function of biodiversity, in this limited context, in managed systems (agroecosystems).

Analyses of indirect interactions have been deepened with the recognition of a distinction between density-mediated and trait-mediated indirect effects (Abrams 1995b; Werner and Anholt 1996). The issue of trait-mediated interactions has mainly addressed animal communities, largely because changes in behavioural traits that can easily be linked to changes in interaction parameters are easily conceptualised. However the effect is most likely far more general, and includes communities with plant species involved as well. For example, the parasitoid *Cotesia glomerata* is less efficient at locating its host, *Pierus brasica* when the cabbage that *P. brassica* is eating is in an intercrop of potato than when it is in a monoculture of cabbage. Thus, the potato has an indirect negative effect on the cabbage by altering a specific trait of the parasitoid that attacks a herbivore of the cabbage (Perfecto and Vet, pers. com.). Even in communities conceptualized as containing only plants the effects of trait-mediated indirect interactions is likely to be important. For example, clonal plants sometimes have a remarkable capacity to change their morphology and spatial arrangement to more efficiently utilise resources (Hutchings and de Kroon 1994), leaving open at least the possibility that the presence of one species will affect the mobility trait of other species in the community. Such effects could be especially important in microbial communities where (1) microhabitats are so clearly modified by component species and (2) species' characteristics are so clearly affected by microhabitat factors.

19.4 Discussion

The above overview emphasized, among other things, the major division between planned versus associated biodiversity, a division having meaning only in the context of a managed system. Many

Table 19.3 Relationship between associated and planned biodiversity in some managed ecosystems

Taxon	Relation to planned biodiversity	References
Vertebrates	Positive	Mellink 1991; Perfecto et al. 1996; Greenberg et al. 1997
Arthropods	Positive	Perfecto et al. 1996, 1997; Perfecto and Vandermeer 1994; Perfecto and Snelling 1997; Hawksworth 1991; Lefeuvre 1992; Crossley et al. 1992; Dennis and Fry 1992; Paoletti and Pimentel 1992; Kim 1994; Burel and Baudry 1995
Non-crop plants	Positive	Van der Maesen 1993; Pimbert and Rajan 1993; Losch et al. 1994; Chacon and Gliessman 1982
Microbes	Positive	Liljeroth et al. 1990; Boehm et al. 1993; Frostegård et al. 1993; Kirchner et al. 1993; Kennedy and Smith 1994; Workneh and ven Bruggen 1994; Zelles et al. 1994; Hawksworth 1991; Olembo and Hawksworth 1990; Anderson et al. 1992; Pimbert and Rajan 1993
	Neutral	Buyer and Kaufman 1996; Doran et al. 1987; Hassink et al. 1991

comparisons of industrial-type agriculture with alternative forms of agriculture have suggested that the alternative forms with their usual incorporation of more planned biodiversity, usually also have higher associated biodiversity (Gall and Orians 1992; Paoletti et al. 1992; Ryan and Starke 1992). It has been established beyond credible doubt that associated biodiversity is positively related to planned species diversity in the case of vertebrates, arthropods, and non-crop plants (Table 19.3). Microbiological biodiversity either is greater in more ecological forms of agriculture or no significant difference is found (Table 19.3). However, it is almost certain that this relationship is non-linear (Fig. 19.1), although the exact form of this nonlinearity remains to be investigated (Pimbert and Rajan 1993; Perfecto et al. 1996, 1997; Swift et al. 1996; Vandermeer 1998). While a certain number of associated species are likely to be incorporated into the system with the addition of each planned species or variety, it is also likely that the planned species combinations will create additional habitats for various associated species (Perfecto and Vandermeer 1996), and possibly change the nature and intensity of interactions among the associated species thus reducing competition or predation coefficients to allow more species coexistence (Perfecto and Vandermeer 1994).

There are two conflicting points of view regarding the function of associated biodiversity. Some

presume that the various species or genotypes perform various functions all of which are necessary for the proper functioning of the ecosystem, and thus a higher biodiversity implies that all of the necessary functions will be served (Wilson 1988; Potter et al. 1990; Hawksworth 1991). An alternative vision is that there is indeed a great amount of redundancy and relatively few functions to be served, and that an efficient system is one in which the 'best' species or genotype dominates in the performance of that function. For example, there is no evidence that two species of biological control agents operate better than a single one (Hawkins 1992), implying that it makes sense then to look for the individual species that performs best. Similarly, Materon et al. (1995) note that native Turkish varieties of Rhizobium leguminosarum generally perform poorly and suggest a programme of active inoculation which is designed to effectively reduce the natural biodiversity of R. leguminosarum. This position is increasingly gaining credibility, especially amongst soil microbiologists (Anderson et al. 1992), although the positive attributes of ecosystem function due to increased biodiversity is still probably the most commonly held position (but see Materon et al. 1995, who suggest a negative function of some biodiversity). Meyer (1994) for example suggests that a large number of different types of bacteria are necessary to break down the various polymers in organic matter.

In contrast to the observation that a single bio-logical control agent is as good as two, Levins (pers. com.) has suggested that in an indirect way the inclusion of a generalist as well as a specialist natural enemy in a biological control system theoretically may be useful in that the generalist may effect control during normal times while the specialist is able to respond quickly to population flushes of the pest. Furthermore, the entire history of the emergence of new pests and resurgence of old ones suggests the need for a variety of potential natural enemies waiting in the wings in anticipation of problems that may not be actualised in a particular moment, but may emerge in the future. Similarly, the various genotypes of R. leguminosarum may perform differently in different environments so that assessments during one or a few years may not encounter that particularly wet (or dry or windy or sunny) year in which one of the poorly performing strains actually performs better. In this sense the biodiversity is seen as a hedge against future changes in environment, be they naturally occurring or human induced—the insurance hypothesis.

The actual functions of associated biodiversity are either well-known or assumed in several cases, including the achievement of greater resource use efficiency, overcoming physical constraints to productivity and sustainability, and the achievement of stable and sustainable resistance to biotic stresses (Spain et al. 1991). The case of biological control implies the existence of a diverse set of natural enemies that can serve this function (Altieri 1991, 1993; Waage 1991; Kogan and Lattin 1993). Other functions of insect biodiversity include pollination (Schwenninger 1992; Pimbert and Rajan 1993; O'Toole and Gauld 1993; Batra 1995), contributions to soil dynamics (Stork and Eggleton 1992) and even as alternative food sources (DeFoliart 1995).

In sum, the arguments, specifically with reference to managed ecosystems are: (1) Biodiversity enhances ecosystem function since different species or genotypes perform slightly different functions (have different niches) and all together function better than some subset. (2) Biodiversity is neutral to negative in that there are many more species than there are ecosystem functions and thus redundancy must be built into the system. (3) Biodiversity

enhances ecosystem function on a long-term basis since those components that appear redundant at one point in time become important when some environmental change happens, which is to say the apparently redundant species have utility as 'buffers' against future change.

The first argument, biodiversity enhances ecosystem function, suggests that society should be concerned with any loss of biodiversity in the ecosystem, since each component has a specific function.

The second argument, in contradistinction, contains an obvious corollary. If one species is likely to be the most efficient at performing the particular ecosystem task, and all the other members of the redundant set are at least slightly less efficient, this suggests that biodiversity should be decreased so as to give preference to the 'best' components. This position has precedence not only in recent technical literature (e.g. Varney et al. (1995) recommend spraying herbicide on conservation headlands to eliminate the highly competitive grasses), but in the literature on traditional agricultural systems (e.g. Chacon and Gliessman (1982)) report that Mayan farmers look at some plant species in their fallow as good components of the fallow and thus deserving of conservation while other components are bad and thus deserving of removal).

The third argument suggests society should have the same concern with biodiversity loss as if the first argument were true. However, it also suggests something more difficult to translate into meaningful practice. Future utility of biodiversity components is difficult to establish beyond speculation. Yet modern ecology has come to recognize the inevitability of surprise in ecosystem dynamics, suggesting that insurance derived from biodiversity, what might be called 'ecosystem buffers', might be the most important function of biodiversity in the end.

Investigations of these hypotheses are really just beginning. The vast majority of literature on biodiversity in managed ecosystems focuses on the elaboration of whatever patterns exists, with little concern for potential biodiversity function. On the other hand, classical work on managed systems has frequently included biodiversity function considerations, well before the debate had crystallized in the formal ecological literature. Intercropping researchers have long studied the mechanisms of overyielding, biocontrol workers have wrestled with the question of classical bio control (which by implication is relatively low biodiversity) as opposed to promoting native natural enemies (which by implication is relatively high biodiversity). And many of the ecological details related to biodiversity function, such as trophic cascades, have long been appreciated in managed systems (e.g. natural enemies 'cascade' their effect to the crop they protect; see Polis 2000).

With this conceptual formulation (first, the notions of planned versus associated biodiversity and second, the three hypotheses of biodiversity function), it is apparent that research agendas may legitimately be focused on several distinct points. For example, little is known about how the planned biodiversity actually creates the associated biodiversity in the first place. Clearly there is a very direct association in that organisms that live only on corn will not live in a bean field and organisms that live only on beans will not live in cornfields, but both sets of organisms very well might live in a corn-bean intercrop. But a more nuanced effect is easily discernible also, in that different degrees of diversity likely imply different degrees of microhabitat availability for organisms, so that a corn-bean intercrop may include organisms that require the specific habitat created by the combination of these two crops (for example the beetle *Diabrotica balteata* eats bean leaves as an adult and corn roots as a larva). There is yet a third level, in that different levels of diversity may imply qualitative modes of interaction of various components of the potential associated biodiversity. For example, as an obligate arboreal nester, the ant *Pheidole punctatissima* is perfectly capable of living in coffee monocultures since it can place its nest in coffee bushes. But the ground-nesting ants that come to dominate the monoculture as distinct from the multiple species traditional system, are competitively dominant and completely eliminate *P. punctatissima* from the modern system. Thus its habitat exists in the less diverse system, but it cannot survive there because of the enhanced competition from other ants that are less able to competitively dominate when the system is more diverse (Perfecto and Vandermeer 1994). Clearly, the question of how the associated biodiversity is determined

by the planned biodiversity is a legitimate and interesting question in its own right, independently of the ultimate function of any of this associated biodiversity.

A second major research agenda and the one most often cited in conjunction with associated biodiversity, is the functional significance of associated biodiversity. The agroecosystem affords an escape from a particular quagmire that unmanaged ecosystem researchers now find themselves—what we referred to above as the normative question. In managed ecosystems, due to their basic anthropogenic nature, function is clear. Biodiversity may increase production (usually a desired goal), it may decrease risk, it may increase sustainability, it may not matter at all for any of these goals, it may matter for some components but not others. Nonetheless, it is clear that function may be clearly defined simply because of the goal-based nature of the managed system.

Yet a different set of questions arise from a consideration of continuing changes in anthropogenic forces. As farmers change their management practices, they may or may not change the planned biodiversity. Clearly as the old multiple cropping systems of native Americans in Mexico are replaced by monocultures of corn, we expect changes in the associated biodiversity also. But the change in corn monocultures in Central America from a high labour/low capital input system to a system involving high capital outlays has a potentially important effect on associated biodiversity. Simply the use of broad-spectrum pesticides implies what may be an enormous loss in biodiversity. Thus, changes in husbandry may force changes in the way planned biodiversity functions, the way it translates into associated biodiversity, and in turn the way the associated biodiversity functions. While a great deal of attention has been accorded the transformation of unmanaged ecosystems into managed ones, sometimes even with the assumption that this transformation is the most critical one for biodiversity loss, we feel that transformation of managed systems from one form to another may involve dramatic changes in biodiversity, the practical consequences of which are not well-understood, not accurately predictable and clearly cause for concern.

PART VI

Synthesis

CHAPTER 20

Perspectives and challenges

M. Loreau, S. Naeem, and P. Inchausti

Few areas of ecology have expanded as fast as biodiversity and ecosystem functioning research during the last few years. Starting from simple intuitive hypotheses, this scientific area has generated a new wave of ambitious experiments using synthesized model ecosystems in both terrestrial and aquatic environments, it has stimulated the emergence of new theoretical approaches linking concepts and perspectives from community ecology and ecosystem ecology, and it has more broadly renewed interest in synthetic approaches in ecology, cutting across increasingly specialized ecological sub-disciplines.

These exciting scientific developments have also been accompanied by an animated debate over the interpretation and implications of recent experiments, as is almost inevitable in any new area (Naeem *et al.*, Chapter 1; Mooney, Chapter 2). This volume, as well as the review paper that we have recently co-authored with the main protagonists in the debate (Loreau *et al.* 2001), are testimony to the fact that such scientific debates can be resolved with three main ingredients: an open mind, an appropriate conceptual and theoretical framework, and strong data. Although the debate is not over, we believe that the work performed during and around the Synthesis Conference does provide the bases for its resolution.

In this concluding chapter, we offer some reflections on where we stand today and how the biodiversity and ecosystem functioning area can profitably be developed in the future.

20.1 Resolving the debate over biodiversity effects in small-scale experiments

The controversy over the interpretation of the biodiversity experiments started with the

realization that their results can be generated by different mechanisms. The mechanisms discussed so far may be grouped into two main classes. First are local deterministic processes, such as niche differentiation and facilitation, which increase the performance of communities above that expected from the performance of individual species grown alone, and which we subsume here under the term 'complementarity' for convenience. Second are local and regional stochastic processes involved in community assembly, which are mimicked in recent experiments by random sampling from a species pool. Random sampling coupled with ecological 'selection' of highly productive species can also lead to increased average primary production with diversity because plots that include many species have a higher probability of containing highly productive species (Huston 1997; Aarssen 1997; Tilman *et al.* 1997b). Two major issues are involved in this controversy: (i) are stochastic community assembly processes relevant? (ii) what is the relative importance of the two classes of mechanisms? A third, related issue concerns the extent to which such mechanisms can be detected in observational studies or in natural systems where extrinsic factors, such as site fertility, disturbance, or climate, covary with biodiversity (see below).

There have been diverging views on the relevance of the sampling component of biodiversity effects. As sampling processes were not an explicit part of the initial hypotheses, they have been viewed by some as 'hidden treatments' (Huston 1997), whereas others have viewed them as the simplest possible mechanism linking diversity and ecosystem functioning (Tilman *et al.* 1997b). Resolving this part of the debate requires increasing knowledge about the patterns and processes of biodiversity loss in nature, which are still poorly known overall. If dominant species control ecosystem processes and mostly rare species go extinct, the vagaries of community assembly or disassembly may have little relevance for the impact of biodiversity loss on natural ecosystems. But environmental changes and landscape fragmentation could prevent recruitment of appropriate dominants (Grime 1998). Also, climate change could lead to gradual losses of species as abiotic conditions begin to exceed species' tolerance limits. Such losses could be random with respect to species effects on any given ecosystem process, leading to patterns of process response to changes in diversity similar to those observed in randomly assembled communities. It should be recalled here that recent experiments were not intended to reproduce any particular sequence of species loss. They were designed to test theoretical hypotheses (Naeem *et al.*, Chapter 1), and thus reflect potential patterns, unaffected by correlations between diversity loss and compositional changes, rather than actual predictions of functional consequences of biodiversity loss under specific global change scenarios.

Recent experiments were also not designed to directly test underlying biological mechanisms. Assessing the relative importance of complementarity and sampling or selection effects has so far been done indirectly, using comparisons between the performances of mixtures and monocultures (Tilman *et al.*, Chapter 3; Hector *et al.*, Chapter 4). Furthermore, it is becoming clear that complementarity and sampling are not mutually exclusive mechanisms as previously thought. Communities with more species have a greater probability of containing a higher phenotypic trait diversity. Ecological 'selection' that leads to dominance of species with particular traits, and complementarity among species with different traits are two ways by which this phenotypic diversity maps onto ecosystem processes. These two mechanisms, however, may be viewed as two poles on a continuum from pure selection to pure complementarity. Intermediate scenarios involve complementarity among particular sets of species or functional groups, or selection of particular subsets of complementary species (Fig. 20.1). Any bias in community assembly that leads to correlations between diversity and community composition may involve both selection and complementarity effects.

Rigorously testing the hypothesis that there is a minimum subset of complementary species that is sufficient to explain diversity effects will often be difficult because it would ideally require testing, with replication, the performance of all species combinations at all diversity levels. Analysis of data from both the Cedar Creek and BIODEPTH

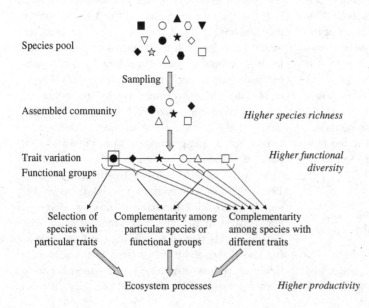

Figure 20.1 Hypothesized mechanisms involved in biodiversity experiments using synthetic communities. Sampling effects are involved in community assembly, such that communities that have more species have a greater probability of containing a higher phenotypic trait diversity. Phenotypic diversity then maps onto ecosystem processes through two main mechanisms: selection of species with particular traits, and complementarity among species with different traits. Intermediate scenarios involve complementarity among particular species or functional groups or, equivalently, selection of particular subsets of complementary species. Modified from Loreau *et al.* (2001).

experiments shows significant effects of species richness on plant biomass even after controlling for the strong effects of certain species, such as legumes (Tilman *et al.*, Chapter 3; Hector *et al.*, Chapter 4). Thus, these results imply that complementarity does occur among at least several species belonging to different functional groups in these experiments. Huston and McBride (Chapter 5) challenge these results, and propose a new hypothesis which combines two-species selection, 'variance reduction' and 'quasi-replication' effects to account for the observed effects. We emphasize, however, that the nature of the debate has changed qualitatively in the process. Previous debate focused on whether single-species selection effects driven by dominant species are sufficient to explain results from biodiversity experiments. Now that the initial hypothesis of single-species selection effects has been rejected in several experiments (Tilman *et al.*, Chapter 3; Hector *et al.*, Chapter 4), the debate is moving to multiple-species selection effects, which, as we discussed above, necessarily involve complementarity. Thus, the borderline between selection and complementarity is vanishing, which, in principle, should facilitate resolution of the debate.

New tests of multiple-species selection effects would be useful to reach the final resolution of this debate. As the many contributions in this volume suggest, however, this debate will no longer be as prominent as it once was. We are reaching a point where it is becoming fruitful to turn to other issues and new experiments. Future experiments should strive to overcome the limitations that led to the recent controversy. Schmid *et al.* (Chapter 6) provide several recommendations and suggestions along this line. Testing biological mechanisms directly and testing more realistic species loss scenarios are two obvious ways to progress in this area.

20.2 Generalizing across ecosystems and trophic levels

Most of the recent experiments that found significant effects of species diversity (Tilman *et al.*, Chapter 3; Hector *et al.*, Chapter 4) have concerned effects of plant diversity on primary production

and nutrient retention in temperate grasslands, both of which are under direct plant control. These experiments have often failed to detect significant effects on below-ground decomposition processes (Naeem *et al.* 1994a; Hector *et al.* 2000a; Knops *et al.* 2001), perhaps because these processes are under microbial control. This questions whether results obtained on primary production in grasslands can be generalized to other processes and ecosystems.

Wardle and van der Putten (Chapter 14) and Mikola *et al.* (Chapter 15) review a large number of studies that addressed effects of species or functional diversity on below-ground soil processes and linkages between above-ground and below-ground processes. The overall conclusion that seems to emerge from this overview is that the complexity of soil biodiversity and the processes they govern defy generality. In contrast, van der Heijden and Cornelissen (Chapter 16) suggest that the diversity of symbiotic mycorrhizal microorganisms is critical for the maintenance of both plant diversity and plant-based ecosystem processes. It would therefore be of interest to revisit the functional role of soil biodiversity using new experiments and new hypotheses based on more solid theory. Of particular importance are the vast areas of biodiversity that involve small organisms such as viruses, bacteria, archaea, protists, microarthropods and nematodes that drive the bulk of ecosystem processes. Modern molecular tools are beginning to make possible the integration of microbial diversity into studies of ecosystem processes (Øvreås 2000).

There have been remarkably few theoretical and experimental investigations into the functional impacts of diversity at higher trophic levels. Yet interesting complex biodiversity effects on ecosystem processes may be expected at these levels. Complementarity and selection effects should tend to improve resource exploitation just as in plants. This should lead to higher secondary productivity if bottom-up control prevails, as in plant–decomposer interactions. But enhanced resource exploitation can also lead to overexploitation, and thus decreased productivity, if top-down control is important, as might be the case with herbivores and predators. Raffaelli *et al.* (Chapter 13) review our current level of knowledge—and ignorance—in this area, and plea eloquently for the establishment of a link between classical food-web theory and modern biodiversity–ecosystem functioning research.

There is also a need to extend our current knowledge to ecosystem types other than temperate grasslands. Aquatic microcosms have proved a particularly valuable tool to test theoretical hypotheses in the biodiversity–ecosystem functioning area (Petchey *et al.*, Chapter 11). But field experimental studies in freshwater and marine ecosystems have been scarce (Emmerson and Huxham, Chapter 12). Forests constitute other important biomes that have been neglected so far. Generalizing across ecosystems is not simply useful for the sake of accumulating knowledge: differences in coexistence mechanisms may lead to differences in biodiversity effects on ecosystem functioning, and thus significant differences may be expected between ecosystem types just as between trophic levels (Bengtsson *et al.*, Chapter 18).

20.3 Scaling-up in space and time

Experiments performed at small spatial and temporal scales are likely to underestimate the functional role of biodiversity because different species have different requirements and typically replace each other along spatial and temporal gradients. Even when high diversity is not critical for maintaining ecosystem processes under constant or benign environmental conditions, it might nevertheless be important for maintaining them under changing conditions. The insurance hypothesis and related hypotheses propose that biodiversity provides a buffer against environmental fluctuations because different species respond differently to these fluctuations, leading to more predictable aggregate community or ecosystem properties (Loreau *et al.*, Chapter 7).

Recent theoretical developments on these issues are relatively solid, and provide new perspectives on the long-standing debate on the relationship between stability and diversity (Loreau *et al.*, Chapter 7; Hughes *et al.*, Chapter 8). The experimental

evidence for these hypotheses, however, is still comparatively thin (Loreau *et al.*, Chapter 7; de Ruiter *et al.*, Chapter 9). An important step forward on this issue would be to design experiments in which long-term environmental variability is directly manipulated to test the potentially important role that environmental fluctuations may play as both the creator and driver of the conditions necessary for the existence of compensatory dynamics among species. Another issue on which greater understanding would be desirable concerns the mechanisms that govern the effects of species diversity on community invasion resistance (Levine *et al.*, Chapter 10). Few theoretical studies have addressed this issue, and explanations are still largely intuitive.

Just as diversity allows functional compensations between species through time, it allows functional compensations through space. The larger the spatial scale, the greater the environmental heterogeneity, and the higher the biological diversity that should be needed to take full advantage of these environmental differences. One of the most potent effects of declining diversity, however, could be the decline in the rate at which appropriate dominant species or combinations of species are recruited during ecosystem assembly. Diversity loss at regional scales and dispersal limitations due to landscape fragmentation may reduce the pool of potential colonists at local scales, and hence the potential for local compositional adjustments to environmental changes. Functional effects of biodiversity changes at landscape to regional scales have been virtually ignored so far, and yet may be particularly important and complex (Bengtsson *et al.*, Chapter 18). These larger scales are all the more critical to investigate since it is at these scales that management and policy decisions usually take place.

20.4 Linking biodiversity dynamics and ecosystem functioning

As one turns to larger spatial and temporal scales, the processes that are responsible for the generation, maintenance and loss of biodiversity are likely to be increasingly important in determining the relationship between diversity and ecosystem functioning. Recent experiments have examined whether diversity alone has a local effect on productivity when all other environmental factors are held constant. This was a historically necessary step to assess and understand the direct causal effect of diversity on ecosystem functioning, which is otherwise hidden by variation in a host of other environmental factors (Lawton *et al.* 1998; Loreau 1998a). Once this causal relationship has been established, we are in a position to return to natural and managed ecosystems, and ask how environmental changes, such as changes in soil, climate and land use, or management regimes affect biodiversity, ecosystem functioning, and the interaction between them. This important step is not only necessary from an applied or management perspective, it also raises new and interesting questions from a basic science perspective. In particular, it allows us to re-establish a much needed link between controlled experiments and natural patterns. Natural patterns of both diversity and ecosystem processes such as productivity and nutrient cycling indicate a variety of relationships between these factors (Bengtsson *et al.*, Chapter 18). What are the respective effects of abiotic factors, such as soil and climate, on diversity, of abiotic factors on productivity, and of diversity on productivity in these patterns? Separation of these effects should become possible, and would allow us to develop a more predictive ecological science. Prediction of future impacts of global change scenarios is another obvious area in which this approach should deliver important results. Experiments have begun to investigate interactions between changes in biodiversity and other components of global change, such as warming, increase in CO_2 concentration and nitrogen deposition (Petchey *et al.* 1999; Reich *et al.* 2001). Understanding the nature of these interactions will be increasingly important in the future.

To understand and predict changes in biodiversity and ecosystem processes at large scales, therefore, we need to move beyond unidirectional causality approaches in which diversity is either cause or effect, and address feedbacks among biodiversity dynamics, ecosystem functioning and environmental factors (Loreau *et al.* 2001). This is a fundamental shift which requires new theoretical, experimental and observational approaches. In recent experimental and theoretical studies, diversity has been manipulated as an independent variable. In reality, biodiversity is itself a dynamic variable, which depends on both extrinsic and intrinsic, abiotic and biotic factors, including the complex interactions which develop among the organisms that make up this biodiversity. Different kinds of interactions between species, and between species and their environment, are known to produce different levels of species diversity, and are equally likely to impose different constraints on the relationship between diversity and ecosystem processes. This simple theoretical observation has far-reaching consequences. Community ecology has been traditionally concerned with the factors that generate and maintain diversity. Merging this perspective with the new perspective on the functional role of biodiversity has the potential to further promote the integration of community ecology and ecosystem ecology, which has already been greatly stimulated by the emergence of the biodiversity–ecosystem functioning area.

In fact, its integration potential goes beyond the borders of ecology, and includes evolutionary biology. Evolutionary constraints have been recognized for a long time in community ecology, but have often been ignored in ecosystem ecology. Yet current ecosystems are the product of past evolution, and ecosystem processes impose constraints on the natural selection of species traits (Loreau 2002). New theoretical and experimental approaches are beginning to tackle the natural selection of ecosystem properties, such as primary productivity, secondary productivity and nutrient cycling efficiency, as the result of evolution of species traits, such as plant and decomposer resource competitive abilities (Loreau 1998b; Swenson *et al.* 2000; de Mazancourt *et al.* 2001; Yamamura *et al.* in press). Incorporating evolutionary constraints and trade-offs responsible for the maintenance of species diversity into such approaches would be an exciting development, which would help bring about a broader synthesis of community, evolutionary and ecosystem ecology. Interestingly, evolutionary concepts, theories, and even equations are already finding striking ecological analogues in the biodiversity–ecosystem functioning area (Loreau and Hector 2001; Norberg *et al.* 2001).

20.5 Conclusion

The biodiversity–ecosystem functioning research programme has been extremely successful in recent years. It has attracted a lot of interest, has made major advances in our understanding of ecological systems, and has pushed towards integration across ecological subdisciplines. The challenges that lie ahead of us, which we have tried to identify in this chapter, are equally exciting from a basic science perspective.

The emergence of this new scientific interface was motivated by a simple but pressing question: Can the current decline in biodiversity alter the functioning and stability of ecosystems and of the Earth system? As evidenced throughout this volume, the answer to this question is 'yes', but there are many intricacies, uncertainties and questioning behind this deceptively simple answer. This answer nevertheless urges adoption of a precautionary approach to biodiversity management and conservation, but we still have some way to go before it can be translated into more concrete, practical recommendations for policy and management decisions from local to global scales.

Despite our current limited ability to deliver quantitative predictions, the social and economic implications of biodiversity–ecosystem functioning research should not be lost from sight. More efforts should be devoted to effects of biodiversity on ecosystem processes that generate 'ecosystem services' of relevance to human societies. The relationship between biodiversity and ecosystem processes and 'services' in managed ecosystems deserves greater attention—a call which seems particularly appropriate since agro-ecology was in several ways a predecessor of the current interest in the productivity of diverse assemblages (Vandermeer *et al.*, Chapter 19). It is perhaps in a management perspective that functional classifications of biodiversity are most important. Specific knowledge of functional types may be critical to predict ecosystem responses under different global change scenarios, or where management priorities seek to manipulate species composition directly, for example in complex agro-ecosystems, forestry, or ecosystem restoration with particular functional goals in mind (Hooper *et al.*, Chapter 17). We have, however, to keep in mind that the predictability of ecological systems is intrinsically limited, and that no functional classification will spare us the need to be prepared to expect the unexpected (Holling 1986).

After nearly a decade of research, it is becoming clear that biodiversity can no longer be regarded as a passive reflection of static abiotic constraints and biotic interactions within communities, but is itself a dynamic component of local and global environmental changes. We are confident that the closer integration of ecological subdisciplines fostered by the biodiversity–ecosystem functioning research programme will help science and society to meet the challenges of human-induced global change.

Bibliography

Aarssen LW (1997). High productivity in grassland eco-systems: affected by species diversity or productive species? *Oikos*, **80**, 183–184.

Aarssen L and Epp GA (1990). Neighbor manipulations in natural vegetation: a review. *Journal of Vegetation Science*, **1**, 13–30.

Aberdeen JEC (1956). Factors influencing the distribution of fungi and plant roots. Part I. Different host species and fungal interactions. *Papers of the Department of Botany of the University of Queensland*, **3**, 113–124.

Abrams PA (1993). Effect of increased productivity on the abundances of trophic levels. *The American Naturalist*, **141**, 351–371.

Abrams PA (1995a). Monotonic or unimodal diversity-productivity gradients: what does competition theory predict? *Ecology*, **76**, 2019–2027.

Abrams PA (1995b). Implications of dynamically variable traits for identifying, classifying, and measuring direct and indirect effecs in ecological communities. *American Naturalist*, **146**, 112–134.

Aerts R (2002). The role of various types of mycorrhizal fungi in nutrient cycling and plant competition. In MGA van der Heijden and IR Sanders, eds *Mycorrhizal ecology*. Ecological Studies 157, Springer Verlag, Heidelberg.

Aerts R and Chapin FS III (2000). The mineral nutrition of wild plants revisited: a re-evaluation of processes and patterns. *Advances in Ecological Research*, **30**, 1–67.

Ågren GI and Fagerström T (1984). Limiting dissimilarity in plants: randomness prevents exclusion of species with similar competitive abilities. *Oikos*, **43**, 369–375.

Aitken M (1995). Comments on J. A. Nelder: The statistics of linear models: back to basics. *Statistics and Computing*, **5**, 85–86.

Allen EB, Allen MF, Helm DJ, Trappe JM, Molina R and Rincon E (1995). Patterns and regulation of mycorrhizal plant and fungal diversity. *Plant and Soil*, **170**, 47–62.

Allen ON and Allen EK (1980). *The leguminosae, A source book of characteristics, uses, and nodulation*. The University of Wisconsin Press, Madison.

Allen-Morley CR and Coleman DC (1989). Resilience of soil biota in various food webs to freezing perturbations. *Ecology*, **70**, 1127–1141.

Aller R (1983). The importance of the diffusive permeability of the animal burrow linings in determining marine sediment chemistry. *Journal of Marine Research*, **41**, 299–322.

Aller RC and Yingst JY (1978). Biogeochemistry of tube dwellings: A study of the sedentary polycheate *Amphitrite ornata* (Leidy). *Journal of Marine Research*, **36**, 201–254.

Aller RC and Yingst JY (1985). Effects of the marine deposit-feeders *Heteromastus filiformis* (Polycheata), *Macoma balthica* (Bivalvia), and *Tellina texana* (Bivalvia) on averaged sedimentary solute transport, reaction rates, and microbial distributions. *Journal of Marine Research*, **43**, 615–645.

Aller RC, Yingst JY and Ullman WJ (1983). Comparative biogeochemistry of water in intertidal *Onuphis* (polychaeta) and *Upogebia* (crustacea) burrows: temporal patterns and causes. *Journal of Marine Research*, **41**, 571–604.

Allison GW (1999). The implications of experimental design for biodiversity manipulations. *American Naturalist*, **153**, 26–45.

Al-Mufti MM, Sydes CL, Furness SB, Grime JP and Band SR (1977). A quantitative analysis of shoot phenology and dominance in herbaceous vegetation. *Journal of Ecology*, **65**, 759–791.

Alphei J, Bonkowski M and Scheu S (1996). Protozoa, Nematoda and Lumbricidae in the rhizosphere of *Hordelymus europaeus* (Poaceae): faunal interactions, response of microorganisms and effects on plant growth. *Oecologia*, **106**, 111–126.

Altierri MA (1983). The question of small development: who teaches whom? *Agriculture Ecosystems and Environment*, **9**, 401–405.

Altierri MA (1990). Why study traditional agriculture? In CR Carroll, JH Vandermeer and P Rosset, eds *Agroecology*, pp. 551–564. McGraw-Hil, New York.

Altierri MA (1991). Traditional farming in Latin America. *The Ecologist*, **21**, 93–96.

Altierri MA (1993). Ethnoscience and biodiversity: key elements in the design of sustainable pest management systems for small farmers in developing countries. *Agriculture, Ecosystems, and Environment*, **46**, 257–272.

Altierri MA (1995). *Agroecology: Creating the synergism for a sustainable agriculture.* UNDP Guidbook Series.

Altierri MA and Letourneau DK (1984). Vegetation diversity and insect pest outbreaks. *CRC Critical Reviews in Plant Sciences*, **2**, 131–169.

Andersen T (1997). *Pelagic nutrient cycles: herbivores as sources and sinks.* Springer-Verlag, New York.

Andersen FØ and Kristensen E (1988). The influence of macrofauna on estuarine benthic community metabolism: a microcosm study. *Journal of Marine Research*, **99**, 591–603.

Anderson JM (1978). Inter- and intra-habitat relationships between woodland Cryptostigmata species diversity and the diversity of soil and litter microhabitats. *Oecologia*, **32**, 341–348.

Anderson JM (1988). Spatiotemporal effects of invertebrates on soil processes. *Biology and Fertility of Soils*, **6**, 216–227.

Anderson J (1995). Soil organisms as engineers: Microsite modulation of macroscale processes. In CG Jones and JH Lawton, eds *Linking species and ecosystems*, pp. 94–106. Chapman and Hall, New York.

Anderson JM, Greenland DJ and Szabolcs I (1992). Functional attributes of biodiversity in land use systems. In DJ Greenland, ed. *Soil resilience and sustainable land use*. Proceedings of a symposium held in Budapest, 28 September–2 October 1992, including the Second Workshop on the Ecological foundations of Sustainable Agriculture (WEFSA II). 2670290.

Anderson RV, Coleman DC, Cole CV and Elliott ET (1981). Effect of the nematodes *Acrobeloides* sp. and *Meso-diplogaster lheritieri* on substrate utilization and nitrogen and phosphorus mineralization in soil. *Ecology*, **62**, 549–555.

Andow DA (1991). Vegetational diversity and arthropod population response. *Annual Review of Entomology*, **36**, 561–586.

André M, Bréchignac F and Thibault P (1994). Biodiversity in model ecosystems. *Nature*, **371**, 565.

Andrén O et al. (1990). Organic carbon and nitrogen flows. *Ecological Bulletin*, **40**, 85–125.

Andrén O, Bengtsson J and Clarholm M (1995). Biodiversity and species redundancy among litter decomposers. In HP Collins, GP Robertson and MJ Klug, eds *The significance and regulation of soil biodiversity*, pp. 141–151. Dordrecht, Kluwer.

Andrén O, Brussaard L and Clarholm M (1999) Soil organism influence on ecosystem-level processes – bypassing the ecological hierarchy? *Applied Soil Ecology* **11**, 177–188.

Aoki I and Mizushima T (2001). Biomass diversity and stability of food webs in aquatic ecosystems. *Ecological Research*, **16**, 65–71.

Arditi R and Ginzburg LR (1989). Coupling in predator prey dynamics: Ratio dependence. *Journal of Theoretical Biology*, **139**, 311–326.

Armstrong RA and McGehee R (1980). Competitive exclusion. *American Naturalist*, **115**, 151–170.

Armsworthy SL, MacDonald BA and Ward J (2001). Feeding activity, absorption efficiency and suspension feeding processes in the ascidian, *Halocynthia pyriformis* (Stolidobranchia: Ascidiacea): responses to variations in diet quantity and quality. *Journal of Experimental Marine Biology and Ecology*, **260**, 41–69.

Asmus H and Asmus R (1985). The importance of grazing food chain for energy flow and production in three intertidal sand bottom communities of the northern Wadden Sea. *Helgolander Meeresuntersuchungen*, **39**, 273–301.

Asmus RM (1986). Nutrient flux in short term enclosures of intertidal sand communities. *Ophelia*, **26**, 1–18.

Asmus RM et al. (1998). The role of water movement and spatial scaling for measurement of dissolved inorganic nitrogen fluxes in intertidal sediments. *Estuarine Coastal and Shelf Science*, **46**, 221–232.

Asmus RM, Sprung M and Asmus H (2000). Nutrient fluxes in intertidal communities of a South European lagoon (Ria Formosa)—similarities and differences with a northern Wadden Sea bay (Sylt-Rømø Bay). *Hydrobiologia*, **436**, 217–235.

Augspurger CK and Kelly CK (1984). Pathogen mortality of tropical tree seedlings: experimental studies of the effects of dispersal distance, seedling density, and light conditions. *Oecologia*, **61**, 211–217.

Austin MP (1982). Use of relative physiological performance value in the prediction of performance in multispecies mixtures from monoculture performance. *Journal of Ecology*, **70**, 559–570.

Austin MP, Fresco LFM, Nicholls AP, Groves RH and Kaye PE (1988). Competition and relative yield: estimation and interpretation at different densities and under various nutrient concentrations using *Silybum marianum* and *Cirsium vulgare*. *Journal of Ecology*, **76**, 157.

Ayensu E et al. (1999). Ecology: International ecosystem assessment. *Science*, **286**, 685–686.

Bååth E et al. (1978). The effect of nitrogen and carbon supply on the development of soil organism populations and pine seedlings: amicrocosm approach. *Oikos*, **31**, 153–163.

Bagwell CE and Lovell CR (2000). Microdiversity of culturable diazotrophs from the rhizoplanes of the salt marsh grasses *Spartina alterniflora* and *Juncus roemerianus*. *Microbial Ecology*, **39**, 128–136.

Bakker JP and Berendse F (1999). Constrains in the restoration of ecological diversity in grassland and heathland communities. *Trends in Ecology & Evolution*, **14**, 63–68.

Bardgett RD and Chan KF (1999). Experimental evidence that soil fauna enhance nutrient mineralization and plant nutrient uptake in montane grassland ecosystems. *Soil Biology and Biochemistry*, **31**, 1007–1014.

Bardgett RD and Cook R (1998). Functional aspects of soil animal diversity in agricultural grasslands. *Applied Soil Ecology*, **10**, 263–276.

Bardgett RD and Shine A (1999). Linkages between plant litter diversity, soil microbial biomass and exosystem function in temperate grasslands. *Soil Biology and Biochemistry*, **31**, 317–321.

Bardgett RD, Leemans DK, Cook R and Hobbs PJ (1997). Seasonality of the soil biota of grazed and ungrazed hill grasslands. *Soil Biology and Biochemistry*, **29**, 1285–1294.

Bardgett RD, Wardle DA and Yeates GW (1998). Linking above-ground and below-ground interactions: how plant responses to foliar herbivory influence soil organisms. *Soil Biology and Biochemistry*, **30**, 1867–1878.

Bardgett RD, Lovell RD, Hobbs PJ and Jarvis SC (1999). Dynamics of below-ground microbial communities in temperate grasslands: influence of management intensity. *Soil Biology and Biochemistry*, **31**, 1021–1030.

Bardgett RD *et al.* (2001). The influence of soil biodiversity on hydrological pathways and the transfer of materials between terrestrial and aquatic ecosystems. *Ecosystems*, **4**, 421–429.

Bascompte J and Rodríguez MA (2001). Habitat patchiness and plant species richness. *Ecology Letters*, **5**, 417–420.

Baskin Y (1997). *The work of nature*. Island Press, Washington DC.

Batra SWT (1995). Bees and pollination in our changing environment. *Apidologie*, **26**, 361–370.

Bauer T (1982). Predation by a carabid beetle specialized for catching Collembola. *Pedobiologia*, **24**, 169–179.

Baxter JW and Dighton J (2001). Ectomycorrhizal diversity alters growth and nutrient acquisition of grey birch (Betula populifolia) seedlings in host-symbiont culture conditions. *New Phytologist*, **152**, 139–149.

Bazzaz FA (1987). Experimental studies on the evolution of niche in successional plant populations. In A Gray,

M Crawley and P Edwards, eds *Colonization, succession and stability*, pp. 245–272. Blackwell Scientific, Oxford.

Bazzaz FA (1996). *Plants in changing environments*. Cambridge University Press, Cambridge.

Bazzaz FA *et al.* (1999). Ecological science and the human predicament. *Science*, **282**, 879.

Beare MH, Neely CL, Coleman DC and Hargrove WL (1991). Characterization of a substrate induced respiration method for measuring fungal, bacterial and total microbial biomass on plant residues. *Agriculture, Ecosystems and Environment*, **34**, 65–73.

Beare MH, Parmelee RW, Hendrix PF, Cheng W, Coleman DC and Crossley DAJ (1992). Microbial and faunal interactions and effects on litter nitrogen and decomposition in agroecosystems. *Ecological Monographs*, **62**, 569–591.

Beare MH, Coleman DC, Crossley DA Jr, Hendrix PF and Odum EP (1995). A hierarchical approach to evaluating the significance of soil biodiversity to biogeochemical cycling. *Plant and Soil*, **170**, 5–22.

Behan-Pelletier V and Newton G (1999). Linking soil biodiversity and ecosystem function—the taxonomic dilemma. *BioScience*, **49**, 149–153.

Bell G (1990). The ecology and genetics of fitness in *Chlamydomonas*. 2. The properties of mixtures of strains. *Proceedings of the Royal Society of London, Series B Biological Sciences*, **240**, 323–350.

Benedetti-Cecchi L (2000). Variance in ecological consumer–resource interactions. *Nature*, **407**, 370–374.

Bengtsson J (1998). Which species? What kind of diversity? Which ecosystem function? Some problems in studies of relations between biodiversity and ecosystem function. *Applied Soil Ecology*, **10**, 191–199.

Bengtsson J (2000). Review of 'ecological scale: Theory and applications' by DL Peterson and VT Parker, eds *Ecological Economics*, **29**, 333–334.

Bengtsson J, Setälä H and Zheng DW (1996). Food webs and nutrient cycling in soils: Interactions and positive feedbacks. In GA Polis and KO Winemiller, eds *Food webs: Integration of patterns and dynamics*, pp. 30–38. Chapman and Hall, New York.

Bengtsson J *et al.* (in press). Reserves, resilience and dynamic landscapes. *Ambio*.

Berendse F (1994a). Competition between plant populations at low and high nutrient supplies. *Oikos*, **71**, 253–260.

Berendse F (1994b). Litter decomposability—a neglected component of plant fitness. *Journal of Ecology*, **82**, 87–190.

Berendse F, Aerts R, Bisby FA and Bobbink R (1993). Atmospheric nitrogen deposition and its impact on terrestrial ecosystems. In CC Vos and P Opdam, eds

Landscape ecology of a stressed environment, pp. 104–121. Chapman and Hall, London.

Berish CW and Ewel JJ (1988). Root development in simple and complex tropical successional ecosystems. *Plant and soil*, **106**, 73–84.

Berry EC and Karlen DL (1993). Comparison of alternative farming systems. II. Earthworm population density and species diversity. *American Journal of Alternative Agriculture*, **8**, 21–26.

Bertolo A, Lacroix G and Lescher-Moutoué F (1999). Scaling food chains in aquatic mesocosms: do the effects of depth override the effects of planktivory? *Oecologia*, **121**, 55–65.

Beukema JJ (1989). Long-term changes in macrozoobenthic abundance on the tidal flats of the western part of the Dutch Wadden Sea. *Helgolander Meeresuntersuchungen*, **43**, 405–415.

Bever JD (1984). Feedback between plants and their soil communities in an old field community. *Ecology*, **75**, 1965–1977.

Bever JD, Morton JB, Antonovics J and Schulz PA (1996). Host-dependent sporulation and species diversity of arbuscular mycorrhizal fungi in mown grassland. *Journal of Ecology*, **84**, 71–82.

Bever JD, Westover KM and Antonovics J (1997). Incorporation of the soil community into plant population dynamics: the utility of the feedback approach. *Journal of Ecology*, **85**, 561–573.

Bezemer TM and Jones TH (1998). Plant-insect herbivore interactions in elevated atmospheric CO_2: quantitative analyses and guild effects. *Oikos*, **82**, 212–222.

Bingham C and Fienberg SE (1982). Textbook analysis of covariance—is it correct? *Biometrics*, **38**, 747–753.

Bisby FA and Coddington J (1995). Biodiversity from a taxonomic and evolutionary perspective. In VH Heywood and RT Watson, eds *Global biodiversity assessment*, pp. 27–57. Cambridge University Press, Cambridge.

Björklund J, Limburg KE and Rydberg T (1999). Impact of production intensity on the ability of the agricultural landscape to generate ecosystem services: and example from Sweden. *Ecological Economics*, **29**, 269–291.

Blair JM, Parmelee RW and Beare MH (1990). Decay rates, nitrogen fluxed and decomposer communities in single and mixed species foliar litter. *Ecology*, **71**, 1976–1985.

Blakemore RJ (1997). Agronomic potential of earthworms in brigalow soils of South-East Queensland. *Soil Biology and Biochemistry*, **29**, 603–608.

Blanchart E, Lavelle P, Braudeau E, Le Bissonnais Y and Valentin C (1997). Regulation of soil structure by geophagous earthworm activities in humid savannas of Côte d'Ivoire. *Soil Biology and Biochemistry*, **29**, 431–439.

Bloem J *et al.* (1994). Dynamics of microorganisms, microbivores and nitrogen mineralisation in winter wheat fields under conventional and integrated management. *Agriculture Ecosystems and Environment*, **51**, 129–143.

Bloemers GF, Hodda M, Lambshead PJD, Lawton JH and Wanless FR (1997). The effects of forest disturbance on diversity of tropical soil nematodes. *Oecologia*, **111**, 575–582.

Boag B and Yeates GW (1998). Soil nematode biodiversity in terrestrial ecosystems. *Biodiversity and Conservation*, **7**, 617–630.

Boddey RM *et al.* (1995). Biological nitrogen-fixation associated with sugar-cane and rice—contributions and prospects for improvement. *Plant and Soil*, **174**, 195–209.

Boddey RM, Peoples MB, Palmer B and Dart PJ (2000). Use of the N-15 natural abundance technique to quantify biological nitrogen fixation by woody perennials. *Nutrient cycling in agroecosystems*, **57**, 235–270.

Boehm MJ, Madden LV and Hoitink HAJ (1993). Effect of organic matter decomposition level on bacterial species diversity and composition in relationship to Pythium damping off severity. *Applied and Environmental Microbiology*, **59**, 4171–4179.

Bohannan BJM and Lenski RE (1999). Effect of prey heterogeneity on the response of a model food chain to resource enrichment. *The American Naturalist*, **153**, 73–82.

Bongers T (1990). The maturity index: an ecological measure of environmental disturbance based on nematode species composition. *Oecologia*, **83**, 14–19.

Bongers T and Bongers M (1998). Functional diversity of nematodes. *Applied Soil Ecology*, **10**, 239–251.

Bonkowski M, Cheng W, Griffiths BS, Alphei J and Scheu S (2000a). Microbial–faunal interactions in the rhizosphere and effects on plant growth. *European Journal of Soil Biology*, **36**, 135–147.

Bonkowski M, Griffiths B and Scrimgeour C (2000b). Substrate heterogeneity and microfauna in soil organic 'hotspots' as determinants of nitrogen capture and growth of ryegrass. *Applied Soil Ecology*, **14**, 37–53.

Borrvall C, Ebenman B and Jonsson T (2000). Biodiversity lessens the risk of cascading extinction in model food webs. *Ecology Letters*, **3**, 131–136.

Both C and Visser ME (2001). Adjustment to climate change is constrained by arrival date in a long-distance migrant bird. *Nature*, **411**, 296–298.

Bouwman LA *et al.* (1994). Short-term and long-term effects of bacterivorous nematodes and nematophagous fungi on carbon and nitrogen mineralization in microcosms. *Biology and Fertility of Soils*, **17**, 249–256.

Bowers MA (1993). Influence of herbivorous mammals on an old-field plant community: years 1–4 after disturbance. *Oikos*, **67**, 129–141.

Box EO (1996). Plant functional types and climate at the global scale. *Journal of Vegetation Science*, **7**, 309–320.

Brand RH and Dunn CP (1998). Diversity and abundance of springtails (Insecta: Collembola) in native and restored tallgrass prairies. *American Midland Naturalist*, **139**, 235–242.

Briand F and McCauley E (1978). Cybernetic mechanisms in lake plankton systems: how to control undesirable algae. *Nature*, **273**, 228–230.

Briggs CJ and Godfray HCJ (1995). Models of intermediate complexity in insect–pathogen interactions: Population dynamics of the microsporidian pathogen, *Nosema pyrausta*, of the European corn borer, *Ostrinia nubilalis*. *Parasitology*, **111**, S71–S89.

Bronmark C, Klosiewski SP and Stein RA (1991). Indirect effects of predation in a freshwater benthic food chain. *Ecology*, **73**, 1662–1674.

Bronmark C, Dhal J and Greenberg LA (1997). Complex trophic interactions in freshwater benthic food chains. In B Streit, T Stadler and CM Lively, eds *Evolutionary ecology of freshwater animals*, pp. 55–88. Birkhauser-Verlag, Basel, Switzerland.

Brooks JL and Dodson SI (1965). Predation, body size, and composition of plankton. *Science*, **150**, 28–35.

Broughton LC, Gross KL and Hector A (submitted). Linking plant community diversity to soil microbial communities: An experimental evaluation from the BIODEPTH experiment. *Journal of Ecology*.

Brown BJ and Ewel JJ (1987). Herbivory in complex and simple tropical successional ecosystems. *Ecology*, **68**, 108–116.

Brown GG (1995). How do earthworms affect microfloral and faunal community diversity? *Plant and Soil*, **170**, 209–231.

Brown JH (1999). The desert granivory experiments at Portal. In J Resetartis and W Bernardo, eds *Issues and perspectives in experimental ecology*. Oxford University Press, Oxford.

Brown VK and Gange AC (1989a). Differential effects of above- and below-ground insect herbivory during early plant succession. *Oikos*, **54**, 67–76.

Brown VK and Gange AC (1989b). Herbivory by soil-dwelling insects depresses plant species richness. *Functional Ecology*, **3**, 667–671.

Brown VK and Gange AC (1990). Insect herbivory below-ground. *Advances in Ecological Research*, **20**, 1–58.

Brown VK and Gange AC (1992). Secondary plant succession—how is it modified by insect herbivory? *Vegetatio*, **101**, 3–13.

Brussaard L *et al.* (1988). The Dutch Programme on Soil Ecology of Arable Farming Systems. I. Objectives, approach and preliminary results. *Ecological Bulletin*, **39**, 35–40.

Brussaard L *et al.* (1990). Biomass, composition and temporal dynamics of soil organisms of a silt loam soil under conventional and integrated management. *Netherlands Journal of Agricultural Science*, **38**, 283–302.

Brussaard L, Bakker JP and Olff H (1996). Biodiversity of soil biota and plants in abandoned arable fields and grasslands under restoration management. *Biodiversity and Conservation*, **5**, 211–221.

Brussaard L *et al.* (1997). Biodiversity and ecosystem functioning in soil. *Ambio*, **26**, 563–570.

Buchman S and Nabhan GP (1996). *The forgotten pollinators*. Island Press, Washington.

Buckland SM, Grime JP, Hogson JG and Thompson KJE (1997). A comparison of plant responses to the extreme drought of 1995 in northern England. *Journal of Ecology*, **85**, 875–882.

Buckling A, Kassen R, Bell G and Rainey PB (2000). Disturbance and diversity in experimental microcosms. *Nature*, **408**, 961–964.

Bullock JM, Pywell RF, Burke MJW and Walker KJ (2001). Restoration of biodiversity enhances agricultural production. *Ecology Letters*, **4**, 185–189.

Burel F and Baudry J (1995). Species biodiversity in changing agricultural landscapes: A case study in the Pays d'Auge, France. *Agriculture, Ecosystems and Environment*, **55**, 3–200.

Burnham KP and Anderson DR (1998). *Model selection and inference: a practical information-theoretic approach*. Springer-Verlag, New York.

Burrows RL and Pfleger FL (2002). Arbuscular-mycorrhizal fungi respond to increasing plant diversity. *Canadian Journal of Botany*, **80**, 120–130.

Butcher SS, Charlson RJ, Orians GH and Wolfe GV (1992). *Global biogeochemical cycles*. Academic Press, London.

Buyer JS and Kaufman DD (1996). Microbial diversity in the rhizosphere of corn grown under conventional and low-input systems. *Applied Soil Ecology*, **5**, 21–27.

Byers RJ and Odum HT (1993). *Ecological microcosms*. Springer-Verlag, New York.

Caldeira MC, Ryel RJ, Lawton JH and Pereira JS (2001). Mechanisms of positive biodiversity–production relationships: insights provided by ^{13}C analysis in experimental Portuguese grassland plots. *Ecology Letters*, **4**, 439–443.

Carpenter SR (1996). Microcosm experiments have limited relevance for community and ecosystem ecology. *Ecology*, **77**, 677–680.

Carpenter SR and Kitchell JF, eds (1993). *The trophic cascade in lakes*. Cambridge University Press, Cambridge.

Carpenter SR *et al.* (2001). Trophic cascades, nutrients and lake productivity: whole lake experiments. *Ecological Monographs*, **71**, 163–187.

Case TJ (1990). Invasion resistance arises in strongly interacting species-rich model competition communities. *Proceedings of the National Academy of Sciences*, **87**, 9610–9614.

Case TJ (1991). Invasion resistance, species build-up and community collapse in metapopulation models with interspecies competition. *Biological Journal of the Linnean Society*, **42**, 239–266.

Cebrian J (1999). Patterns in the fate of production in plant communities. *American Naturalist*, **154**, 449–468.

Chacon JC and Gliessman SR (1982). Use of the 'non-weed' concept in traditional tropical agroecosystems of southeastern Mexico. *Agroecosystems*, **8**, 1–11.

Chan KY (2001). An overview of some tillage impacts on earthworm population abundance and diversity—implications for functioning in soils. *Soil and Tillage Research*, **57**, 179–191.

Chapin FS III (1980). The mineral nutrition of wild plants. *Annual Review of Ecology and Systematics*, **11**, 233–260.

Chapin FS III, Vitousek PM and Van Cleve K (1986). The nature of nutrient limitation in plant communities. *American Naturalist*, **127**, 48–58.

Chapin FS III, Autumn K and Pugnaire F (1993). Evolution of suites of traits in response to environmental stress. *The American Naturalist*, **142**, S78–S92.

Chapin FS III, Bret-Harte S, Hobbie S and Zhong H (1996a). Plant functional types as predictors of transient responses of arctic vegetation to global change. *Journal of Vegetation Science*, **7**, 347–358.

Chapin FS III, Reynolds H, D'Antonio C and Eckhart V (1996b). The functional role of species in terrestrial ecosystems. In B Walker and W Steffen, eds *Global change in terrestrial ecosystems*, pp. 403–428. Cambridge University Press, Cambridge.

Chapin FS III *et al.* (1997). Biotic control over the functioning of ecosystems. *Science*, **277**, 500–504.

Chapin FS III *et al.* (1998). Ecosystem consequences of changing biodiversity. *Bioscience*, **48**, 45–51.

Chapin FS III *et al.* (2000). Consequences of changing biodiversity. *Nature*, **405**, 234–242.

Chapman K, Whittaker JB and Heal OW (1988). Metabolic and faunal activity in litters of tree mixtures compared with pure stands. *Agriculture, Ecosystems and Environment*, **24(1–3)**, 33–40.

Chase JM (1999). Food web effects of prey size-refugia: Variable interactions and alternative stable equilibria. *American Naturalist*, **154**, 559–570.

Chen J and Ferris H (2000). Growth and nitrogen mineralization of selected fungi and fungal-feeding nematodes on sand amended with organic matter. *Plant and Soil*, **218**, 91–101.

Chen B and Wise DH (1999). Bottom-up limitation of predaceous arthropods in a detritus-based terrestrial food web. *Ecology*, **80**, 761–772.

Chesson PL and Warner RR (1981). Environmental variability promotes coexistence in lottery competitive systems. *American Naturalist*, **117**, 923–943.

Christensen B, Vedel A and Kristensen E (2000). Carbon and nitrogen fluxes in sediment inhabited by suspension feeding (*Nereis diversicolor*) and non-suspension feeding (*N. virens*) polycheates. *Marine Ecology-Progress Series*, **192**, 203–217.

Christie P, Newman EI and Campbell R (1974). Grassland plant species can influence the abundance of microbes on each other's roots. *Nature*, **250**, 570–571.

Christie P, Newman EI and Campbell R (1978). The influence of neighbouring grassland plants on each other's endomicorrhizas and root-surface microorganisms. *Soil Biology and Biochemistry*, **10**, 521–527.

Clapp JP, Helgason T, Daniell TJ and Young JPW (2002). Genetic studies of the structure and diversity of arbuscular mycorrhizal fungal communities. In MGA van der Heijden and IR Sanders, eds *Mycorrhizal ecology*. Ecological Studies 157. Springer-Verlag, Heidelberg.

Clarholm M (1985). Interactions of bacteria, protozoa and plants leading to mineralization of soil nitrogen. *Soil Biology and Biochemistry*, **17**, 181–187.

Clarke KR and Warwick RM (1998). A taxonomic distinctness index and its statistical properties. *Journal of Applied Ecology*, **35**, 523–531.

Clawson ML and Benson DR (1999). Natural diversity of Frankia strains in actinorhizal root nodules from promiscuous hosts in the family Myricaceae. *Applied and Environmental Microbiology*, **65**, 4521–4527.

Clay K and Holah J (1999). Fungal endophyte symbiosis and plant diversity in successional fields. *Science*, **285**, 1742–1744.

Clegg CD, Ritz K and Griffiths BS (1998). Broad-scale analysis of soil microbial community DNA from upland grasslands. *Antonie van Leeuwenhoek*, **73**, 9–14.

Clements FE (1916). *Plant succession*. Carnegie Institution, Washington.

Cloern JE (1996). Phytoplankton bloom dynamics in coastal ecosystems. A review with some general lessons from sustained investigation of San Francisco Bay, California. *Reviews of Geophysics*, **34**, 127–168.

Cohen JE (1978). *Food webs in niche space*. Princeton University Press, Princeton.

Cohen JE and Newman CM (1985). A stochastic theory of community food web. I. Models and aggregated data. *Proceedings of the Royal Society of London, Series B*, **224**, 421–448.

Cohen JE *et al.* (1990). *Community food webs: data and theory*. Springer-Verlag, New York.

Cohen AE, Gonzalez A, Lawton JH, Petchey OL, Wildman D and Cohen JE (1998). A novel experimental apparatus to study the impact of white and 1/f noise on animal populations. *Proceedings of the Royal Society of London, Series B*, **265**, 11–15.

Cole JJ (1982). Interactions between bacteria and algae in aquatic ecosystems. *Annual Review of Ecology and Systematics*, **13**, 291–314.

Cole L, Bardgett RD and Ineson P (2000). Enchytraeid worms (Oligochaeta) enhance mineralization of carbon in organic upland soils. *European Journal of Soil Science*, **51**, 185–192.

Coleman DC, Anderson RV, Cole CV, Elliott ET, Woods LE and Campion MK (1978). Trophic interactions in soils as they affect energy and nutrient dynamics. IV. Flows of metabolic and biomass carbon. *Microbial Ecology*, **4**, 373–380.

Coleman DC, Reid CPP and Cole CV (1983). Biological strategies of nutrient cycling in soil ecosystems. *Advances in Ecological Research*, **13**, 1–55.

Coley PD (1983). Herbivory and defensive characteristics of tree species in a lowland tropical forest. *Ecological Monographs*, **53**, 209–233.

Coley PD (1988). Effects of plant growth rate and leaf lifetime on the amount and type of antiherbivore defense. *Oecologia*, **74**, 531–536.

Collins WW and Qualset CO (1999). *Biodiversity in agroecosystems*. CRC Press, Boca Raton, Florida.

Coma R, Josep-Maria J and Rikel Z (1995). Trophic ecology of a benthic marine hydroid, *Campanularia everta*. *Marine Ecology-Progress Series*, **119**, 211–220.

Connell JH (1975). Some mechanisms producing structure in natural communities: a model and evidence from field experiments. In ML Cody and J Diamond, eds *Ecology and evolution of communities*. pp. 460–490, Belknap, Cambridge.

Connell JH (1978). Diversity in tropical rainforests and coral reefs. *Science*, **199**, 1302–1310.

Connolly JH, Wayne P and Bazzaz FA (2001). Interspecific competition in plants: How well do current methods answer fundamental questions? *The American Naturalist*, **157(2)**, 107–125.

Cornelissen JHC, Castro-Diez P and Carnelli AL (1998). Variation in relative growth rate among woody species. In H Lambers, H Poorter and MMI van Vuuren, eds *Inherent variation in plant growth: physiological mechanisms and ecological consequences*, pp. 363–392. Backhuys Publications, Leiden.

Cornelissen JHC *et al.* (1999). Leaf structure and defense control litter decomposition rate across species, life forms and continents. *New Phytologist*, **143**, 191–200.

Cornelissen JHC, Aerts R, Cerabolini B, Werger MJA and van der Heijden MGA (2001). Carbon cycling traits of plant species are linked with mycorrhizal strategy. *Oecologia*, **129**, 611–619.

Cottingham KL and Schindler DE (2000). Effects of grazer community structure on phytoplankton response to nutrients pulses. *Ecology*, **81**, 183–200.

Cottingham KL, Rusak JA and Leavitt PR (2000). Increased ecosystem variability and reduced predictability following fertilisation: Evidence from paleolimnology. *Ecology Letters*, **3**, 340–348.

Cottingham KL, Brown BL and Lennon JT (2001). Biodiversity may regulate the temporal variability of ecological systems. *Ecology Letters*, **4**, 72–85.

Cragg RG and Bardgett RD (2001). How changes in soil faunal diversity and composition within a trophic group influence decomposition processes. *Soil Biology and Biochemistry*, **33**, 2073–2081.

Cramer W (1997). Using plant functional types in a global vegetation model. In TM Smith, HH Shugart and FI Woodward, eds *Plant functional types: Their relevance to ecosystem properties and global change*, pp. 271–288. Cambridge University Press, Cambridge.

Crawley MJ, Brown SL, Heard MS and Edwards GR (1999). Invasion-resistance in experimental grassland communities: species richness or species identity? *Ecology Letters*, **2**, 140–148.

Crittenden PD (1983). The role of lichens in the nitrogen economy of subarctic woodlands: nitrogen loss from the nitrogen-fixing lichen Stereocaulon paschale during rainfall. In JA Lee, McNeill S and IH Rorison, eds *Nitrogen as an ecological factor*, pp. 43–68. Blackwell Science, Oxford.

Crossley DA Jr, Mueller BR and Perdue JC (1992). Biodiversity of microarthropods in agricultural soils: relations to processes. *Agriculture, Ecosystems and Environment*, **40(1–4)**, 37–46.

Cuenca G, De Andrade Z and Escalante G (1998). Diversity of Glomalean spores from natural, disturbed and revegetated communities growing on nutrient-poor tropical soils. *Soil Biology and Biochemistry*, **30**, 711–719.

Cummins KW (1974). Structure and function of stream ecosystems. *BioScience*, **24**, 631–641.

Cummins KW and Klug MJ (1979). Feeding ecology of stream invertebrates. *Annual Review of Ecology and Systematics*, **10**, 147–172.

Cutler A (1991) Nested faunas and extinctions in fragmented habitats. *Conservation Biology*, **5**, 496–504.

Damascos MA and Gallopin GG (1992). Ecology of an introduced shrub (*Rosa rubiginosa* L. = *Rosa eglanteria* L.)—invasion risks and the effects on the plant communities of the Andean-Patagonian region of Argentina. *Revista Chilena de Historia Natural*, **65**, 395–407.

Dame RF, Bushek D and Prins TC (2001). Benthic suspension feeders as determinants of ecosystem structure and function in shallow coastal waters. In K Reise, ed. *Ecological comparisons of sedimentary shores*, pp. 11–37. Springer-Verlag, Berlin, Heidelberg.

Daily G, ed. (1997). *Nature's services: societal dependence on natural ecosystems*. Island Press, Washington DC.

Daily GC and Ellison K (2002). *The new economy of nature: the quest to make conservation profitable*. Island Press, Washington DC.

Daily G et al. (1997). Ecosystem services: benefits supplied to human societies by natural ecosystems. *Issues in Ecology*, **2**, 1–16.

Dauber J and Wolters V (2000). Microbial activity and functional diversity in the mounds of three different ant species. *Soil Biology and Biochemistry*, **32**, 93–99.

Daufresne T and Loreau M (2001). Ecological stoichiometry, primary producers–decomposer interactions, and ecosystem persistence. *Ecology*, **82**, 3069–3082.

Dauwe B, Herman PMJ and Heip CHR (1998). Community structure and bioturbation potential of macrofauna at four North Sea stations with contrasting food supply. *Marine Ecology-Progress Series*, **173**, 67–83.

Dawkins HC (1959). The volume increment of natural tropical high-forest and limitations on its improvement. *Empire Forestry Review*, **38**, 175–180.

Dawkins HC (1964). The productivity of lowland tropical high forest and some comparisons with its competitors. *Oxford University Forestry Society Journala, Fifth Series*, **12**, 1–8.

DeAngelis DL (1975). Stability and connectance in food web models. *Ecology*, **56**, 238–243.

DeAngelis DL (1992). *Dynamics of nutrient cycling and food webs*. Chapman and Hall, London.

de Faria SM, Lewis GP, Sprent JI and Sutherland JM (1989). Occurrence of nodulation in the Leguminosae. *New Phytologist*, **111**, 607–619.

De Mazancourt C and Loreau M (2000). Effects of herbivory on primary production, and plant species replacement. *American Naturalist*, **155**, 734–754.

De Mazancourt C, Loreau M and Dieckmann U (2001). Can the evolution of plant defense lead to plant–herbivore mutualism? *The American Naturalist*, **158**, 109–123.

De Rooij-Van der Goes PCEM (1995). The role of plant-parasitic nematodes and soil-borne fungi in the decline of *Ammophila arenaria* L. Link. *New Phytologist*, **129**, 661–669.

de Ruiter PC et al. (1993a). Simulation of nitrogen dynamics in the belowground food webs of two winter-wheat fields. *Journal of Applied Ecology*, **30**, 95–106.

de Ruiter PC et al. (1993b). Calculation of nitrogen mineralization in soil food webs. *Plant and Soil*, **157**, 263–273.

de Ruiter PC, Neutel A-M and Moore JC (1994). Modelling food webs and nutrient cycling in agro-ecosystems. *Trends in Ecology and Evolution*, **9**, 378–383.

de Ruiter PC, Neutel A-M and Moore JC (1995). Energetics, patterns of interaction strengths, and stability in real ecosystems. *Science*, **269**, 1257–1260.

de Ruiter PC et al. (1998). Biodiversity in soil ecosystems: the role of energy flow and community stability. *Applied Soil Ecology*, **10**, 217–228.

DeFoliart GR (1995). Edible insects as minilivestock. *Biodiversity and Conservation*, **4**, 306–321.

Degens BP (1998). Decreases in microbial functional diversity do not result in corresponding changes in decomposition under different moisture conditions. *Soil Biology and Biochemistry*, **30**, 1989–2000.

Degrange V, Lensi R and Bardin R (1997). Activity, size and structure of a Nitrobacter community as affected by organic carbon and nitrite in sterile soil. *FEMS Microbiology Ecology*, **24**, 173–180.

Degrange V, Coûteaux MM, Anderson JM, Berg MP and Lensi R (1998). Nitrification and occurrence of Nitrobacter in low and high coniferous forest soils. *Plant and Soil*, **198**, 201–208.

Dennis P and Fry GLA (1992). Field margins: can they enhance natural enemy population densities and general arthropod diversity on farmland? *Agriculture, Ecosystems and Environment*, **40**, 1–4, 95–115.

deWit CT, Tow PG and Ennik GC (1966). Competition between legumes and grasses. *Verslegen Landbouwkundig Onderzoek*, **687**, 3–30.

Díaz S and Cabido M (1997). Plant functional types and ecosystem function in relation to global change: a multiscale approach. *Journal of Vegetation Science*, **8**, 463–474.

Díaz S and Cabido M (2001). Vive la différence: Plant functional diversity matters to ecosystem processes. *Trends in Ecology and Evolution*, 16, 646–655.

Díaz S, Cabido M and Casanoves F (1998). Plant functional traits and environmental filters at a regional scale. *Journal of Vegetation Science*, 9, 113–122.

Díaz S *et al.* (1999). Plant traits as links between ecosystem structure and functioning. In D Eldridge and D Freudenberger, eds Proceedings of the VIth International Rangeland Congress, pp. 896–901. VI International Rangeland Congress, Townsville.

Diehl S and Feissel M (2000). Effects of enrichment on three-level food chains with omnivory. *American Naturalist*, 155, 200–218.

Diehl S and Kornijow R (1997). The influence of submerged macrophytes on trophic interactions between fish and macroinvertebrates. In E Jeppesen, MA Sondergaard, MO Sondergaard and K Christoffersen, eds *The structuring role of submerged macrophytes in lakes*, pp. 24–46. Springer-Verlag, New York.

Diemer M and Schmid B (2001). Effects of biodiversity loss and disturbance on the survival and performance of two *Ranunculus* species with differing clonal architectures. *Ecography*, 24, 59–67.

Diemer M, Joshi J, Körner C, Schmid B and Spehn EM (1997). An experimental protocol to assess the effects of plant diversity on ecosystem functioning utilized in a European research network. *Bulletin of the Geobotanical Institute ETH*, 63, 95–107.

Diggle PJ (1983). *Statistical analysis of spatial point patterns*. Academic Press, London.

Doak DF, Bigger D, Harding EK, Marvier MA, O'Malley RE and Thomson D (1998). The statistical inevitability of stability–diversity relationships in community ecology. *American Naturalist*, 151, 264–276.

Dodd ME, Silvertown J, McConway K, Potts J and Crawley M (1994). Stability in the plant communities of the Park Grass Experiment: The relationships between species richness, soil pH and biomass variability. *Philosophical Transactions of the Royal Society of London B Biological Sciences*, 346, 185–193.

Dodson SI (1974). Adaptative change in plankton morphology in response to size-selective predation: a new hypothesis of cyclomorphosis. *Limnology and Oceanography*, 19, 721–729.

Dollar SJ *et al.* (1991). Annual cycle of benthic nutrient fluxes in Tomales Bay, California, and contribution of the benthos to total ecosystem metabolism. *Marine Ecology-Progress Series*, 79, 115–125.

Doran JW, Fraser DG, Culik MN and Liebhardt WC (1987). Influence of alternative and conventional agricultural management on soil microbial processes and nitrogen availability. *American Journal of Alternative Agriculture*, 2, 99–106.

Downing AL (2001). *The role of biological diversity for the functioning and stability of pond ecosystems. Dissertation.* University of Chicago, Chicago.

Drake JA (1988). Models of community assembly and the structure of ecological landscapes. In TG Hallam, LJ Gross and SA Levin, eds *Mathematical ecology* pp. 585–605. World Press, Singapore.

Drake JA (1990). The mechanics of community assembly and succession. *Journal of Theoretical Biology*, 147, 213–233.

Drake JA and Mooney HA, eds (1989). *Biological invasions: a global perspective*. John Wiley and Sons, Chichester.

Duarte CM (2000). Marine biodiversity and ecosystem services: an elusive link. *Journal of Experimental Marine Biology and Ecology*, 250, 117–131.

Dublin HT, Sinclair ARE and McGlade J (1990). Elephants and fire as causes of multiple stable states in the Serengeti-Mara woodlands. *Journal of Animal Ecology*, 59, 1147–1164.

Duffy JE *et al.* (2001). Grazer diversity, functional redundancy, and productivity in seagrass beds: an experimental test. *Ecology*, 82, 2417–2434.

Dukes JS (2001a). Biodiversity and invasibility in grassland microcosms. *Oecologia*, 126, 563–568.

Dukes JS (2001b). Productivity and complementarity in grassland microcosms of varying diversity. *Oikos*, 94, 468–480.

Dunwoody S (1999). Scientists, journalists, and the meaning of uncertainty. In SM Friedman, S Dunwoody and CL Rogers, eds *Communicating uncertainty: Media coverage of new and controversial science*, pp. 59–79. Lawrence Erlbaum Associates, Mahwah.

Du Toit JT and Cumming DHM (1999). Functional significance of ungulate diversity in African savannas and the ecological implications of the spread of pastoralism. *Biodiversity and Conservation*, 8, 1643–1661.

Eddison JC and Ollason JG (1978). Diversity in constant and fluctuating environments. *Nature*, 275, 309–310.

Egerton FN (2001). A history of the ecological sciences. *Bulletin of the Ecological Society of America*, 82, 93–97.

Egerton-Warburton LM and Allen EB (2000). Shifts in arbuscular mycorrhizal communities along an anthropogenic nitrogen deposition gradient. *Ecological Applications*, 10, 484–496.

Ehrlich PR (1988). The loss of biodiversity: causes and consequences. In EO Wilson, ed. *Biodiversity*, pp. 21–27. National Academy Press, Washington, DC.

Ehrlich PR and Wilson EO (1991). Biodiversity studies: Science and Policy. *Science*, 253, 758–762.

Ekschmitt K and Griffiths BS (1998). Soil biodiversity and its implications for ecosystem functioning in a heterogeneous and variable environment. *Applied Soil Ecology*, **10**, 201–215.

Elliott ET, Anderson RV, Coleman DC and Cole CV (1980). Habitable pore space and microbial trophic interactions. *Oikos*, **35**, 327–335.

Elser J, Dobberfuhl DR, MacKay NA and Schampel JH (1996). Organism size, life history and N:P stoichiometry. *BioScience*, **46**, 674–684.

Elser J *et al.* (2000). Biological stoichiometry from genes to ecosystems. *Ecology Letters*, **3**, 540–550.

Elton C (1927). *Animal ecology*. McMillan, New York.

Elton CS (1958). *The ecology of invasions by animals and plants*. Methuen, London.

Emmerson MC and Raffaelli D (2000). Detecting the effects of diversity on measures of ecosystem function: experimental design, null models and empirical observations. *Oikos*, **91**, 195–203.

Emmerson MC, Solan M, Emes C, Paterson DM and Raffaelli D (2001). Consistent patterns and the idiosyncratic effects of biodiversity in marine ecosystems. *Nature*, **411**, 73–77.

Engelhardt KAM and Ritchie ME (2001). Effects of macrophyte species richness on wetland ecosystem functioning and services. *Nature*, **411**, 687–689.

Enserink M (1999). Biological invaders sweep in. *Science*, **285**, 1834–1836.

Ericson L and Wennstrom A (1997). The effect of herbivory on the interaction between the clonal plant *Trientalis europea* and its smut fungus *Urocystis trientalis*. *Oikos*, **80**, 107–111.

Erlinge S (1987). Predation and noncyclicity in a microtine rodent population in southern Sweden. *Oikos*, **50**, 347–352.

Ernst WG (2000). The Earth's place in the solar system. In WG Ernst, ed. *Earth systems: Processes and issues*, pp. 45–58. Cambridge University Press, Cambridge.

Ewel JJ (1986). Designing agroecosystems for the humid tropics. *Annual Review of Ecology and Systematics*, **17**, 245–271.

Ewel JJ (1991). Yes, we got some bananas. *Conservation Biology*, **5**, 423–425.

Ewel JJ (1999). Natural systems as models for the design of sustainable systems of land use. *Agroforestry Systems*, **45**, 1–21.

Faber JH (1991). Functional classification of soil fauna: a new approach. *Oikos*, **62**, 110–117.

Faber JH and Verhoef HA (1991). Functional differences between closely-related soil arthropods with respect to decomposition processes in the presence or absence of pine tree roots. *Soil Biology and Biochemistry*, **23**, 15–23.

Felton GW and Korth KL (2000). Trade offs between pathogen and herbivore resistance. *Current Opinion in Plant Biology*, **3**, 309–314.

Fenchel T, Esteban GF and Finlay BJ (1997). Local versus global diversity of microorganisms: cryptic diversity of ciliated protozoa. *Oikos*, **80**, 220–225.

Ferris H, Lau S and Venette R (1995). Population energetics of bacterial-feeding nematodes: respiration and metabolic rates based on CO_2 production. *Soil Biology and Biochemistry*, **27**, 319–330.

Ferber D (2001). Keeping the Stygian waters at bay. *Science*, **291**, 968–973.

Findlay S, Carreiro M, Krischic V and Jones CJ (1996). Effects of damage to living plants on leaf litter quality. *Ecological Applications*, **6**, 269–275.

Finlay BJ and Clarke KJ (1999a). Apparent global ubiquity of species in the protist genus *Paraphysomonas*. *Protist*, **150**, 419–430.

Finlay BJ and Clarke KJ (1999b). Ubiquitous dispersal of microbial species. *Nature*, **400**, 828.

Finlay BJ and Fenchel T (1999). Divergent perspectives on protist species richness. *Protist*, **150**, 229–233.

Finlay BJ, Maberly SC and Cooper JI (1997). Microbial diversity and ecosystem function. *Oikos*, **80**, 209–213.

Fitter AH (1990). The role and ecological significance of vesicular-arbuscular mycorrhizas in temperate ecosystems. *Agriculture, Ecosystems and Environment*, **29**, 257–265.

Fitter AH, Graves JD, Watkins NK, Robinson D and Scrimgeour C (1998). Carbon transfer between plants and its control in networks of arbuscular mycorrhizas. *Functional Ecology*, **12**, 406–412.

Flanagan PW and van Cleve K (1983). Nutrient cycling in relation to decomposition and litter quality in Taiga ecosystems. *Canadian Journal of Forest Research*, **13**, 795–817.

Foissner W (1999). Protist diversity: Estimates of the near-imponderable. *Protist*, **150**, 363–368.

Folke C, Holling CS and Perrings C (1996). Biological diversity, ecosystems and the human scale. *Ecological Applications*, **6**, 1018–1024.

Forman RTT (1975). Canopy lichens with blue-green algae: a nitrogen source in a Colombian rain forest. *Ecology*, **56**, 1176–1184.

Fox BJ and Brown JH (1993). Assembly rules for functional groups in North American desert rodent communities. *Oikos*, **67**, 358–370.

Fragoso C *et al.* (1997). Agricultural intensification, soil biodiversity and agroecosystem function in the tropics: the role of earthworms. *Applied Soil Ecology*, **6**, 17–35.

Francis R and Read DJ (1995). Mutualism and antagonism in the mycorrhizal symbiosis, with special reference to impact on plant community structure. *Canadian Journal of Botany,* **73,** 1301–1309.

François F, Poggiale J-C, Durbec J-P and Stora G (1997). A new approach for the modelling of sediment reworking induced by a macrobenthic community. *Acta Biotheoretica,* **45,** 295–319.

Frane JW (1986). BMD and BMDP approaches to unbalanced data. *Technical report* No. 41. Los Angeles, BMDP Statistical Software.

Frank SA (1997). The Price equation, Fisher's fundamental theorem, kin selection and causal analysis. *Evolution,* **51,** 1712–1729.

Frank DA and Groffman PM (1998). Ungulate vs. landscape control of soil C and N processes in grasslands of Yellowstone National Park. *Ecology,* **79,** 2229–2241.

Frank DA and McNaughton SJ (1991). Stability increases with diversity in plant communities: empirical evidence from the 1988 Yellowstone drought. *Oikos,* **62,** 360–362.

Franklin J (2001). *The science of conjecture: evidence and probability before Pascal.* Johns Hopkins University Press, Baltimore.

Franklin JF and MacMahon JA (2000). Messages from a mountain. *Science,* **288,** 1183–1184.

Freckman DW and Ettema CM (1993). Assessing nematode community structure in agroecosystems of varying intervention. *Agriculture, Ecosystems and Environment,* **45,** 239–261.

Freckman DW, Blackburn TH, Brussaard L, Hutchings P, Palmer MA and Snelgrove PVR (1997). Linking biodiversity and ecosystem functioning of soils and sediments. *Ambio,* **26,** 556–562.

Fridley JD (2001). The influence of species diversity on ecosystem productivity: how, where, and why? *Oikos,* **93,** 514–526.

Fridley JD (2001). Relative and interactive effects of resource availability and species diversity on ecosystem productivity in experimental plant communities. *Oikos,* **93,** 514–526.

Frost TM, Carpenter SR, Ives AR and Kratz TK (1995). Species compensation and complementarity in ecosystem function. In CG Jones and JH Lawton, eds *Linking species and ecosystems.* Chapman and Hall, New York.

Frostegård A, Bååth E and Tunlid A (1993). Shifts in the structure of soil microbial communities in limed forests as revealed by phospholipid fatty acid analysis. *Soil Biology and Biochemistry,* **25,** 723–730.

Fukami T, Naeem S and Wardle DA (2001). On similarity among local communities in biodiversity experiments. *Oikos,* **95,** 340–348.

Funes G, Basconcelo S, Díaz S and Cabido M (1999). Seed size and shape predict seed persistence in the soil bank in grasslands of central Argentina. *Seed Science Research,* **9,** 341–345.

Gall GAE and Orians GH (1992). Agricultural and biological conservation. *Agriculture, Ecosystems, and Environment,* **42,** 1–8.

Gamble JC (1991). Mesocosms, statistical and experimental design considerations. In CM Lalli, ed. *Enclosed experimental marine ecosystems, a review and recommendations,* pp. 188–196. Springer-Verlag, New York.

Gange A and Brown VK, eds (1996). *Multitrophic interactions in terrestrial systems.* Blackwell Science, Oxford.

Gange AC and West HM (1994). Interactions between arbuscular mycorrhizal fungi and foliar-feeding insects in *Plantago lanceolata* L. *New Phytologist,* **128,** 79–87.

Gange AC, Brown VK and Farmer LM (1990). A test of mycorrhizal benefit in an early successional plant community. *New Phytologist,* **115,** 85–91.

Gange AC, Brown VK and Sinclair GS (1993). Vesicular-arbuscular mycorrhizal fungi: a determinant of plant community structure in early succession. *Functional Ecology,* **7,** 616–622.

Gardner MR and Ashby WR (1970). Connectance of large (cybernetic) systems: critical values for stability. *Nature,* **228,** 784.

Gardner RH, Kemp WM, Kennedy VS and Petersen JE (2001). *Scaling relations in experimental ecology.* Columbia University Press, New York.

Gardner LR, Sharma P and Moore WS (1987). A regeneration model for the effect of bioturbation by fiddler crabs on 210Pb profiles in salt marsh sediments. *Journal of Environmental Radioactivity,* **5,** 25–36.

Garnier E, Navas M-L, Austin MP, Lilley JM and Gifford RM (1997). A problem for biodiversity–productivity studies: how to compare the productivity of multispecific plant mixtures to that of monocultures? *Acta Oecologica,* **18,** 657–670.

Giblin AE, Foreman KH and Banta GT (1995). Biogeochemical processes and marine benthic community structure: which follows which? In CG Jones and JH Lawton, eds *Linking species and ecosystems,* pp. 37–44. Chapman and Hall, San Diego.

Gilbert F, Gonzalez A and Evans-Freke I (1998). Corridors maintain species richness in the fragmented landscape of a microecosystem. *Proceedings of the Royal Society of London, Series B, Biological Sciences,* **265,** 577–582.

Giller KE, Beare MH, Lavelle P, Izac A-MN and Swift MJ (1997). Agricultural intensification, soil biodiversity and agroecosystem function. *Applied Soil Ecology,* **6,** 3–16.

Giller KE, Witter E and McGrath S (1998). Toxicity of heavy metals to microoorganisms and microbial processes in agricultural soils: a review. *Soil Biology and Biochemistry*, **30**, 1389–1414.

Giller PS (1996). The diversity of soil communities, the 'poor man's tropical rainforest'. *Biodiversity and Conservation*, **5**, 135–168.

Giller PS, Hildrew AG and Raffaelli DG (1994). *Aquatic ecology: scale, pattern and process*. Blackwell Science, Oxford.

Gitay H and Noble IR (1997). What are functional types and how should we seek them? In TM Smith, HH Shugart and FI Woodward, eds *Plant functional types*, pp. 3–19. University Press, Cambridge.

Gitay H, Wilson JB and Lee WG (1996). Species redundancy: a reduntant concept? *Journal of Ecology*, **84**, 121–124.

Givnish TJ (1994). Does diversity beget stability? *Nature*, **371**, 113–114.

Gleason HA (1926). The individualistic concept of the plant association. *Bulletin of the Torrey Botanical Club*, **53**, 7–26.

Glick BR (1995). The enhancement of plant growth by free living bacteria. *Canadian Journal of Microbiology*, **41**, 109–117.

Goldberg DE and Miller TE (1990). Effects of different resource additions on species diversity in an annual plant community. *Ecology*, **71**, 213–225.

Goldberg DE and Werner PA (1983). Equivalence of competitors in plant communities: a null hypothesis and a field experimental approach. *American Journal of Botany*, **70**, 1098–1104.

Gonzalez et al. (1999). Metapopulation dynamics, abundance, and distribution in a microecosystem. *Science*, **281**, 2045–2047.

Gotelli NJ and Çolwell RK (2001). Quantifying biodiversity: procedures and pitfalls in the measurement and comparison of species richness. *Ecology Letters*, **4**, 349–391.

Gough L, Osenberg CW, Gross KL and Collins SL (2000). Fertilisation effects on species density and primary productivity in herbaceous plant communities. *Oikos*, **89**, 428–439.

Goverde M, Van der Heijden M, Wiemken A, Sanders IR and Erhardt A (2000). Arbuscular mycorrhizal fungi influence life history traits of a lepidopteran larvae. *Oecologia*, **125**, 362–369.

Grayston S, Griffiths G, Mawdsley JL, Campbell C and Bardgett RD (2001). Accounting for variability in soil microbial communities of temperate upland grassland ecosystems. *Soil Biology and Biochemistry*, **33**, 533–551.

Greenberg R, Bichier P, Cruz Angon A and Reitsma Rl (1997). Bird populations in shade and sun coffee plantations in central Guatemala. *Conservation Biology*, **11**, 448–459.

Griffiths BS (1994). Microbial-feeding nematodes and protozoa in soil: their effects on microbial activity and nitrogen mineralization in decomposition hotspots and the rhizosphere. *Plant and Soil*, **164**, 25–33.

Griffiths BS, Bonkowski M, Dobson G and Caul S (1999). Changes in soil microbial community structure in the presence of microbial-feeding nematodes and protozoa. *Pedobiologia*, **43**, 297–304.

Griffiths BS et al. (2000). Ecosystem response of pasture soil communities to fumigation-induced microbial diversity reductions: an examination of the biodiversity—ecosystem function relationship. *Oikos*, **90**, 279–294.

Griffiths BS, Bonkowski M, Roy J and Ritz K (2001a) Functional stability, substrate utilisation and biological indicators of soils following environmental impacts. *Applied Soil Ecology*, **16**, 49–61.

Griffiths BS et al. (2001b). An examination of the biodiversity–ecosystem function relationship in arable soil microbial communities. *Soil Biology and Biochemistry*, **33**, 1713–1722.

Grime JP (1973a). Competitive exclusion in herbaceous vegetation. *Nature*, **242**, 344–347.

Grime JP (1973b). Controls of species density in herbaceous vegetation. *Journal of Environmental Management*, **1**, 151–167.

Grime JP (1979). *Plant strategies and vegetation processes*. John Wiley and Sons, Chichester.

Grime JP (1987). Dominant and subordinate components of plant communities: implications for succession, stability and diversity. In AJ Gray, MJ Crawley and PJ Edwards, eds *Colonization, succession and stability*, pp. 413–428. Blackwell Science, Oxford.

Grime JP (1988). The CSR model of primary plant strategies—origins, implications and tests. In LD Gottlieb and S Jain, eds *Evolutionary plant biology*, pp. 371–393. Chapman and Hall, London.

Grime JP (1997). Biodiversity and ecosystem function: the debate deepens. *Science*, **277**, 1260–1261.

Grime JP (1998). Benefits of plant diversity to ecosystems: immediate, filter and founder effects. *Journal of Ecology*, **86**, 902–910.

Grime JP (2001). *Plant strategies, vegetation processes and ecosystem properties*. John Wiley and Sons, Chichester.

Grime JP and Campbell BD (1991). Growth rate, habitat productivity, and plant strategy as predictors of stress response. In HA Mooney, WE Winner and EJ Pell, eds

Response of plants to multiple stresses, pp. 143–161. Academic Press, San Diego.

Grime JP and Hunt R (1975). Relative growth-rate: its range and significance in a local flora. *Journal of Ecology*, **63**, 393–422.

Grime JP, Mackey JML, Hillier SH and Read DJ (1987). Floristic diversity in a model system using experimental microcosms. *Nature*, **328**, 420–422.

Grime JP *et al.* (1997a). Functional types: testing the concept in Northern England. In TM Smith, HH Shugart and FI Woodward, eds *Plant functional types: Their relevance to ecosystem properties and global change*, pp. 122–152. Cambridge University Press, Cambridge.

Grime JP *et al.* (1997b). Integrated screening validates primary axes of specialization in plants. *Oikos*, **79**, 259–281.

Grimm NB (1995). Why link species and ecosystems? A perspective from ecosystem ecology. In CG Jones and JH Lawton, eds *Linking species and ecosystems*, pp. 5–15. International Thomson Publishing, New York.

Groffman PM and Bohlen PJ (1999). Soil and sediment biodiversity. *BioScience*, **49**, 139–148.

Groffman PM, Eagan P, Sullivan WM and Lemunyon JL (1996). Grass species and soil type effects on microbial biomass and activity. *Plant and Soil*, **183**, 61–67.

Groppe K, Steinger T, Schmid B, Baur B and Boller T (2001). Effects of small-scale habitat fragmentation on choke disease (*Epichloë bromicola*) in the grass *Bromus erectus*. *Journal of Ecology*, **89**, 247–255.

Gross KL, Willig MR, Gough L, Inouye R and Cox SB (2000). Patterns of species density and productivity at different spatial scales in herbaceous plant communities. *Oikos*, **89**, 417–427.

Gunderson L (2000). Ecological resilience—in theory and application. *Annual Review of Ecology and Systemetics*, **31**, 425–439.

Gunnarsson T and Tunlid A (1986). Recycling of fecal pellets in isopods: microorganisms and nitrogen compounds as potential food for *Oniscus asellus* L. *Soil Biology and Biochemistry*, **18**, 595–600.

Gurney WSG, Ross AH and Broekhuizen N (1995). Coupling the dynamics of species and materials. In CG Jones and JH Lawton, eds *Linking species and ecosystems*, pp. 176–193. Chapman and Hall, New York.

Guterman L (2000). *Have ecologists oversold biodiversity? The chronicle of higher education*, Issue of 13 Oct. 2000.

Haggar J and Ewel J (1997). Primary productivity and resource partitioning in model tropical ecosystems. *Ecology*, **78**, 1211–1221.

Haimi J, Huhta V and Boucelham M (1992) Growth increase of birch seedlings under the influence of

earthworms – a laboratory study. *Soil Biology and Biochemistry* **24**, 1525–1528.

Hairston NG (1989). Hard choices in ecological experimentation. *Herpetologica*, **45**, 119–122.

Hairston NGJ and Hairston NGS (1993). Cause-effect relationships in energy flow, trophic structure, and interspecific interactions. *American Naturalist*, **142**, 379–411.

Hairston NG, Smith FE and Slobodkin LB (1960). Community structure, population control and competition. *American Naturalist*, **94**, 421–425.

Hale SS (1975). The roles of benthic communities in the nitrogen and phosphorous cycles of an estuary. *Recent Advances in Estuary Research*, **1**, 291–308.

Hall DO, Scurlock JMO, Bolhar-Nordenkampf HR, Leegood RC and Long SP (1993). *Photosynthesis and production in a changing environment*. Chapman and Hall, London.

Hall SJ and Raffaelli DG (1993). Food webs: theory and reality. *Advances in Ecological Research*, **24**, 187–239.

Hall SJ and Raffaelli DG (1997). Food web patterns: what do we really know? In A Gange and VK Brown, eds *Multitrophic interactions in terrestrial systems*, pp. 395–419. Blackwell Science, Oxford.

Halley JM (1996). Ecology, evolution and 1/f noise. *Trends in Ecology and Evolution*, **11**, 33–37.

Handley WRC (1954). *Mull and mor in relation to forest soils*. Her Majesty's Stationery Office, London.

Hanlon RDG (1981). Influence of grazing by Collembola on the activity of senescent fungal colonies grown on media of different nutrient concentration. *Oikos*, **36**, 362–367.

Hanlon RDG and Anderson JM (1979). The effects of collembola grazing on microbial activity in decomposing leaf litter. *Oecologia*, **38**, 93–99.

Hansen RA (2000). Effect of habitat complexity and composition on a diverse litter microarthropod assemblage. *Ecology*, **81**, 1120–1132.

Hanski I, ed. (1988). Ecological significance of spatial and temporal variability. *Annales Zoologici Fennici* (Special Issue).

Hanski I (1999a). Habitat connectivity, habitat continuity, and metapopulations in dynamic landscapes. *Oikos*, **87**, 209–219.

Hanski I (1999b). *Metapopulation ecology*. Oxford University Press.

Hanski I and Ranta E (1983). Co-existence in a patchy environment: Three species of *Daphnia* in rock pools. *Journal of Animal Ecology*, **52**, 263–279.

Hardin G (1960). The competitive exclusion principle. *Science*, **131**, 1292–1297.

Harley JL and Harley EL (1987). A check list of mycorrhiza in the British Flora. *New Phytologist*, **105**, 1–102.

Harper JL (1977). *Population biology of plants*. Academic Press, London.

Harper JL and Hawksworth DL (1994). Biodiversity: measurement and estimation. *Philosophical Transactions of the Royal Society of London B*, **345**, 5–12.

Harrison GW (1979). Stability under environmental stress: resistance, resilience, persistence, and variability. *American Naturalist*, **113**, 659–669.

Harte J and Kinzig AP (1993). Mutualism and competition between plants and decomposers: implications for nutrient allocation in ecosystems. *The American Naturalist*, **141**, 829–846.

Hartnett DC and Wilson WT (1999). Mycorrhizae influence plant community structure and diversity in tall grass prairie. *Ecology*, **80**, 1187–1195.

Hassall M, Visser S and Parkinson D (1986). Vertical migration of *Onychiurus subtenuis* (Collembola) in relation to rainfall and microbial activity. *Pedobiologia*, **29**, 175–182.

Hassall M, Turner JG and Rands MRW (1987). Effects of terrestrial isopods on the decomposition of woodland leaf litter. *Oecologia*, **72**, 597–604.

Hassink J, Oude Voshaar JH, Nijhuis EH and van Veen JA (1991). Dynamics of the microbial populations of a reclaimed-polder soil under a conventional and a reduced-input farming system. *Soil Biology and Biochemistry*, **23**, 515–524.

Hawkins BA (1992). Parasite-host food webs and donor control. *Oikos*, **65**, 159–162.

Hawksworth DL, ed. (1991). *The biodiversity of microorganisms and invertebrates: its role in sustainable agriculture*. CAB International, Wallingford.

Hayman DS and Tavares M (1985). Plant growth responses to vesicular-arbuscular mycorrhiza: XV. Influence of soil pH on the symbiotic efficiency of different endophytes. *New Phytologist*, **100**, 367–377.

Hector A (1998). The effect of diversity on productivity: detecting the role of species complementarity. *Oikos*, **82**, 597–599.

Hector A (2002). Biodiversity and the Functioning of Grassland Ecosystems: Multi-Site Comparisons. In AP Kinzig, D Tilman and SW Pacala, eds *The functional consequences of biodiversity: Empiric progress and theoretical extensions*, pp. 71–95. Princeton University Press, Princeton.

Hector A and Hooper RE (2002). Darwin and the first ecological experiment. *Science*, **295**, 639–640.

Hector A *et al.* (1999). Plant diversity and productivity experiments in European grasslands. *Science*, **286**, 1123–1127.

Hector A, Beale AJ, Minns A, Otway SJ and Lawton JH (2000a). Consequences of the reduction of plant diversity for litter decomposition: effects through litter quality and microenvironment. *Oikos*, **90**, 357–371.

Hector A *et al.* (2000b). No consistent effect of plant diversity on productivity? Response to Huston *et al. Science*, **289**, 1255.

Hector A, Dobson K, Minns A, Bazely-White E and Lawton JH (2001a, in press). Community diversity and invasion resistance: an experimental test in a grassland ecosystem and a review of comparable studies. *Ecological Research*, **16**(5) (supplement).

Hector A, Joshi J, Lawler SP, Spehn EM and Wilby A (2001b). Conservation implications of the link between biodiversity and ecosystem functioning. *Oecologia*, **129**, 624–628.

Hector A, Schmid B, Beierkuhnlein C, Caldeira MC, Diemer M, Dimitrakopoulos PG, Finn J, Freitas H, Giller PS, Good J, Harris R, Högberg P, Huss-Danell K, Joshi J, Jumpponen A, Körner C, Leadley PW, Loreau M, Minns, A, Mulder CPH, O'Donovan G, Otway SJ, Pereira JS, Prinz A, Read DJ, Scherer-Lorenzen M, Schulze E-D, Siamantziouras A-SD, Spehn, EM, Terry AC, Troumbis AY, Woodward FI, Yachi S and Lawton JH (2002a). Biodiversity and the Functioning of Grassland Ecosystems: Multisite Studies. *Functional Consequences of Biodiversity: Experimental Progress and Theoretical Extensions* (eds A. Kinzig, D. Tilman & S.P. Pacala), pp. Chapter 4. Princeton University Press, Princeton.

Hedlund K and Sjögren Öhrn M (2000). Tritrophic interactions in a soil community enhance decomposition rates. *Oikos*, **88**, 585–591.

Hedlund K, Boddy L and Preston CM (1991). Mycelial responses of the soil fungus, *Mortierella isabellina*, to grazing by *Onychiurus armatus* (Collembola). *Soil Biology and Biochemistry*, **23**, 361–366.

Hedlund K, Bengtsson G and Rundgren S (1995). Fungal odour discrimination in two sympatric species of fungivorous collembolans. *Functional Ecology*, **9**, 869–875.

Helgason T, Donnel TJ, Husband R, Fitter AH and Young JPY (1998). Ploughing up the wood-wide web? *Nature*, **394**, 431.

Hellman H (1998). *Great feuds in science: Ten of the liveliest disputes ever*. John Wiley and Sons, New York.

Hendrix LJ, Carter MW and Scott DT (1982). Covariance analyses with heterogeneity of slopes in fixed models. *Biometrics*, **38**, 641–650.

Hendrix PF *et al.* (1986). Detritus food webs in conventional and no-tillage agroecosystems. *Bioscience*, **36**, 374–380.

Heneghan L, Coleman DC, Zou X, Crossley DAJ and Haines BL (1999). Soil microarthropod contributions to decomposition dynamics: tropical-temperate comparisons of a single substrate. *Ecology*, **80**, 1873–1882.

Henry J (1997). *The scientific revolution and the origins of modern science*. Macmillam Press, Houndmills.

Herman PMJ *et al.* (1999). Ecology of estuarine macrobenthos. *Advances in Ecological Research*, **29**, 195–240.

Herms A and Mattson WJ (1992). The dilemma of plants: to grow or defend. *Quarterly Review of Biology*, **67**, 293.

Hey J (2001). The mind of the species problem. *Trends in Ecology and Evolution*, **16**, 326–329.

Higgins SI, Richardson DM, Cowling RM and Trinder-Smith TH (1999). Predicting the landscape-scale distribution of alien plants and their threat to diversity. *Conservation Biology*, **13**, 303–313.

Hobbie SE (1992). Effects of plant species on nutrient cycling. *Trends in Ecology and Evolution*, **7**, 336–339.

Hobbie SE (1996). Temperature and plant species control over litter decomposition in Alaskan tundra. *Ecological Monographs*, **66**, 503–522.

Hobbie SE and Chapin FSI (1998). The response of tundra plant biomass, aboveground production, nitrogen, and CO_2 flux to experimental warming. *Ecology*, **79**, 1526–1544.

Hobbie SE, Jensen DB and Chapin FSI (1993). Resource supply and disturbance as controls over present and future plant diversity. In E-D Schulze and HA Mooney, eds *Biodiversity and ecosystem function*, pp. 385–408. Springer-Verlag, Berlin.

Hobbs RJ and Mooney HA (1991). Effects of rainfall variability and gopher disturbance on serpentine annual grassland dynamics. *Ecology*, **72**, 59–68.

Hobbs RJ and Mooney HA (1998). Broadening the extinction debate: population deletions and additions in California and Western Australia. *Conservation Biology*, **12**, 271–283.

Hochberg ME, Clobert J and Barbault R (1996). *Aspects of the genesis and maintenance of biological diversity*. Oxford University Press, Oxford.

Hodge A, Campbell CD and Fitter AH (2001). An arbuscular mycorrhizal fungus accelerates decomposition and acquires nitrogen directly from organic material. *Nature*, **413**, 297–299.

Hodgson J, Thompson K, Wilson P and Bogaard A (1998). Does biodiversity determine ecosystem function? The Ecotron experiment reconsidered. *Functional Ecology*, **12**, 843–848.

Hodgson JG *et al.* (1999). Allocating CSR plant functional types: a soft approach to a hard problem. *Oikos*, **85**, 282–294.

Holah JC and Alexander HM (1999). Soil pathogenic fungi have the potential to affect the co-existence of two tallgrass prairie species. *Journal of Ecology*, **87**, 598–608.

Holland J and Fahrig L (2000). Effect of woody borders on insect density and diversity in crop field: a landscape-scale analysis. *Agriculture, Ecosystems and Environment*, **78**, 1125–1129.

Hollibaugh JT *et al.* (1988). Tomales Bay, California: a macrocosm for examining biogeochemical coupling at the land-sea interface. *Eos*, **69**, 843–845.

Holling CS (1973). Resilience and stability of ecological systems. *Annual Review of Ecology and Systematics*, **4**, 1–23.

Holling CS (1986). The resilience of ecosystems: local surprise and global change. In WC Clark and RE Munn, eds *Sustainable development of the biosphere*, pp. 292–317. Cambridge University Press, Cambridge.

Holling CS, Schindler DW, Walker BW and Roughgarden J (1995). Biodiversity in the functioning of ecosystems: an ecological synthesis. In CA Perrings, K-G Mäler, C Folke, CS Holling and B–O Jansson, eds *Biodiversity loss. Economic and ecological issues*, pp. 44–83. Cambridge University Press, Cambridge.

Holmstrup M (2001). Sensitivity of life history parameters in the earthworm *Aporrectodea caliginosa* to small changes in soil water potential. *Soil Biology and Biochemistry*, **33**, 1217–1223.

Holt RD (1997). Community modules. In A Gange and VK Brown, eds *Multitrophic interactions in terrestrial systems*, pp. 333–350. Blackwell Science, Oxford.

Holt RD and Loreau M (2002). Biodiversity and ecosystem functioning: the role of trophic interactions, and the importance of system openness. In A Kinzig, D Tilman and SW Pacala, eds *Biodiversity and ecosystem functioning: empirical and theoretical analyses*. Princeton University Press, Princeton.

Holt RD, Grover J and Tilman D (1994). Simple rules for interspecific dominance in systems with exploitative and apparent competition. *The American Naturalist*, **144**, 741–771.

Holway DA (1998). Factors governing rate of invasions: a natural experiment using Argentine ants. *Oecologica*, **115**, 206–212.

Holyoak M (2000). Habitat subdivision causes changes in food web structure. *Ecology Letters*, **3**, 509–515.

Holyoak M and Lawler SP (1996a). Persistence of an extinction prone predator–prey interaction through metapopulation dynamics. *Ecology*, **77**, 1867–1879.

Holyoak M and Lawler SP (1996b). The role of dispersal in predator–prey metapopulation dynamics. *Journal of Animal Ecology*, **65**, 640–652.

Hooper DU (1998). The role of complementarity and competition in ecosystem responses to variation in plant diversity. *Ecology*, **79**, 704–719.

Hooper DU and Vitousek PM (1997). The effects of plant composition and diversity on ecosystem processes. *Science*, **277**, 1302–1305.

Hooper DU and Vitousek PM (1998). Effects of plant composition and diversity on nutrient cycling. *Ecological Monographs*, **68**, 121–149.

Hooper D, Hawksworth D and Dhillion S (1995). Microbial diversity and ecosystem processes. In United Nations Environment Programme, ed. *Global biodiversity assessment*, pp. 433–443. Cambridge University Press, Cambridge.

Hooper DU *et al.* (2000). Interactions between above- and below-ground biodiversity in terrestrial ecosystems: patterns, mechanisms and feedbacks. *BioScience*, **50**, 1049–1061.

Houghton J (1997). *Global warming: The complete briefing.* Cambridge University Press, Cambridge.

Hrbácek JM, Dvorakova M and Korínek VEA (1961). Demonstration of the effect of the fish stock on the species composition of zooplankton and the intensity of metabolism of the whole plankton association. *Verhandlungen der Internationale Vereinigung für Theoretische und Angewandte Limnologie*, **14**, 192–195.

Hubbell SP (1999). Tropical tree richness and resource-based niches. *Science*, **285**, 554.

Hubbell SP (2001). *The unified neutral theory of biodiversity and biogeography.* Princeton Monographs, Princeton.

Hubbell SP and Foster RB (1986). Biology, chance, and history and the structure of tropical rain forest tree communities. In J Diamond and TJ Case, eds *Community ecology*, pp. 314–329. Harper and Row, New York.

Hubbell SP *et al.* (1990). Light-gap disturbances, recruitment limitation, and tree diversity in a neotropical forest. *Science*, **283**, 554–557.

Hughes DJ, Atkinson RJA and Ansell AD (2000). A field test of the effects of megafaunal burrows on benthic chamber measurements of sediment-water solute fluxes. *Marine Ecology-Progress Series*, **195**, 189–199.

Hughes JB and Petchey OL (2001). Merging perspectings on biodiversity and ecosystem functioning. *Trends in Ecology and Evolution*, **16**, 222–223.

Hughes JB and Roughgarden J (1998). Aggregate community properties and the strength of species' interactions. *Proceedings of the National Academy of Sciences of the United States of America*, **95**, 6837–6842.

Hughes JB and Roughgarden J (2000). Species diversity and biomass stability. *American Naturalist*, **155**, 618–627.

Hughes JB, Daily GC and Ehrlich PR (1997). Population diversity: Its extent and extinction. *Science*, **278**, 689–691.

Huhta V and Viberg K (1999). Competitive interactions between the earthworm *Dendrobaena octaedra* and the enchytraeid *Cognettia sphagnetorum*. *Pedobiologia*, **43**, 886–890.

Hulot FD, Lacroix G, Lescher-Moutoue F and Loreau M (2000). Functional diversity governs ecosystem response to nutrient enrichment. *Nature*, **405**, 340–344.

Hunt HW *et al.* (1987). The detrital food web in a short-grass prairie. *Biology and Fertility of Soils*, **3**, 57–68.

Hunter MD and Price PW (1992). Playing chutes and ladders: heterogeneity and the relative roles of bottom-up and top-down forces in natural communities. *Ecology*, **73**, 724–731.

Huntly N (1991). Herbivores and the dyanmics of communities and ecosystems. *Annual Review of Ecology and Systematics*, **22**, 477–503.

Hurd LE and Wolf LL (1974). Stability in relation to nutrient enrichment in arthropod consumers of old-field successional ecosystems. *Ecological Monographs*, **44**, 465–482.

Hurlbert SH (1984). Pseudoreplication and the design of ecological field experiments. *Ecological Monographs*, **54**, 187–211.

Huston MA (1979). A general hypothesis of species diversity. *American Naturalist*, **113**, 81–101.

Huston MA (1980). Soil nutrients and tree species richness in Costa Rican forests. *Journal of Biogeography*, **7**, 147–157.

Huston MA (1993). Biological diversity, soils, and economics. *Science*, **262**, 1676–1680.

Huston MA (1994). *Biological diversity. The coexistence of species on changing landscapes.* Cambridge University Press, Cambridge.

Huston MA (1997). Hidden treatments in ecological experiments: re-evaluating the ecosystem function of biodiversity. *Oecologia*, **110**, 449–460.

Huston MA and DeAngelis DL (1994). Competition and coexistence: the effects of resource transport and supply rates. *American Naturalist*, **144**, 954–977.

Huston MA *et al.* (2000). No consistent effect of plant diversity on productivity. *Science*, **289**, 1255.

Hutchings MJ and de Kroon H (1994). Foraging in plants: the role of morphological plasticity in resource acquisition. *Advances in Ecological Research*, **25**, 159–238.

Hutchings P (1998). Biodiversity and functioning of polycheates in benthic sediments. *Biological Conservation*, **7**, 1133–1145.

Huxham M, Raffaelli DG and Pike AW (1995). Parasites and food web patterns. *Journal of Animal Ecology*, **64**, 164–176.

Huxham M, Roberts I and Bremner J (2000). A field test of the intermediate disturbance hypothesis in the soft-bottom intertidal. *International Review of Hydrobiology*, **8**, 379–394.

Ingham RE, Trofymow JA, Ingham ER and Coleman DC (1985). Interactions of bacteria, fungi and their nematode grazers: effects on nutrient cycling and plant growth. *Ecological Monographs*, **55**, 119–140.

Iverson LR and Prasad A (1998). Estimating regional plant biodiversity with GIS modelling. *Diversity and Distributions*, **4**, 49–61.

Ives AR (1995a). Measuring resilience in stochastic systems. *Ecological Monographs*, **65**, 217–233.

Ives AR (1995b). Predicting the response of populations to environmental-change. *Ecology*, **76**, 926–941.

Ives AR, Foufopoulos J, Klopper ED, Klug JL and Palmer TM (1996). Bottle or big-scale studies: how do we do ecology. *Ecology*, **77**, 681–685.

Ives AR, Gross K and Klug JL (1999). Stability and variability in competitive communities. *Science*, **286**, 542–544.

Ives AR, Klug JL and Gross K (2000). Stability and species richness in complex communities. *Ecology Letters*, **3**, 399–411.

Ives AR and Hughes JB (in press). General relationships between species diversity and stability in competitive systems. *American Naturalist*.

Jacot KA, Luscher A, Nosberger J and Hartwig UA (2000). The relative contribution of symbiotic N_2 fixation and other nitrogen sources to grassland ecosystems along an altitudinal gradient in the Alps. *Plant and Soil*, **225**, 201–211.

Jakobsen I, Abbott LK and Robson AD (1992). External hyphae of vesicular-arbuscular mycorrhizal fungi associated with *Trifolium subterraneum* L. II: Hyphal transport of ^{32}P over defined distances. *New Phytologist*, **120**, 509–516.

Jentschke G, Bonkowski M, Godbold DL and Scheu S (1995). Soil protozoa and forest tree growth: non-nutritional effects and interaction with mycorrhizae. *Biology and Fertility of Soils*, **20**, 263–269.

Johnson KH (2000). Trophic-dynamic considerations in relating species diversity to ecosystem resilience. *Biological Reviews*, **75**, 347–376.

Johnson NC, Zak DR, Tilman D and Pfleger FL (1991). Dynamics of vesicular arbuscular mycorrhizae during old field succession. *Oecologia*, **86**, 349–358.

Jolliffe PA (2000). The replacement series. *Journal of Ecology*, **88**, 371–385.

Jonasson S, Michelson A, Schmidt IK, Nielsen E and Callaghan TV (1986). Microbial biomass carbon nitrogen and phosphorus in two arctic soils and responses to NPK

fertiliser and sugar: implications for plant nutrient uptake. *Oecologia*, **106**, 507–515.

Jones CG and Lawton JH, eds (1995). *Linking species and ecosystems*. Chapman and Hall, New York.

Jones CG, Lawton JG and Shachak M (1994). Organisms as ecosystem engineers. *Oikos*, **69**, 373–386.

Jonsson L, Nilsson M-C, Wardle DA and Zackrisson O (2001). Context dependent effects of ectomycorrhizal species richness on tree seedling productivity. *Oikos*, **93**, 353–364.

Jonsson M and Malqvist B (2000). Ecosystem process rate increases with animal species richness: evidence from leaf-eating, aquatic insects. *Oikos*, **89**, 519–523.

Joshi J, Matthies D and Schmid B (2000). Root hemi-parasites and plant diversity in experimental grassland communities. *Journal of Ecology*, **88**, 634–644.

Joshi J et al. (2001). Local adaptation enhances performance of common plant species. *Ecology Letters*, **4**, 1–9.

Jumars PA and Nowell ARM (1984). Effects of benthos on sediment transport: difficulties with functional grouping. *Continental Shelf Research*, **3**, 115–130.

Kaiser J (2000). Rift over biodiversity divides ecologists. *Science*, **289**, 1282–1283.

Kaneko N and Salamanca N (1999). Mixed leaf litter effects on decomposition rates and soil arthropod communities in an oak-pine forest stand in Japan. *Ecological Research*, **14**, 131–138.

Karban R and Baldwin IT (1997). *Induced response to herbivory*. Chicago University Press, Chicago.

Kareiva P (1989). Renewing the dialogue between theory and experiments in population ecology. In J Roughgarden, RM May and SA Levin, eds *perspectives in ecological theory*, pp. 68–88. Princeton University Press, Princeton.

Kass DC (1978). Polyculture cropping systems: review and analysis. *Cornell International Agriculture Bulletin*, **32**, 1–69.

Kassen R, Buckling A, Bell G and Rainey PB (2000). Diversity peaks at intermediate productivity in a laboratory microcosm. *Nature*, **406**, 508–512.

Kaunzinger CMK and Morin PJ (1998). Productivity controls food-chain properties in microbial communities. *Nature*, **395**, 495–497.

Kayang H, Sharma GD and Mishra RR (1994). Effect of an isopod grazing (*Burmoniscus* sp.) upon microbes and nutrient release from the decomposing leaf litter of *Alnus nepalensis* D. Don. *European Journal of Soil Biology*, **30**, 11–15.

Kaye JP and Hart SC (1997). Competition for nitrogen between plants and soil microorganisms. *Trends in Ecology and Evolution*, **12**, 139–143.

Keddy P, Twolan-Strutt L and Shipley B (1999). Experimental evidence that interspecific competitive asymmetry increases with soil productivity. *Oikos*, **80**, 253–256.

Kelt D and Brown J (1999). Community structure and assembly rules: confronting conceptual and statistical issues with data on desert rodents. In E Weiher and P Keddy, eds *Ecological assembly rules: Perspectives, advances, retreats*, pp. 75–107. Cambridge University Press, Cambridge.

Kelt D, Taper M and Meserve P (1995). Assessing the impact of competition on the assembly of communities: a case study using small mammals. *Ecology*, **76**, 1283–1296.

Kempton RA and Lockwood G (1984). Inter–plot competition in variety trials of field beans (*Vicia faba* L.). *Journal of Agricultural Science, Cambridge*, **103**, 293–302.

Kennedy AC and Smith KL (1994). Soil microbial diversity and ecosystem functioning. In GP Robertson and H Collins, eds *The functional significance and regulation of soil biodiversity*. Kluwer, The Netherlands.

Kenny DA (1979). *Correlation and causality*. John Wiley and Sons, New York.

Kielland K (1994). Amino acid absorption by arctic plants: implications for plant nutrition and nitrogen cycling. *Ecology*, **75**, 2373–2383.

Kielland K, Bryant JP and Ruess RW (1997). Moose herbivory and carbon turnover of early successional stands in interior Alaska. *Oikos*, **80**, 25–30.

Kim KC (1994). Entomology in the changing world: biodiversity and sustainable agriculture. *Korean Journal of Entomology*, **24**, 145–153.

King AW and Pimm SL (1983). Complexity, diversity, and stability: a reconciliation of theoretical and empirical results. *The American Naturalist*, **122**, 229–239.

Kingsolver JG and RT Paine (1991). Theses, antitheses, and syntheses: conversational biology and ecological debate. In LA Real and JH Brown, eds *Foundations of ecology: classic papers with commentaries*, pp. 309–317. University of Chicago Press, Chicago.

Kirchner MJ, Wollum II AG and King LD (1993). Soil microbial populations and activities in reduced chemical input agroecosystems. *Soil Science Society of America Journal*, **57**, 1289–1295.

Klironomos JN (2000). Host-specificity and functional diversity among arbuscular mycorrhizal fungi. In CR Bell, M Brylinsky and P Johnson-Green, eds *Microbial biosystems: New Frontiers*. Proceedings of the 8th International Symposium on Microbial Ecology. Halifax, Canada.

Klironomos JN and Kendrick WB (1996). Palatability of microfungi to soil arthropods in relation to the functioning of arbuscular mycorrhizae. *Biology and Fertility of Soils*, **21**, 43–52.

Klironomos JN, Widden P and Deslandes I (1992). Feeding preferences of the collembolan *Folsomia candida* in relation to microfungal succession on decaying litter. *Soil Biology and Biochemistry*, **24**, 685–692.

Klironomos JN, McCune J, Hart M and Neville J (2000). The influence of arbuscular mycorrhizae on the relationship between plant diversity and productivity. *Ecology letters*, **3**, 137–141.

Klug JL, Fischer JM, Ives AR and Dennis B (2000). Compensatory dynamics in planktonic community responses to pH perturbations. *Ecology*, **81**, 387–398.

Knapp AK and Smith MD (2001). Variation among biomes in temporal dynamics of aboveground primary production. *Science*, **291**, 481–484.

Knops JMH, Griffin JR and Royalty AC (1995). Introduced and native plants of the Hastings reservation, central coastal California: a comparison. *Biological Conservation*, **71**, 115–123.

Knops JMH *et al.* (1999). Effects of plant species richness on invasion dynamics, disease outbreaks, insect abundances and diversity. *Ecology Letters*, **2**, 286–293.

Knops J, Wedin D and Tilman D (2001). Biodiversity and decomposition in experimental grassland communities. *Oecologia*, **126**, 429–433.

Koehler HH (1997). Mesostigmata (Gamasina, Uropodina), efficient predators in agroecosystems. *Agriculture, Ecosystems and Environment*, **62**, 105–117.

Kogan M and Lattin JD (1993). Insect conservation and pest management. *Biodiversity and Conservation*, **2,3**, 242–257.

Koide R (1991). Nutrient supply, nutrient demand and plant response to mycorrhizal infection. *New Phytologist*, **117**, 365–386.

Kokkoris GD, Troumbis AY and Lawton JH (1999). Patterns of species interaction strength in assembled competition communities. *Ecology Letters*, **2**, 70–74.

Koricheva J (1999). Interpreting phenotypic variation in plant allelochemistry: problems with the use of concentrations. *Oecologia*, **119**, 467–473.

Koricheva J, Mulder CPH, Schmid B, Joshi J and Huss-Danell K (2000). Numerical responses of different trophic groups of invertebrates to manipulations of plant diversity in grasslands. *Oecologia*, **126**, 310–320.

Körner C (2000). Biosphere responses to CO_2 enrichment. *Ecological Applications*, **10**, 1590–1619.

Korthals GW, Smilauer P, Van Dijk C and Van der Putten WH (2001). Linking above- and belowground biodiversity: abundance and trophic complexity in soil as a response to experimental plant communities on abandoned arable land. *Functional Ecology*, **15**, 506–514.

Krantzberg G (1985). The influence of bioturbation on physical, chemical and biological parameters in aquatic environments: a review. *Environmental Pollution (Series A)*, **39**, 99–122.

Kranz R (2000). Crossing the moat: using ecosystem services to communicate ecological ideas beyond the ivory tower. *Bulletin of the Ecological Society of America*, **81**, 95–96.

Krebs CJ *et al.* (1995) Impact of food and predation on the snowshoe hare cycle. *Science*, **269**, 1112–1115.

Krieg NR and Holt JG, eds (1984). *Bergey's manual of systematic bacteriology*. Williams and Wilkins, Baltimore.

Kristensen E (1984). Effect of natural concentrations on nutrient exchange between a polycheate burrow in estuarine sediment and the overlying water. *Journal Experimental Marine and Biological Ecology*, **75**, 171–190.

Kristensen E (1993). Seasonal variations in benthic community metabolism and nitrogen dynamics in a shallow, organic poor Danish lagoon. *Estuarine and Coastal Shelf Science*, **36**, 565–586.

Kristensen K and Hansen K (1999). Transport of carbon dioxide and ammonium in bioturbated (*Nereis diversicolor*) coastal, marine sediments. *Biogeochemistry*, **45**, 147–168.

Kruess A and Tscharntke T (1994). Habitat fragmentation, species loss, and biological control. *Science*, **264**, 1581–1584.

Kruger FJ, Breytenbach GJ, MacDonald IAW and Richardson DM (1989). In JA Drake, HA Mooney, F DiCastri, RH Groves, FJ Kruger, M Rejmanek and M Williamson, eds *Biological invasions: a global perspective*, pp. 181–213. John Wiley, Chichester.

Kück U, ed (1995). *The mycota, vol. 2: Genetics and biotechnology*. Springer-Verlag, Berlin.

Kuhn TS (1962). *The structure of scientific revolutions*. The University of Chicago Press, Chicago.

Kuikman PJ, Jansen AG, van Veen JA and Zehnder AJB (1990). Protozoan predation and the turnover of soil organic carbon and nitrogen in the presence of plants. *Biology and Fertility of Soils*, **10**, 22–28.

Laakso J and Setälä H (1999a). Population- and ecosystem-level effects of predation on microbial-feeding nematodes. *Oecologia*, **120**, 279–286.

Laakso J and Setälä H (1999b). Sensitivity of primary production to changes in the architecture of belowground food webs. *Oikos*, **87**, 57–64.

Laakso J, Setälä H and Palojarvi A (2000). Influence of decomposer food web structure and nitrogen availability on plant growth. *Plant and Soil*, **225**, 153–165.

Lacroix G and Lescher-Moutoué F (1991). Interaction effects of nutrient loading and density of young-of-the-year cyprinids on eutrophication in a shallow lake: an experimental mesocosm study. *Memorie dell'Istituto Italiano di Idrobiologia*, **48**, 53–74.

Lalli CM, ed (1991). Enclosed experimental marine ecosystems, a review and recommendations. Springer-Verlag, New York.

Lamont BB (1995). Testing the effects of composition/structure on its functioning. *Oikos*, **74**, 283–295.

Landeweert R, Hoffland E, Finlay RD, Kuyper ThW and van Breemen N (2001). Linking plants to rocks: Ectomycorrhizal fungi mobilize nutrients from minerals. *Trends in Ecology and Evolution*, **16(5)**, 248–254.

Landsberg J (1999). Response and effect—different reasons for classifying plant functional types under grazing. In D Eldridge and D Freudenberger, eds *People and rangelands: Building the future*, pp. 911–915. Proceedings of the VI International Rangeland Congress. VI International Rangeland Congress, Townsville, Australia.

Lavelle P and Pashanasi B (1989) Soil macrofauna and land management in Peruvian Amazonia (Yurimaguas, Loreto). *Pedobiologia* **33**, 283–291.

Lavelle P (1994). Faunal activities and soil processes: adaptive strategies that determine ecosystem function. In XV ISSS Congress Proceedings, Acapulco, Mexico, Vol. 1: Introductory Conferences, pp. 189–220.

Lavelle P, Bignell D and Lepage M (1997). Soil function in a changing world: the role of invertebrate ecosystem engineers. *European Journal of Soil Biology*, **33**, 159–193.

Lavelle P *et al.* (1994). The relationship between soil macro fauna and tropical soil fertility in PL Woomer and MJ Swift, eds. *The Biological Management of Tropical Soil Fertility*. John Wiley & Sons, Chichester.

Lavorel S and Garnier E (2001). Aardvarck to Zyzyxia—functional groups across kingdoms. *New Phytologist*, **149**, 1–4.

Lavorel S, McIntyre S, Landsberg J and Forbes TDA (1997). Plant functional classifications: from general groups to specific groups based on response to disturbance. *Trends in Ecology and Evolution*, **12**, 474–478.

Lavorel S, Prieur-Richard A-H and Grigulis K (1999). Invasibility and diversity of plant communities: from patterns to processes. *Diversity and Distributions*, **5**, 41–49.

Law R (1988). Some ecological properties of intimate mutualisms involving plants. In AJ Davy, MJ Hutchings and AR Watkinson, eds *Plant population ecology*. pp. 315–341. Blackwell Science, Oxford.

Law R and Blackford JC (1992). Self assembling food webs: a global viewpoint of coexisting species in Lotka-Volterra communities. *Ecology*, **73**, 567–578.

Law R and Morton RD (1996). Permanence and the assembly of ecological communities. *Ecology*, **77**, 762–775.

Lawes JB, Gilbert JH and Masters MT (1882). Agricultural, chemical, and botanical results of experiments on the mixed herbage of permanent grasslands, conducted for more than twenty years in succession on the same land. Part II. The botanical results. *Philosophical Transactions of the Royal Society of London, Series A and B*, **173**, 1181–1423.

Lawler SP (1998). Ecology in a bottle: using microcosms to test theory. In WJJ Resetarits and J Bernardo, eds *Experimental ecology. Issues and perspectives*, pp. 236–253. Oxford University Press, Oxford.

Lawler SP and Morin PJ (1993). Food web architecture and population dynamics in laboratory microcosms of protists. *The American Naturalist*, **141**, 675–686.

Lawler S, Armesto JJ and Kareiva P (2002). How relevant are studies of biodiversity and ecosystem functioning to conservation? In A Kinzig, S Pacala and D Tilman, eds *Functional consequences of biodiversity: experimental progress and theoretical extension*. Princeton University Press.

Lawrence DC (1996). Trade-offs between rubber production and maintenance of diversity: the structure of rubber gardens in West Kalimantan, Indonesia. *Agroforestry Systems*, **34**, 83–100.

Lawrence DC, Leighton M and Peart DR (1995). Availability and extraction of forest products in managed and primary forest around a Dayak village in West Falimantau, Indonesia. *Conservation Biology*, **9**, 76–88.

Lawrence KL and Wise DH (2000). Spider predation on forest-floor Collembola and evidence for indirect effects on decomposition. *Pedobiologia*, **44**, 33–39.

Lawton JH (1994). What do species do in ecosystems? *Oikos*, **71**, 367–374.

Lawton JH (1995). Ecological experiments with model systems. *Science*, **269**, 328–331.

Lawton JH (1998). Ecological experiments with model systems: the ecotron facility in context. In WJJ Resetarits and J Bernardo, eds *Experimental ecology: Issues and perspectives*. Oxford University Press, Oxford.

Lawton JH (1999). Size matters. *Oikos*, **85**, 19–21.

Lawton JH (2000). *Community ecology in a changing world*. Ecology Institute, Oldendorf, Germany.

Lawton JH and Brown VK (1993). Redundancy in ecosystems. In ED Schulze and HA Mooney, eds. *Biodiversity and ecosystem function*, pp. 255–270. Springer Verlag, New York.

Lawton JH and May RM, eds (1995). *Extinction rates*. Oxford University Press, Oxford.

Lawton JH *et al.* (1993). The Ecotron: a controlled environmental facility for the investigation of populations and ecosystem processes. *Philosophical Transactions of the Royal Society of London, Series B*, **341**, 181–194.

Lawton JH *et al.* (1996). Carbon flux and diversity of nematodes and termites in Cameroon forest soils. *Biodiversity and Conservation*, **5**, 261–273.

Lawton JH, Naeem S, Thompson LJ, Hector A and Crawley MJ (1998). Biodiversity and ecosystem functioning: Getting the Ecotron experiment in its correct context. *Functional Ecology*, **12**, 843–856.

Leadley PW and Körner C (1996). Effects of elevated CO_2 on plant species dominance in a highly diverse calcareous grassland. In C Körner and FA Bazzaz, eds *Carbon dioxide, populations and communities*, pp. 159–175. Academic Press, New York.

Leake JR, Donnelly DP and Boddy L (2002). Interactions between ecto-mycorrhizal fungi and saprotrophic fungi. In MGA van der Heijden and IR Sanders, eds *Mycorrhizal ecology*. Ecological Studies 157. Springer-Verlag, Heidelberg.

Leemans R (1997). The use of plant functional type classifications to model global land cover and simulate the interactions between the terrestrial biosphere and the atmosphere. In TM Smith, HH Shugart and FI Woodward, eds *Plant functional types: Their relevance to ecosystem properties and global change*, pp. 289–316. Cambridge University Press, Cambridge.

Lefeuvre JC (1992). L'agriculture et la gestion des ressources renouelables. *Economie-Rurale*, **208–209**, 79–84.

Lehman CL and Tilman D (2000). Biodiversity, stability, and productivity in competitive communities. *American Naturalist*, **156**, 534–552.

Leibold MA (1989). Resource edibility and the effects of predators and productivity on the outcome of trophic interactions. *American Naturalist*, **134**, 922–949.

Leibold MA (1995). The niche concept revisited: mechanistic models and community context. *Ecology*, **76**, 1371–1382.

Leibold MA (1996). A graphical model of keystone predators in food webs: trophic regulation of abundance, incidence, and diversity patterns in communities. *American Naturalist*, **147**, 784–812.

Leibold MA and Wilbur HM (1992). Interactions between food web structure and nutrients on pond organisms. *Nature*, **360**, 341–343.

Leibold MA, Chase JM, Shurin JB and Downing AL (1997). Species turnover and the regulation of trophic structure. *Annual Review of Ecology and Systematics*, **28**, 467–494.

Lemmens J *et al.* (1996). Filtering capacity of seagrass meadows and other habitats of Cockburn Sound, Western Australia. *Marine Ecology-Progress Series*, **143**, 187–200.

Lepš J, Osbornova-Kosinoa J and Rejmanek K (1982). Community stability, complexity and species life-history strategies. *Vegetatio*, **50**, 53–63.

Lepš J *et al.* (2001). Separating the chance effect from the other diversity effects in the functioning of plant communities. *Oikos*, **92**, 123–134.

Lesser MP, Shumway SE, Cucci T and Smith J (1992). The impact of fouling organisms on mussel rope culture: Interspecific competition for food among suspension-feeding invertebrates. *Journal Experimental Marine and Biological Ecology*, **165**, 91–102.

Levin SA (1981). The role of theoretical ecology in the description and understanding of populations in heterogeneous environments. *American Zoologist*, **21**, 865–875.

Levin SA (1992). The problem of pattern and scale in ecology. *Ecology*, **73**, 1943–1967.

Levin SA (1999). *Fragile dominion: complexity and the commons.* Perseus Books, Reading, Massachusetts.

Levin SA and Pacala S (1997). Theories of simplification and scaling of spatially distributed processes. In D Tilman and P Kareiva, eds *Spatial ecology.* Princeton University Press, Princeton.

Levine JM (2000a). Species diversity and biological invasions: Relating local process to community pattern. *Science*, **288**, 852–854.

Levine JM (2000b). Complex interactions in a streamside plant community. *Ecology*, **81**, 3431–3444.

Levine JM (2001). Local interactions, dispersal, and native and exotic plant diversity along a California stream. *Oikos*, **95**, 397–408.

Levine JM and D'Antonio CM (1999). Elton revisited: a review of evidence linking diversity and invasibility. *Oikos*, **87**, 15–26.

Levins R (1969). The effect of random variations on different types of population growth. *Proceedings of the National Academy of Sciences*, **62**, 1061–1065.

Levins R (1970). Complex systems. In CH Waddington, ed *Towards a theoretical biology*, pp. 73–88. Edinburgh University Press, Edinburgh.

Levins R (1979). Coexistence in a variable environment. *American Naturalist*, **114**, 765–783.

Levins R and Culver D (1971). Regional coexistence of species and competition between rare species. *Proceedings of the National Academy of Sciences*, **68**, 1246–1248.

Levinton J (1995). Bioturbators as ecosystem engineers: control of the sediment fabric, inter-individual interactions, and material fluxes. In CG Jones and JH Lawton, eds *Linking species and ecosystems*, pp. 29–36. Chapman and Hall, San Diego.

Liebman M (1995). Polyculture cropping systems. In MA Altieri, *Agroecology.* Westview Press, Boulder, Colorado.

Liebman M and Dyck E (1993). Crop rotation and inter-cropping strategies for weed management. *Ecological Applications*, **3**, 92–122.

Liiri M, Setälä H, Haimi J, Pennanen T and Fritze H (2002). Relationship between soil microarthropod species diversity and plant growth does not change when the system is disturbed. *Oikos*, **96**, 138–150.

Likens GE (1992). *The ecosystem approach: Its use and abuse.* Ecology Institute, Oldendorf/Luhe, Germany.

Liljeroth E, Scheling GC and van Veen JA (1990). Influence of different application rates of nitrogen to soil on rizosphere bacteria. *Netherlands Journal of Agricultural Science*, **38**, 255–264.

Linusson AC, Berlin GAI and Olsson EGA (1998). Reduced community diversity in semi-natural meadows in southern Sweden 1965–1990. *Plant Ecology*, **136**, 77–94.

Lonsdale WM (1999). Global patterns of plant invasions and the concept of invasibility. *Ecology*, **80**, 1522–1536.

Loreau M (1996). Coexistence of multiple food chains in a heterogeneous environment: interactions among community structure, ecosystem functioning, and nutrient dynamics. *Mathematical Biosciences*, **134**, 153–188.

Loreau M (1998a). Biodiversity and ecosystem functioning: a mechanistic model. *Proceedings of the National Academy of Sciences of the United States of America*, **95**, 5632–5636.

Loreau M (1998b). Separating sampling and other effects in biodiversity experiments. *Oikos*, **82**, 600–602.

Loreau M (1998c). Ecosystem development explained by competition within and between material cycles. *Proceedings of the Royal Society of London, Series B*, **265**, 33–38.

Loreau M (2000a). Biodiversity and ecosystem function: recent theoretical advances. *Oikos*, **91**, 3–17.

Loreau M (2000b). Are communities saturated? On the relationship between a, b and g diversity. *Ecology Letters*, **3**, 73–76.

Loreau M (2001). Microbial diversity, producer–decomposer interactions and ecosystem processes: a theoretical model. *Proceedings of the Royal Society of London, Series B*, **268**, 1–7.

Loreau M (2002). Evolutionary processes in ecosystems. In HA Mooney and J Canadell, eds *Encyclopedia of global environmental change*, Vol. 2, *The Earth system: biological and ecological dimensions of global environmental change*, pp. 292–297. John Wiley and Sons, London.

Loreau M and Behera N (1999). Phenotypic diversity and stability of ecosystem processes. *Theoretical Population Biology*, **56**, 29–47.

Loreau M and Hector A (2001). Partitioning selection and complementarity in biodiversity experiments. *Nature*, **412**, 72–76 and **413**, 548.

Loreau M and Mouquet N (1999). Immigration and maintenance of local species diversity. *American Naturalist*, **154**, 427–440.

Loreau M et al. (2001). Biodiversity and ecosystem functioning: current knowledge and future challenges. *Science*, **294**, 804–808.

Losch R, Thomas D, Kaib U, Peters R and Toatman N (1994). Resource use of crops and weeds on extensively managed field margins. In *Field margins: integrating agriculture and conservation*. Proceedings of a symposium held at Coventry, UK 18–20 April 1994, pp. 203–208. BCPC Monograph No. 58, British Crop Protection Council: Farnham, UK.

Lovelock J (1979). *Gaia*. Oxford University Press, Oxford.

Lubchenco JEA (1991). The sustainable biosphere initiative: an ecological research agenda. *Ecology*, **72**, 371–412.

Lubchenco J (1998). Entering the century of the environment: a new social contract for science. *Science*, **279**, 491–497.

Luckinbill LS (1974). The effects of space and enrichment on a predator–prey system. *Ecology*, **55**, 1142–1147.

Lyons KG and Schwartz MW (2001). Rare species loss alters ecosystem function—invasion resistance. *Ecology Letters*, **4**, 358–365.

MacArthur RH (1955). Fluctuations of animal populations and a measure of community stability. *Ecology*, **36**, 533–536.

MacArthur RH (1960). On the relative abundance of species. *American Naturalist*, **94**, 25–36.

MacArthur RH (1968). The theory of the niche. In RC Lewontin, ed. *Population Biology and Evolution*, pp. 159–176. Syracuse University Press, Syracuse, New York.

MacArthur RH (1970). Species-packing and competitive equilibrium for many species. *Theoretical Population Biology*, **1**, 1–11.

MacArthur RH (1972). *Geographical ecology: Patterns in the distribution of species*. Harper and Row, New York.

MacArthur RH and Wilson EO (1967). *The theory of island biogeography*. Princeton University Press, Princeton.

MacDonald D et al. (2000). Agricultural abandonment in mountain areas of Europe: Environmental consequences and policy response. *Journal of Environmental Management*, **59**, 47–69.

MacGillivray CW et al. (1995). Testing predictions of the resistance and resilience of vegetation subjected to extreme events. *Functional Ecology*, **9**, 640–649.

Magurran AE (1988). *Ecological diversity and its measurement*. Chapman and Hall, London.

Mahdi A, Law R and Willis AJ (1989). Large niche overlaps among coexisting plant species in a limestone grassland community. *Journal of Ecology*, **77**, 386–400.

Malý S, Korthals GW, van Dijk C, van der Putten WH and de Boer W (2000). Effect of vegetation manipulation of abandoned land on soil microbial properties. *Biology and Fertility of Soils*, **31**, 121–127.

Malta EJ, Draisma SGA and Kamermans P (1999). Free-floating Ulva in the southwest Netherlands: species or morphotypes? A morphological, molecular and ecological comparison. *European Journal of Phycology*, **34**, 443–454.

Manly B (1992). *The design and analysis of research studies*. Cambridge University Press, Cambridge.

Maraun M, Migge S, Schaefer M and Scheu S (1998). Selection of microfungal food by six oribatid mite species (Oribatida, Acari) from two different beech forests. *Pedobiologia*, **42**, 232–240.

Marcogliese DJ and Cone DK 1997. Food webs: a plea for parasites. *Trends in Ecology and Evolution*, **12**, 320–325.

Margulis L (1993). *Symbiosis in cell evolution*. WH Freeman, New Work.

Marschner H (1995). *Mineral nutrition of higher plants*. Academic Press, London.

Martin TH, Crowder LB, Dumas CF and Burkholder JM (1992). Indirect effects of fish on macrohytes in Bays Mountain Lake: evidence for a littoral trophic cascade. *Oecologia*, **89**, 476–481.

Martinez ND (1996). Defining and measuring functional aspects of biodiversity. In KJ Gaston, ed *Biodiversity. A biology of numbers and difference*, pp. 115–148. Blackwell Science, Oxford.

Marx DH (1991). The practical significance of ectomycorrhizae in forest establishment. In *Ecophysiology of ectomycorrhizae of forest trees*. Marcus Wallenberg Foundation Symposia Proceedings, Vol. 7, pp. 54–90. M. Wallenberg Foundation, Stockholm, Sweden.

Masters GJ (1995). The impact of root herbivory on aphid performance: field and laboratory evidence. *Acta Œcologica*, **16**, 135–142.

Masters GJ and Brown VK (1992). Plant-mediated interactions between two spatially separated insects. *Functional Ecology*, **6**, 175–179.

Masters GJ, Jones TH and Rogers M (2001). Host-plant mediated effects of root herbivory on insect predators and their parasitoids. *Oecologia*, **127**, 246–250.

Materon LA, Keatinge JDH, Bek DP, Yurtsever N, Karuc K and Altuntas S (1995). The role of rhizobial biodiversity in legume crop productivity in the west asian highlands II. *Rhizobium leguminosarum. Experimental Agriculture*, **31**, 485–491.

Mather K and Jinks JL (1982). *Biometrical genetics, 3rd Edition*. Chapman and Hall, London.

May RM (1972). Will large and complex systems be stable? *Nature*, **238**, 413–414.

May RM (1973). *Stability and complexity in model ecosystems, 2nd Edition*. Princeton University Press, Princeton.

May RM (1975). Patterns of species abundance and diversity. In ML Cody and JM Diamond, eds *Ecology and evolution of communities*. pp. 81–120. Belknap Press, Cambridge, MA.

May RM (1984). *Exploitation of marine communities*. Springer-Verlag, Berlin.

May RM (1997). Concluding remarks. In A Gange, VK Brown, eds *Multitrophic interactions in terrestrial systems*, pp. 419–422. Blackwell Science, Oxford.

May RM and MacArthur RH (1972). Niche overlap as a function of environmental variability. *Proceedings of the National Academy of Sciences*, **69**, 1109–1113.

Mayer MS, Schaffner L and Kemp WM (1995). Nitrification potentials of benthic macrofaunal tubes and burrow walls: effects of sediment NH_4^+ and animal irrigation behaviour. *Marine Ecology-Progress Series*, **121**, 157–169.

McCaig AE, Glover LA and Prosser JI (1999). Molecular analysis of bacterial community structure and diversity in unimproved and improved upland grass pastures. *Applied and Environmental Microbiology*, **65**, 1721–1730.

McCann KS (2000). The diversity stability debate. *Nature*, **405**, 228–233.

McCann K, Hastings A and Strong DR (1998a). Trophic cascades and trophic trickles in pelagic food webs. *Proceedings of the Royal Society of London, Series B, Biological Sciences*, **265**, 205–209.

McCann K, Hastings A and Huxel GR (1998b). Weak trophic interactions and the balance of nature. *Nature*, **395**, 794–798.

McCullagh P and Nelder JA (1989). *Generalized linear models. 2nd Edition*. Chapman and Hall, London.

McGilchrist CA (1965). Analysis of competition experiments. *Biometrics*, **27**, 975–986.

McGrady-Steed J, Harris PM and Morin PJ (1997). Biodiversity regulates ecosystem predictability. *Nature*, **390**, 162–165.

McGrady-Steed J and Morin PJ (2000). Biodiversity, density compensation, and the dynamics of populations and functional groups. *Ecology*, **81**, 361–373.

McIntyre S and S Lavorel (2001). Livestock grazing in subtropical pastures: steps in the analysis of attribute response and plant functional types. *Journal of Ecology*, **89**, 209–226.

McIntyre S, Lavorel S and Tremont RM (1995). Plant life-history attributes: their relationship to disturbance response in herbaceous vegetation. *Journal of Ecology*, **83**, 31–44.

McKee KL and Faulkner PL (2000). Restoration of biogeochemical function in mangrove forests. *Restoration Ecology*, **8**, 247–259.

McNaughton SJ (1977). Diversity and stability of ecological communities: a comment on the role of empiricism in ecology. *American Naturalist*, **111**, 515–525.

McNaughton SJ (1979). Grazing as an optimization process: grass-ungulate relationships in the Serengeti. *American Naturalist*, **113**, 691–703.

McNaughton SJ (1985). Ecology of a grazing system: the Serengeti. *Ecological Monographs*, **55**, 259–294.

McNaughton SJ (1993). Biodiversity and function of grazing ecosystems. In ED Schulze and HA Mooney, eds *Biodiversity and ecosystem function*, pp. 361–383. Springer-Verlag, Berlin.

McNaughton SJ, Ruess RW and Seagle SW (1988). Large mammals and process dynamics in African ecosystems. *BioScience*, **38**, 794–800.

McNaughton SJ, Oesterheld M, Frank DA and Williams KJ (1989). Ecosystem-level patterns of primary productivity and herbivory in terrestrial habitats. *Nature*, **341**, 142–144.

Mead R and Riley J (1981). A review of statistical ideas relevant to intercropping research. *Journal of the Royal Statistical Society, Series A*, **144**, 462–509.

Mellinger MV and McNaughton SJ (1975). Structure and function of successional vascular plant communities in central New York. *Ecological Monographs*, **45**, 161–182.

Mellink E (1991). Bird communities associated with three traditional agroecosystems in the San Luis Potosi Plateau, Mexico. *Agriculture Ecosystems and Environment*, **36**, 37–50.

Mermillod F, Creuzé de Châtelliers M, Gérino M and Gaudet J-P (2000). Testing the effect of Limnodrilus sp. (Oligochaeta, Tubificidae) on organic matter and nutrient processing in the hyporheic zone: a microcosm method. *Archiv für Hydrobiologie*, **149**, 467–487.

Meyer O (1994). Functional groups of microorganisms. In ED Schultze and HA Mooney, eds *Biodiversity and ecosystem function*. Springer-Verlag, Berlin.

Micheli F et al. (1999). The dual nature of community variability. *Oikos*, **85**, 161–169.

Michelsen A, Schmidt IK, Jonasson S, Quarmby C and Sleep D (1996). Leaf ^{15}N abundance of subarctic plants provides field evidence that ericoid, ectomycorrhizal and non- and arbuscular mycorrhizal species access different sources of soil nitrogen. *Oecologia*, **105**, 53–63.

Michon G, Mary F and Bolmpard J (1986). Multistoried agroforestry garden system in West Sumatra, Indonesia. *Agroforestry Systems*, **4**, 315–338.

Mihail JD, Alexander HM and Taylor SJ (1998). Interactions between root-infecting fungi and plant density in an annual legume, *Kummerowia stipulacea*. *Journal of Ecology*, **86**, 739–748.

Mikola J and Setälä H (1998a). Relating species diversity to ecosystem functioning: mechanistic backgrounds and experimental approach with a decomposer food web. *Oikos*, **83**, 180–194.

Mikola J and Setälä H (1998b). No evidence of trophic cascades in an experimental microbial-based soil food web. *Ecology*, **79**, 153–164.

Mikola J and Setälä H (1998c). Productivity and trophic-level biomasses in a microbial-based soil food web. *Oikos*, **82**, 158–168.

Mikola J and Sulkava P (2001). Responses of microbial-feeding nematodes to organic matter distribution and predation in experimental soil habitat. *Soil Biology and Biochemistry*, **33**, 811–817.

Miller RM and Jastrow JD (2000). Mycorrhizal fungi influence soil structure. In Y Kapulnik and DD Douds, eds *Arbuscular mycorrhizae: Molecular biology and physiology*. Kluwer Academic Press. The Netherlands.

Mills KE and Bever JD (1998). Maintenance of diversity within plant communities: soil pathogens as agents of negative feedback. *Ecology*, **79**, 1595–1601.

Minerdi D, Fani R, Gallo R, Boarino A and Bonfante P (2001). Nitrogen fixation genes in an endosymbiotic *Burholderia* Strain. *Applied and Environmental Microbiology*, **67**, 725–732.

Mittelbach GG and Osenberg CW (1993). Stage-structured interactions in bluegill: consequences of adult resource variation. *Ecology*, **74**, 2381–2394.

Molina R, Massicotte H and Trappe JM (1992). Specificity phenomena in mycorrhizal symbioses: Community-ecological consequences and practical implications. In MF Allen, ed. *Mycorrhizal functioning*, pp. 357–423. Chapman and Hall, New York.

Mook DH (1981). Removal of suspended particles by fouling communities. *Marine Ecology-Progress Series*, **5**, 279–281.

Mooney HA (1990). Toward the study of the earth's metabolism. *Bulletin of the Ecological Society of America*, **71**, 221–228.

Mooney HA (1991). Emergence of the study of global ecology. Is terrestrial ecology an impediment to progress. *Ecological Applications*, **1**, 2–5.

Mooney HA (1999). On the road to global ecology. *Annual Review of Energy and Environment*, **24**, 1–31.

Mooney HA and Drake JA, eds. (1986). *Ecology of biological invasions of North America and Hawaii*. Springer-Verlag, New York.

Mooney HA et al., eds. (1996). *Functional roles of biodiversity: A global perspective*. John Wiley, Chichester.

Moora M and Zobel M (1996). Effect of arbuscular mycorrhiza on inter- and intraspecific competition of two grassland species. *Oecologia*, **108**, 79–84.

Moore JC and de Ruiter PC, eds (1997). *Compartmentalization of resource utilization within soil ecosystems. Multitrophic interactions in terrestrial ecosystems*. Blackwell Science, Oxford.

Moore JC and Hunt HW (1988). Resource compartmentation and the stability of real ecosystems. *Nature*, **333**, 261–263.

Moore JC et al. (1993). Influence of productivity on the stability of real and model ecosystems. *Science*, **261**, 906–908.

Moore JC et al. (1988). Arthropod regulation of micro- and mesobiota in belowground food webs. *Annual Review of Entomology*, **33**, 419–439.

Moore J, Mouquet N, Lawton JH and Loreau M (2001). Coexistence, saturation and invasion resistance in simulated plant assemblages. *Oikos*, **94**, 303–314.

Moore PD (1996). Biodiversity and climate change dominate environmental research. *The Scientist*, **10**, 14.

Morgan Ernest SK and Brown JH (2001). Homeostasis and compensation: the role of species and resources in ecosystem stability. *Ecology*, **82**, 2118–2132.

Morgan JW (1998). Patterns of invasion of an urban remnant of a species-rich grassland in southeastern Australia by non-native plant species. *Journal of Vegetation Science*, **9**, 181–190.

Morin PJ (1998). Realism, precision, and generality in experimental ecology. In WJJ Resetarits and J Bernardo, eds *Experimental ecology. Issues and perspectives*, pp. 236–253. Oxford University Press, Oxford.

Mortimer SR, Booth RG, Harris SJ and Brown VK (2001). Effects of initial site management on the Coleoptera assemblages colonising newly-established chalk grasslands on ex-arable land. *Biological Conservation*, **99**, 29–46.

Mulder CPH, Koricheva J, Huss-Danell K, Högberg P and Joshi J (1999). Insects affect relationships between plant species richness and ecosystem processes. *Ecology Letters*, **2**, 237–246.

Mulder CPH, Uliassi DD and Doak DF (2001). Physical stress and diversity-productivity relationships: The role of positive interactions. *Proceedings of the National Academy of Sciences of the United States of America*, **98**, 6704–6708.

Muller PE (1884). Studier over skovjord, som bidrag til skovdyrkningens theori. II. Om muld og mor i egeskove og paa heder. *Tidsskrift for Skovbrug*, **7**, 1–232.

Müller-Dombois D and Ellenberg H (1974). *Aims and methods of vegetation ecology*. John Wiley and Sons, New York.

Murdoch WW and Briggs CJ (1996). Theory for biological control. *Ecology*, **77**, 2001–2013.

Murdoch WW, Chesson J and Chesson PLO (1985). Biological control in theory and practice. *American Naturalist*, **125**, 344–366.

Myllius SD, Klumpers K, De Roos AM and Persson L (2001). Impact of intraguild predation and stage structure on simple communities along a productivity gradient. *American Naturalist*, **158**, 259–276.

Myrold DD, Matson PA and Peterson DL (1989). Relationships between soil microbial properties and aboveground stand characteristics of coniferous forests in Oregon. *Biogeochemistry*, **8**, 265–281.

Naeem S (1998). Species redundancy and ecosystem reliability. *Conservation Biology*, **12**, 39–45.

Naeem S (1999). Power behind Nature's throne. *Nature*, **401**, 653–654.

Naeem S (2000). Reply to Wardle *et al. Bulletin of the Ecological Society of America*, **81**, 241–246.

Naeem S (2001). Experimental validity and ecological scale as criteria for evaluating research programs. In RH Gardner *et al.*, eds *Scaling relations in experimental ecology*, pp. 223–250. Columbia Universtiy Press.

Naeem S (in press). In S Levin, ed. *Functioning of biodiversity. Encyclopedia of global environmental change*. John Wiley and Sons.

Naeem S and Li S (1997). Biodiversity enhances ecosystem reliability. *Nature*, **390**, 507–509.

Naeem S, Thompson LJ, Lawler SP, Lawton JH and Woodfin RM (1994a). Declining biodiversity can alter the performance of ecosystems. *Nature*, **368**, 734–736.

Naeem S, Thompson LJ, Lawler SP, Lawton JH and Woodfin RM (1994b). Biodiversity in model ecosystems (response). *Nature*, **371**, 565.

Naeem S, Thompson LJ, Lawler SP, Lawton JH and Woodfin RM (1995). Empirical evidence that declining species diversity may alter the performance of terrestrial ecosystems. *Transactions of the Royal Society of London B*, **347**, 249–262.

Naeem S, Håkansson K, Lawton JH, Crawley MJ and Thompson LJ (1996). Biodiversity and plant productivity in a model assemblage of plant species. *Oikos*, **76**, 259–264.

Naeem S, Hahn DR and Schuurman G (2000a). Producer-decomposer co-dependency influences biodiversity effects. *Nature*, **403**, 762–764.

Naeem S, Knops JMH, Tilman D, Howe KM, Kennedy T and Gale S (2000b). Plant diversity increases resistance to invasion in the absence of covarying extrinsic factors. *Oikos*, **91**, 97–108.

Nee S and May RM (1992). Dynamics of metapopulations: habitat destruction and competitive coexistence. *Journal of Animal Ecology*, **61**, 37–40.

Neter J, Kutner MH, Nachtsheim CJ and Wasserman W (1996). *Applied linear statistical models, 4th Edition*. Irwin, Chigaco.

Netting R and Stone MP (1996). Agro-diversity on a farming frontier: Kofyar smallholders on the Benue plains of central Nigera. *Africa London*, **66**, 52–70.

Neubert MG and Caswell H (1997). Alternatives to resilience for measuring the responses of ecological systems to perturbations. *Ecology*, **78(3)**, 653–665.

Neutel AM, Heesterbeek JAP, de Ruiter PC (2002). Stability in real food webs: weak links in long loops. *Science*, **296**, 1120–1123.

Newell K (1984a). Interaction between two decomposer basidiomycetes and a collembolan under Sitka spruce: grazing and its potential effects on fungal distribution and litter decomposition. *Soil Biology and Biochemistry*, **16**, 235–239.

Newell K (1984b). Interaction between two decomposer basidiomycetes and a collembolan under Sitka spruce: distribution, abundance and selective grazing. *Soil Biology and Biochemistry*, **16**, 227–233.

Newman EI (1988). Mycorrhizal links between plants: their functioning and ecological significance. *Advances in Ecological Research*, **18**, 243–270.

Newsham KK, Fitter AH and Watkinson AR (1995a). Arbuscular mycorrhizae protect an annual grass from root pathogenic fungi in the field. *Journal of Ecology*, **83**, 991–1000.

Newsham KK, Fitter AH and Watkinson AR (1995b). Multi-functionality and biodiversity in arbuscular mycorrhizas. *Trends in Ecology and Evolution*, **10**, 407–411.

Newton PCD, Clark H, Edwards GR and Ross DJ (2001). Experimental confirmation of ecosystem model predictions comparing transient and equilibrium plant responses to elevated CO_2. *Ecology Letters*, **4**, 344–347.

Nijs I and Impens I (2000). Underlying effects of resource use efficiency in diversity–productivity relationships. *Oikos*, **91(1)**, 204–208.

Nijs I and Roy J (2000). How important are species richness, species evenness and interspecific differences to productivity? A mathematical model. *Oikos*, **88(1)**, 57–66.

Niklaus PA, Kandeler E, Leadley PW, Schmid B, Tscherko D and Körner C (2001). A link between plant diversity, elevated CO_2 and soil nitrate. *Oecologia*, **127**, 540–548.

Nilsson MC, Wardle DA and Dahlberg A (1999). Effects of plant litter species composition and diversity on the boreal forest plant-soil system. *Oikos*, **86**, 16–26.

Nixon SW, Oviatt CA and Hale SS (1976). Nitrogen regeneration and the metabolism of coastal marine bottom communities. In JM Anderson and A Macfadyen, eds *The role of terrestrial and aquatic organisms in decomposition processes*, pp. 269–283. Blackwell Science, London.

Noble IR and Slatyer RO (1980). The use of vital attributes to predict successional changes in plant communities subject to recurrent disturbances. *Vegetatio*, **43**, 5–21.

Norberg J (2000). Resource-niche complementarity and autotrophic compensation determines ecosystem level responses to increased cladoceran species richness. *Oecologia*, **122**, 264–272.

Norberg J, Swaney DP, Dushoff J, Lin J, Casagrandi R and Levin SA (2001). Phenotypic diversity and ecosystem functioning in changing environments: a theoretical framework. *Proceedings of the National Academy of Sciences of the United States of America*, **98**, 11376–11381.

Nyström M and Folke C (2001). Spatial resilience in coral reefs. *Ecosystems*, **4**, 406–417.

Nyström M, Folke C and Moberg F (2000). Coral reef disturbance and resilience in a human-dominated environment. *Trends in Ecology and Evolution*, 413–417.

Odum EP (1953). *Fundamentals of ecology*. Saunders, Philadelphia.

Odum EP (1963). *Ecology*. Holt, Rinehart, and Winston, New York.

Odum EO (1969). The strategy of ecosystem development. *Science*, **164**, 262–270.

Okano S, Sato K and Inoue E (1991). Negative relationship between microbial biomass and root amount in topsoil of a renovated grassland. *Soil Science and Plant Nutrition*, **37**, 47–53.

Oksanen J (1996). Is the humped relationship between species richness and biomass an artifact due to plot size? *Journal of Ecology*, **84**, 293–296.

Oksanen L and Oksanen T (2000). The logic and realism of the hypothesis of exploitation ecosystems. *American Naturalist*, **155**, 703–723.

Oksanen L, Fretwell SD, Arruda J and Niemelä P (1981). Exploitation ecosystems in gradients of primary productivity. *American Naturalist*, **118**, 240–261.

Olembo R and Hawksworth DL (1991). Importance of microorganisms and invertebrates as components of biodiversity. In DL Hawksworth, ed. *The biodiversity of microorganisms and invertebrates: its role in sustainable agriculture*, pp. 7–15. CAB International, Wallingford, UK.

Olff H and Ritchie ME (1998). Effects of herbivores on grassland plant diversity. *Trends in Ecology and Evolution*, **13**, 61–265.

Olff H, Hoorens B, de Goede RGM, Van der Putten WH and Gleichman JM (2000). Small-scale shifting mosaics of two dominant grassland species: the possible role of soil pathogens. *Oecologia*, **125**, 45–54.

Ollason JG (1977). Freshwater microcosms in fluctuating environments. *Oikos*, **28**, 262–269.

O'Neill RV (1969). Indirect estimation of energy fluxes in animal food webs. *Journal of Theoretical Biology*, **22**, 284–290.

O'Neill RV and King AW (1998). Homage to St. Michael; or, why are there so many books on scale. In DL Peterson and VT Parker, eds *Ecological scale. Theory and applications*, pp. 3–15. Columbia University Press, New York.

Ong CK (1994). Alley cropping, ecological pie in the sky? *Agroforestry Today*, **6**, 8–10.

Ong CK (1995). The 'dark side' of intercropping: manipulation of soil resources. In H Sinoqut and P Cruz, eds Proceedings of the International Conference on Ecophysiology of Intercropping, 6–11 Dec. 1993, Guadeloupe, pp. 45–46. INRA, Science Update.

Ong CK (1997). A framework for quantifying the varius effects of tree-crop interactions. In CK Ong and P Huxley. *Tree-crop interactions: A physiological approach*. CAB International, Wallingford, UK.

Osenberg CW and Mittlebach GG (1996). The realtive importance of resource limitation and predator limitation in food chains In GA Polis and KO Winemiller, eds *Food webs: integration of patterns and dynamics*, pp. 134–148. Chapman and Hall, New York.

Östman Ö, Ekbom B and Bengtsson J (2001). Natural enemy impacts on a pest aphid varies with landscape structure and farming practice. *Basic and Applied Ecology*, **2**, 365–371.

O'Toole C and Gauld ID (1993). Diversity of native bees and agroedosystems. In La J Salle, ed. *Hymenoptera and biodiversity*, pp. 169–196. CAB International, Wallingford, UK.

Øvreås L (2000). Population and community level approaches for analysing microbial diversity in natural environments. *Ecology Letters*, **3**, 236–251.

Pacala SW and Crawley MJ (1992). Herbivores and plant diversity. *American Naturalist*, **140**, 243–260.

Pacala SW and Silander JA Jr (1985). Neighborhood models of plant population dynamics. I. Single-species models of annuals. *American Naturalist*, **125**, 385–411.

Pacala SW and Silander JA (1987). Neighbourhood interference among velvet leaf, *Abutilon theophrasti* and pigweed, *Amaranthus retroflexus*. *Oikos*, **48**, 217–224.

Pacala SW and Silander JA Jr (1990). Field tests of neighborhood population dynamic models of two annual weed species. *Ecological Monographs*, **60**, 113–134.

Pacala S and Tilman D (2002). The transition from sampling to complementarity. In A Kinzig, S Pacala and D Tilman, eds *Functional consequences of biodiversity: Experimental progress and theoretical extensions*. Princeton University Press, New Jersey.

Pace ML, Cole JJ and Carpenter SR (1998). Trophic cascades and compensation: differential responses of microzooplankton in whole-lake experiments. *Ecology*, **79**, 138–152.

Pace ML, Cole JJ, Carpenter SR and Kitchell JF (1999). Trophic cascades revealed in diverse ecosystems. *Trends in Ecology and Evolution*, **14**, 483–488.

Pace NR (1997). A molecular view of microbial diversity and the biosphere. *Science*, **276**, 734–740.

Packer A and Clay K (2000). Soil pathogens and spatial patterns of seedling mortality in a temperate tree. *Nature*, **404**, 278–281.

Paine RT (1965). Food web complexity and species diversity. *The American Naturalist*, **100**, 65–75.

Paine RT (1966). Food web complexity and community stability. *American Naturalist*, **100**, 65–75.

Paine RT (1969). A note on trophic complexity and community stability. *American Naturalist*, **103**, 91–93.

Paine RT (1974). Intertidal community structue: experimental studies on the relationship between a dominant competitor and its principal predator. *Oecologia*, **15**, 93–120.

Paine RT (1980). Food webs: linkage, interaction strength and community infrastructure. *Journal of Animal Ecology*, **49**, 667–685.

Paine RT (1988). On food webs: road maps of interactions or the grist for theoretcial development? *Ecology*, **69**, 1648–1654.

Paine RT (1992). Food-web analysis through field measurement of per capita interaction strength. *Nature*, **355**, 73–75.

Palmer MW (1994). Variation in species richness: Towards a unification of hypothese. *Folia Geobotentice et Phytotoxonomicie, Praha*, **29**, 511–530.

Palmer MW and Maurer TA (1997). Does diversity beget diversity? A case study of crops and weeds. *Journal of Vegetation Science*, **8**, 235–240.

Paoletti MG and Pimentel D, eds (1992). *Biotic diversity on agroecosystems*. Elsevier, Amsterdam, The Netherlands.

Paoletti MG, Pimentel D, Stinner BR and Stinner D (1992). Agroecosytem biodiversity: matching production and conservation biology. *Agriculture Ecosystems and Environment*, **40**, 3–23.

Paoletti MG, Foissner W and Coleman DC (1993). *Soil biota, nutrient cycling and farming systems*. Lewis, Boca Raton, Florida.

Paracer S and Ahmadjian V (2000). *Symbiosis; an introduction to biological associations*. Oxford University Press, New York.

Parker JD, Duffy JE and Orth RJ (2001). Experimental tests of plant diversity effects on epifaunal diversity and production in a temperate seagrass bed. *Marine Ecology-Progress Series*, **224**, 55–67.

Parkinson D, Visser S and Whittaker JB (1979). Effects of collembolan grazing on fungal colonization of leaf litter. *Soil Biology and Biochemistry*, **11**, 529–535.

Pastor J, Aber JD, McClaugherty CA and Melillo JM (1984). Aboveground productivity and N and P cycling along a nitrogen mineralization gradient on Blackhawk. *Ecology*, **65**, 256–268.

Pastor J, Naiman RJ, Dewey B and McInnes P (1988). Moose, microbes and the boreal forest. *BioScience*, **38**, 770–777.

Patten BC (1975). Ecosystem linearization: an evolutionary design problem. *American Naturalist*, **109**, 529–539.

Paul ND, Hatcher PE and Taylor JE (2000). Coping with multiple enemies: an integration of molecular and ecological perspectives. *Trends in Plant Science*, **5**, 220–225.

Payne RW et al. (1993). *Genstat 5, release 3 reference manual*. Clarendon Press, Oxford.

Pearson TH and Rosenberg R (1978). Macrobenthic succession in relation to organic enrichment and pollution of the marine environment. *Oceanography and Marine Biology: an Annual Review*, **16**, 229–311.

Peart DR and Foin TC (1985). Analysis and prediction of population and community change: a grassland case study. In J White, ed. *The populations structure of vegetation*, pp. 313–339. Dr W. Junk Publishers, Dordecht.

Perfecto I and Snelling R (1995). Biodiversity and tropical ecosystem transformation: ant diverstiy in the coffee agroecosystem in Costa Rica. *Ecological Applications*, **5**, 1084–1097.

Perfecto I and Vandermeer JH (1994). The ant fauna of a transforming agroecosystem in Central America. *Trends in Agricultural Science*, **2**, 7–13.

Perfecto I, Rice R, Greenberg R and Van der Voolt M (1996). Shade coffee as refuge of biodiversity. *BioScience*, **46**, 598–608.

Perfecto I, Vandermeer J, Hanson P and Cartín V (1997). Arthropod biodiversity loss and the transformation of a tropical agro-ecosystem. *Biodiversity and Conservation*, **6**, 935–945.

Persson A *et al.* (2001). Effects of enrichment on simple aquatic food webs. *American Naturalist*, **157**, 654–669.

Persson L (1999). Trophic cascades: abiding heterogeneity and the trophic level concept at the end of the road. *Oikos*, **85**, 385–397.

Persson L, Diehl S, Johansson L, Andersson G and Hamrin SF (1992). Trophic interactions in temperate lake ecosystems: a test of food chain theory. *American Naturalist*, **140**, 59–84.

Petchey OL *et al.* (1999). Environmental warming alters food web structure and ecosystem function. *Nature*, **402**, 69–72.

Petchey OL (2000). Species diversity, species extinction, and ecosystem function. *American Naturalist*, **155**, 696–702.

Petersen JE and Hastings A (2001). Dimensional approaches to scaling experimental ecosystems: Designing mousetraps to catch elephants. *The American Naturalist*, **157**, 324–333.

Petersen JK and Riisgard HU (1992). Filtration capacity of the ascidian Ciona intestinalis and its grazing impact in a shallow fjord. *Marine Ecology-Progress Series*, **88**, 9–17.

Peterson DL and Parker VT, eds (1998). *Ecological scale. Theory and applications*. Columbia University Press, New York.

Peterson G, Allen CR and Holling CS (1998). Ecological resilience, biodiversity, and scale. *Ecosystems*, **1**, 6–18.

Pfisterer AB and Schmid B (2002). Diversity-dependent production can decrease the stability of ecosystem functioning. *Nature*, **416**, 84–86.

Pickard J (1984). Exotic plant distribution on Lord Howe Island: distribution in space and time. 1853–1981. *Journal of Biogeography*, **11**, 181–208.

Pielou EC (1975). *Ecological diversity*. Wiley, New York.

Pimbert M and Rajan V (1993). *The community-nature relationship. The making of agricultural biodiversity in Europe. Rebuilding commuities: experiences and experiments in Europe*. Green Books Ltd. Totnes, Devon.

Pimentel D *et al.* (1992). Conserving biological diversity in agricultural/forestry systems. *BioScience*, **42**, 354–362.

Pimm SL (1981). *The balance of nature*. The University of Chicago Press, Chicago.

Pimm SL (1982). *Food webs*. Chapman and Hall, London.

Pimm SL (1984). The complexity and stability of ecosystems. *Nature*, **307**, 321–326.

Pimm SL (1991). *The balance of nature? Ecological issues in the conservation of species and communities*. University of Chicago Press, Chicago.

Pimm SL and Lawton JH (1977). Number of trophic levels in ecological communities. *Nature*, **268**, 329–331.

Pimm SL, Jones HL and Diamond J (1988). On the risk of extinction. *American Naturalist*, **132**, 757–785.

Pimm SL *et al.* (1991). Food web patterns and their consequences. *Nature*, **350**, 669–674.

Planty-Tabacchi A, Tabacchi E, Naiman RJ, DeFerrari C and Decamps H (1996). Invasibility of species rich communities in riparian zones. *Conservation Biology*, **10**, 598–607.

Polis GA (1991). Complex trophic interactions in deserts: An empirical critique of food-web theory. *American Naturalist*, **138**, 123–155.

Polis GA (1998). Stability is woven by complex food webs. *Nature*, **395**, 744.

Polis GA and Strong DR (1996). Food web complexity and community dynamics. *American Naturalist*, **147**, 813–846.

Polis GA and Winemiller KO, eds (1996). *Food webs: integration of patterns and dynamics*. Chapman and Hall, New York.

Polis GA, Andersen WB and Holt RD (1997). Towards an integration of landscape and food web ecology: the dynamics of spatially subsidised food webs. *Annual Review of Ecology and Systematics*, **28**, 289–316.

Polis GA *et al.* (2000). When is a trophic cascade a trophic cascade? *Trends in Ecology and Evolution*, **15(11)**, 473–475.

Poly F, Ranjard L, Nazaret S, Gourbière F and Monrozier LJ (2001). Comparison of *nif* H gene pools in soils and soil microenvironments with contrasting properties. *Applied and Environmental Microbiology*, **67**, 2255–2262.

Poorter H and Bergkotte M (1992). Chemical composition of 24 wild species differing in relative growth rate. *Plant Cell and Environment*, **15**, 221–229.

Post WM and Pimm SL (1983). Community assembly and food web stability. *Mathematical Biosciences*, **64**, 169–192.

Potter CS, Reyer RE, Singh RP, Par JF and Stewart BA (1990). The role of soil biodiversity in sustainable dryland farming systems. *Advances in Soil Science*, **13**, 241–251.

Power ME (1990). Effects of fish in river food webs. *Science*, **250**, 811–814.

Power ME (1992). Top-down and bottom-up forces in food webs: do plants have primacy? *Ecology*, **73**, 733–746.

Price GR (1970). Selection and covariance. *Nature*, **227**, 520–521.

Price GR (1995). The nature of selection. *Journal of Theoretical Biology*, **175**, 389–396.

Prieur-Richard A-H, Lavorel S, Grigulis K and Dos Santos A (2000). Plant community diversity and invasibility by exotics: invasion of Mediterranean old fields by *Conyza bonariensis* and *Conyza canadensis*. *Ecology Letters*, **3**, 412–422.

Proulx M and Mazumder A (1998). Reversal of grazing impact on plant species richness in nutrient poor vs. nutrient rich ecosystems. *Ecology*, **79**, 2581–2592.

Pugh GJF (1980). Strategies in fungal ecology. *Transactions of the British Mycological Society*, **75**, 1–14.

Pysek P and Pysek A (1995). Invasion by *Heracleum mantagazzianum* in different habitats in the Czech Republic. *Journal of Vegetation Science*, **6**, 711–718.

Raffaelli DG (2000). Trends in research on shallow water food webs. *Journal of Experimental Marine Biology and Ecology*, **250**, 223–232.

Raffaelli DG and Burslem D (2000). Ecological contrasts—different concepts or different ecologists? *Bulletin of the British Ecological Society*, **31**, 22–23.

Raffaelli DG and Hall SJ (1996). Assessing the relative importance of trophic links in food webs. In GA Polis and KO Winemiller, eds *Food webs: integration of patterns and dynamics*, pp. 185–191. Chapman and Hall, New York.

Raffaelli D and Hawkins S (1996). *Intertidal Ecology*. Chapman and Hall, London.

Raffaelli DG and Moller H (2000). Manipulative experiments in animal ecology—do they promise more than they can deliver? *Advances in Ecological Research*, **30**, 229–338.

Rapport DJ (1995). Ecosystem services and management options as blanket indicators or ecosystem health. *Journal of Aquatic Ecosystem Health*, **4**, 97–105.

Rashit E and Bazin M (1987). Environmental fluctuations, productivity, and species diversity: an experimental study. *Microbial Ecology*, **14**, 101–112.

Rastetter EB, Gough L, Hartley AE, Herbert DA, Nadelhoffer KJ and Williams M (1999). A revised assessment of species redundancy and ecosystem reliability. *Conservation Biology*, **13**, 440–443.

Raunkiaer C (1934). *Life forms of plants and statistical plant geography*. Clarendon Press, Oxford.

Ravnskov S and Jakobsen I (1995). Functional compatibility in arbuscular mycorrhizas measured as hyphal P transport to the plant. *New Phytologist*, **129**, 611–618.

Read DJ (1991). Mycorrhizas in ecosystems. *Experientia*, **47**, 376–391.

Reader RJ *et al.* (1994). Plant competition in relation to neighbor biomass—an intercontinental study with *Poa Pratensis*. *Ecology*, **75**, 1753–1760.

Reaka-Kudla ML, Wilson DE and Wilson EO, eds (1997). *Biodiversity II*. Island Press, Washington DC.

Redecker D, Morton JB and Bruns TD (2000). Ancestral lineages of arbuscular mycorrhizal fungi (glomales). *Molecular phylogenetics and evolution*, **14**, 276–284.

Reich PB, Walters MB and Ellsworth DS (1992). Leaf-life span in relation to leaf plant and stand characteristics among diverse ecosystems. *Ecological Monographs*, **62**, 365–392.

Reich PB, Walters MB and Ellsworth DS (1997). From tropics to tundra: global convergence in plant functioning. *Proceedings of the National Academy of Sciences*, **94**, 13730–13734.

Reich PB, Knops J and Tilman D (2001). Plant diversity enhances ecosystem responses to elevated CO_2 and nitrogen deposition. *Nature*, **410**, 809–812.

Reinhold-Hurk B and Hurek T (1998). Life in grasses: diazotrophic endophytes. *Trends in Microbiology*, **6**, 139–144.

Remy W, Taylor TN, Haas H and Kerp H (1994). Four hundred-million-year-old vesicular-arbuscular mycorrhizae. *Proceedings of the National Academy of Science*, **91**, 11841–11843.

Ricciardi A and Bourget E (1999). Global patterns of macroinvertebrate biomass in marine intertidal communities. *Marine Ecology-Progress Series*, **185**, 21–35.

Rieman B and Christoffersen K (1993). Microbial trophodynamics in temperate lakes. Marine Microbiology and Food Webs, **7**, 69–100.

Rieman B, Sondergaard M, Persson L and Johansson L (1986). Carbon metabolism and community regulation in eutrophic temperate lakes. In B Rieman and M Sondergaard, eds *Carbon dynamics in eutrophic temperate lakes*, pp. 267–280. Elsevier Publishers.

Risch SJ, Andow D and Altieri M (1983). Agroecosystem diversity and pest control: data, tentative conclusions, and new research directions. *Environmental Entomology*, **12**, 625–629.

Ritchie M and Olff H (1999). Spatial scaling laws yield a synthetic theory of biodiversity. *Nature*, **400**, 557–560.

Robertson GP, Huston MA, Evans FC and Tiedje JM (1988). Spatial variability in a successional plant community: Patterns of nitrogen mineralization, nitrification, and denitrification. *Ecology*, **69**, 1517–1524.

Robinson CH, Dighton J, Frankland JC and Coward PA (1993). Nutrient and carbon dioxide release by interacting species of straw-decomposing fungi. *Plant and Soil*, **151**, 139–142.

Robinson D (2001). delta ^{15}N as an integrator of the nitrogen cycle. *Trends in Ecology and Evolution*, **16**, 153–162.

Robinson GR, Quinn JF and Stanton ML (1995). Invasibility of experimental habitat islands in a California winter annual grassland. *Ecology*, **76**, 786–794.

Robinson JV and Dickerson J (1984). Testing the invulnerability of laboratory island communities to invasion. *Oecologia*, **61**, 169–174.

Robinson JV and Valentine WD (1979). The concepts of elasticity, invulnerability, and invadibility. *Journal of Theoretical Biology*, **81**, 91–104.

Rodríguez MA (1994). Stability may decrease with diversity in grassland communities: empirical evidence from the 1986 Cantabrian Mountains (Spain) drought. *Oikos*, **71**, 177–180.

Root RB (1967). The niche exploitation pattern of the bluegray gnatcatcher. *Ecological Monographs*, **37**, 317–350.

Root R (1973). Organization of a plant-arthropod association in simple and diverse habitats: The fauna of collards (*Brassica oleracea*). *Ecological Monographs*, **43**, 95–124.

Rosemund AD (1996). Indirect effects of herbivores modify predicted effects of resources and consumption on plant biomass. In GA Polis and KO Winemiller, eds *Food webs: integration of patterns and dynamics*, pp. 149–159. Chapman and Hall, New York.

Rosenthal R and Rosnow RL (1985). *Contrast analysis: Focused comparisons in the analysis of variance*. Cambridge University Press, Cambridge.

Rosenzweig ML (1971). The paradox of enrichment: destabilitization of exploitation ecosystems in ecological time. *Science*, **171**, 385–387.

Rosenzweig ML and Abramsky Z (1993). How are diversity and productivity related? In RE Ricklefs and D Schluter, eds *Species diversity in ecological communities: Historical and geographical perspectives*, pp. 52–65. University of Chicago Press, Chicago.

Roughgarden J (1974). Niche width: biogeographic patterns among *Anolis* lizard populations. *American Naturalist*, **108**, 429–442.

Rowe GT, Clifford CH and Smith KL (1975). Benthic nutrient regeneration and its coupling to primary productivity in coastal waters. *Nature*, **255**, 215–217.

Roy J, Saugier B and Mooney HA, eds (2001). *Terrestrial productivity*. Academic Press, San Diego.

Ruesnik JL and Srivastava DS (2001). Numerical and per capita responses to species loss: mechanisms maintaining ecosystem function in a community of stream insect detrivores. *Oikos*, **93**, 221–234.

Russell EP (1989). Enemies hypothesis: A review of the effect of vegetational diversity on predatory insects and parasitoids. *Environmental Entomology*, **18**, 590–599.

Ryan JC and Stark L (1992). *Conserving biolotical diversity. State of the world 1992: a world watch institute report on progress, towards a sustainable society*, pp. 9–26. Earthsdian Publications, London.

Ryan MR, Burger LW and Kurzejeski EW (1998). The impact of CRP on avian wildlife: A review. *Journal of Production Agriculture*, **11**, 61–66.

Rygiewicz PT and Andersen CP (1994). Mycorrhizae alter quality and quantity of carbon allocated below ground. *Nature*, **369**, 58–60.

Saetre P (1998). Decomposition, microbial community structure, and earthworm effects along a birch-spruce soil gradient. *Ecology*, **79**, 834–846.

Sala OE *et al.* (2000). Global biodiversity scenarios for the year 2100. *Science*, **287**, 1770–1774.

Salonius PO (1981). Metabolic capabilities of forest soil microbial populations with reduced species diversity. *Soil Biology and Biochemistry*, **13**, 1–10.

Sanchez PA (1995). Science in agroforestry. *Agroforestry Systems*, **30**, 5–55.

Sanders IR (2002). Specificity in the arbuscular mycorrhizal symbiosis. In MGA van der Heijden and IR Sanders, eds *Mycorrhizal ecology*. Ecological Studies 157. Springer-Verlag, Heidelberg.

Santos PF, Phillips J and Whitford WG (1981). The role of mites and nematodes in early stages of buried litter decomposition in a desert. *Ecology*, **63**, 664–669.

Sardar MA and Murphy PW (1987). Feeding test of grassland soil-inhabiting gamasine predators. *Acarologia*, **28**, 117–122.

SAS (1990). SAS/STAT user's guide, version 6, vol. 1, 4th Edition. SAS Institute Inc. Cary, NC, USA.

SAS (2000). JMP® statistics and graphics guide, version 4. SAS Institute Inc., Cary, NC, USA.

Scheffer M, Hosper SH, Meijer M-L, Moss B and Jeppesen E (1993). Alternative equilibria in shallow lakes. *Trends in Ecology and Evolution*, **8**, 275–279.

Scheffer M *et al.* (2001). Catastrophic shifts in ecosystems. *Nature*, **413**, 591–96.

Scherer-Lorenzen M (1999). Effects of plant diversity on ecosystem processes in experimental grassland communities. *Bayreuther Forum Ökologie*, **75**, 1–195.

Scheu S and Schaefer M (1998). Bottom-up control of the soil macrofauna community in a beechwood on limestone—manipulation of food resources. *Ecology*, **79**, 1573–1584.

Scheu S and Schulz E (1996). Secondary succession, soil formation and development of a diverse community of oribatids and saprophagous soil macro-invertebrates. *Biodiversity and Conservation*, **5**, 235–250.

Scheu S and Setälä H (2001). Multitrophic interactions in decomposer food webs. In BA Hawkins, ed. *Multitrophic interactions in terrestrial systems*. Cambridge University Press, Cambridge.

Scheu S, Theenhaus A and Jones TH (1999). Links between the detritivore and herbivore system: effects of earthworms and Collembola on plant growth and aphid development. *Oecologia*, **119**, 541–551.

Shigesada N, Kawasaki K and Teramoto E (1984). The effects of interference competition on stability, structure, and invasion of a multispecies system. *Journal of Mathematical Biology*, **21**, 97–113.

Schindler DE *et al.*(1996). Food web structure and littoral zone oupling to pelagic trophic cascades. In GA Polis and KO Winemiller, eds *Food webs: integration of patterns and dynamics*, pp. 96–105. Chapman and Hall, New York.

Schindler DE, Carpenter SR, Cole JJ, Kitchell JF and Pace ML (1997). Influence of food web structure on carbon exchange between lakes and the atmosphere. *Science*, **277**, 248–251.

Schippers B (1992). Prospects for management of natural suppressiveness to control soilborne pathogens. In EC Tjamos, GC Papavizas and RJ Cook, eds *Biological control of plant disease. Progress and challenges for the future*. Plenum Press, New York.

Schläpfer F and Erickson JD (2001). A biotic control perspective on nitrate contamination of groundwater from agricultural production. *Agricultural and Resource Economics Review*.

Schläpfer F and Schmid B (1999). Ecosystem effects of biodiversity: A classification of hypotheses and cross-system exploration of empirical results. *Ecological Applications*, **9**, 893–912.

Schlesinger WH (1997). *Biogeochemistry, 2nd Edition*. Academic Press, San Diego.

Schmid B (2002). The species richness–productivity controversy. *Trends in Ecology and Evolution*, **17**, 113–114.

Schmid B and Harper JL (1985). Clonal growth in grassland perennials. I. Density and pattern dependent competition between plants with different growth form. *Journal of Ecology*, **73**, 793–808.

Schmid B, Joshi J and Schläpfer F (2002). Empirical evidence for biodiversity-ecosystem functioning relationships. In AP Kinzig, SW Pacala and D Tilman, eds *Functional consequences of biodiversity: empiric progress and theoretical extensions*. Princeton University Press, Princeton.

Schmitz OJ, Hambäck PA and Beckerman AP (2000). Trophic cascades in terrestrial systems: a review of the effects of carnivore removal on plants. *American Naturalist*, **155**, 141–153.

Schneider DC (2001). The rise of the concept of scale in ecology. *BioScience*, **51**, 545–553.

Schoener TW (1976). The species–area relation within archipelagoes: models and evidence from island land birds. Proceedings of the 16th International Ornothological Congress (1974), Canberra, pp. 629–642.

Schultz PA (1991). Grazing preferences of two collembolan species, *Folsomia candida* and *Proisotoma minuta*, for ectomycorrhizal fungi. *Pedobiologia*, **35**, 313–325.

Schulze ED and Mooney HA (1994). *Biodiversity and Ecosystem Function*. Springer-Verlag, Berlin.

Schwartz MW, Brigham CA, Hoeksema JD, Lyons KG, Mills MH and van Mantgem PJ (2000). Linking biodiversity to ecosystem function: implications for conservation ecology. *Oecologia*, **122**, 297–305.

Schwenninger HR (1992). Untersuchungen zum Einfuss der Bewirschaftungsintensitat auf das orkommen von Insektenarten in der Agrarlàndschaft, dargestellt am Beispiel der Wildbienen (Hymenoptera: Apoidea). *Zoologische Jahrbucher, Abteilung fur Systematik, Okologie und Geographie der Tiere*, **119**, 543–561.

Searle SR (1971). *Linear models*. John Wiley and Sons, New York.

Searle SR (1995). Comments on J. A. Nelder: The statistics of linear models: back to basics. *Statistics and Computing*, **5**, 103–107.

Seastedt TR (1984). The role of microarthropods in decomposition and mineralization procresses. *Annual Review of Entomology*, **29**, 25–46.

Seastedt TR and Crossley DA (1980). Effects of microarthropods on the seasonal dynamics of nutrients in forest litter. *Soil Biology and Biochemistry*, **12**, 337–342.

Setälä H (2000). Reciprocal interactions between Scots pine and soil food web structure in the presence and absence of ectomycorrhiza. *Oecologia*, **125**, 109–118.

Setälä H and Huhta V (1991). Soil fauna increases *Betula pendula* growth: laboratory experiments with coniferous forest floor. *Ecology*, **72**, 665–671.

Setälä H, Kulmala P, Mikola J and Markkola AM (1999). Influence of ectomycorrhiza on the structure of detrital food webs in pine rhizosphere. *Oikos*, **87**, 113–122.

Shaver GR *et al.* (2000). Global warming and terrestrial ecosystems: a conceptual framework for analysis. *BioScience*, **50**, 871–882.

Shaw C and Pawluk S (1986). The development of soil structure by *Octolasion tyrtaeum, Aporrectodea turgida* and *Lumbricus terrestris* in parent materials belonging to different textural classes. *Pedobiologia*, **29**, 327–339.

Shaw RG and Mitchell-Olds T (1993). ANOVA for unbalanced data: an overview. *Ecology*, **74(6)**, 1638–1645.

Shigesada N, Kawasaki K and Teramoto E (1984). The effects of interference competition on stability, structure, and invasion of a multispecies system. *Journal of Mathematical Biology*, **21**, 97–113.

Shmida A and Ellner S (1984). Coexistence of plant species with similar niches. *Vegetatio*, **58**, 29–55.

Shurin JB (2000). Dispersal limitation, invasion resistance, and the structure of pond zooplankton communities. *Ecology*, **81**, 3074–3086.

Siemann E (1998). Experimental tests of plant productivity and diversity on grassland arthropod diversity. *Ecology*, **79**, 2057–2060.

Siemann E, Tilman D, Haarstad J and Ritchie M (1998). Experimental tests of the dependence of arthropod diversity on plant diversity. *American Naturalist*, **152**, 738–750.

Siepel H (1996). Biodiversity of soil microathropods: the filtering of species. *Biodiversity and Conservation*, **5**, 251–260.

Silander JA and Antonovics J (1982). Analysis of interspecific interactions in a coastal plant community—a perturbation approach. *Nature*, **298**, 557–560.

Silver WL, Brown S, Scatena FN and Ewel JJ (2001). Managed ecosystems deserve greater attention. *Bulletin of the Ecological Society of America*, **82**, 91–93.

Silvertown J and Law R (1987). Do plants need niches? Some recent developments in plant community ecology. *Trends in Ecology and Evolution*, **2**, 24–26.

Silvertown J, Dodd ME, Gowing DJG and Mountford JO (1999). Hydrologically defined niches reveal a basis for species richness in plant communities. *Nature*, **400**, 61–63.

Simard SW, Perry DA, Jones MD, Myrold DD, Durall DM and Molina R (1997). Net transfer of carbon between tree species with shared ecto-mycorrhizal fungi. *Nature*, **388**, 579–582.

Simard SW, Durall D and Jones M (2002). Carbon and nutrient fluxes within and between mycorrhizal plants. In MGA van der Heijden & IR Sanders, eds *Mycorrhizal ecology*. Ecological Studies 157. Springer-Verlag, Heidelberg.

Simberloff D and Dayan T (1991). The guild concept and the structure of ecological communities. *Annual Review of Ecology and Systematics*, **22**, 115–143.

Sinclair ARE (1973). Population increases of buffalo and wildebeast in the Serengeti. *East African Wildlife Journal*, **11**, 93–107.

Smallwood KS (1994). Site invasibility by exotic birds and mammals. *Biological Conservation*, **69**, 251–259.

Smedes GW and Hurd LE (1981). An empirical test of community stability: resistance of a fouling community to a biological patch-forming disturbance. *Ecology*, **62**, 1561–1572.

Smedling FW and Joenje W (1999). Farm-Nature Plan: Landscape ecology based farm planning. *Landscape and Urban Planning*, **46**, 109–115.

Smith FA, Jakobsen I and Smith SE (2000). Spatial differences in acquisition of soil phosphate between two arbuscular mycorrhizal fungi in symbiosis with *Medicago truncatula. New Phytologist*, **147**, 357–366.

Smith JG (1976). Influence of crop background on aphids and other phytophagous insects on brussels sprouts. *Annals of Applied Biology*, **83**, 1–13.

Smith MD and Knapp AK (1999). Exotic plant species in a C_4-dominated grassland: invasibility, disturbance, and community structure. *Oecologia*, **120**, 605–612.

Smith MD, Hartnett DC and Wilson GWT (1999). Interacting influence of mycorrhizal symbiosis and competition on plant diversity in tallgrass prairie. *Oecologia*, **121**, 574–582.

Smith SE and Read DJ (1997). *Mycorrhizal symbiosis*. Academic Press, London.

Smith SV et al. (1987). Stoichiometry of C, N, P, and Si fluxes in a temperate-climate embayment. *Journal of Marine Research*, **45**, 427–460.

Smith SV et al. (1989). Tomales Bay, California: A case for carbon-controlled nitrogen cycling. *Limnology and Oceanography*, **34**, 37–52.

Smith TM, Shugart HH and Woodward FI, eds (1997). *Plant functional types: Their relevance to ecosystem properties and global change*. University Press, Cambridge.

Snaydon RW and Harris PM (1979). Interactions belowground—the use of nutrients and water. In RW Willey, ed. pp. 188–201. Proceedings of the international workshop on intercropping. ICRISAT, Hyderabad, India.

Snedecor GW and Cochran WG (1980). *Statistical methods. 7th Edition*. Iowa State University Press, Ames (Iowa).

Snelgrove P (1998). The biodiversity of macrofaunal organisms in marine sediments. *Biological Conservation*, **7**, 1123–1132.

Snelgrove P (1999). Getting to the bottom of marine biodiversity: sedimentary habitats. *BioScience*, **49**, 129–138.

Snelgrove PVR et al. (1997). The importance of marine sediment biodiversity in ecosystem processes. *Ambio*, **26**, 578–583.

Snelgrove P et al. (2000). Linking biodiversity above and below the marine sediment-water interface. *BioScience*, **50**, 1076–1088.

Soemarwoto O, Soemarwoto I, Karyano EM, Soekartadiredja and Famlan A (1985). The Javanese home gardens as an integrated ecosystem. *Food and Nutrition Bulletin*, **7**, 85–101.

Sohlenius B (1985). Influence of climatic conditions on nematode coexistence: a laboratory experiment with a coniferous forest soil. *Oikos*, **44**, 430–438.

Sokal RR and Rohlf FJ (1981). *Biometry, the principles and practices of statistics in biological research, 2nd Edition*, WH Freeman and Company, San Francisco.

Solan M (2000). *The concerted use of 'traditional' and Sediment Profile Imagery (SPI) methodologies in marine benthic characterisation and monitoring.* Ph.D. National University of Ireland, Galway.

Solan M and Kennedy R (2002). Observation and quantification of in-situ animal-sediment relations using time lapse sediment profile imagery (t-SPI). *Marine Ecology Progress Series,* **228,** 179–191.

Soulé ME (1991). Conservation: tactics for a constant crisis. *Science,* **253,** 744–750.

Southwood TRE, Brown VK and Reader PM (1979). The relationships of plant and insect diversities in succession. *Biological Journal of the Linnean Society,* **12,** 327–348.

Spaekova I and Leps J (2001). Procedure for separating the selection effect from other effects in diversity–productivity relationship. *Ecology Letters,* **4,** 585–594.

Spain JM, Dumanski J, Pushparajah E, Latham M and Meyers R (1991). Genetic resources and biodiversity as tools in sustainable land management. In J Dumanski, E Pushparajah, M Latham, eds *Evaluation for sustainable land management in the developing world, Vol. 2: gechnical papers.* IBSRAM proceedings no. **12(2)** 157–171.

Spaink HP, Kondorosi A and Hooykass PJJ, eds (1998). *The rhizobiaceae; Molecular Biology of model plant-associated bacteria.* Kluwer Academic Publisher, Dordrecht, The Netherlands.

Spehn EM, Joshi J, Schmid B, Alphei J and Körner C (2000a). Plant diversity on soil heterotrophic activity in experimental grassland systems. *Plant and Soil,* **224,** 217–230.

Spehn EM, Joshi J, Schmid B, Diemer M and Körner C (2000b). Above-ground resource use increases with plant species richness in experimental grassland ecosystems. *Functional Ecology,* **14,** 326–337.

Spehn EM *et al.* (2002, in press). The role of legumes as a component of biodiversity in a cross-European study of grassland biomass nitrogen. *Oikos.*

Spencer M and Warren PH (1996). The effects of energy input, immigration and habitat size on food web structure: a microcosm experiment. *Oecologia,* **108,** 764–770.

Sprent JI and Sprent P (1990). *Nitrogen fixing organisms: pure and applied aspects.* Chapman and Hall, London.

Springett JA (1992). Distribution of lumbricid earthworms in New Zealand. *Soil Biology and Biochemistry,* **24,** 1377–1381.

Sprung M (1994). Macrobenthic secondary production in the intertidal zone of the Ria Formosa—a lagoon in Southern Portugal. *Estuarine and Coastal Shelf Science,* **38,** 539–558.

Squartini A (2000). Functional ecology of the rhizobium-legume symbiosis. In R Pinton, Z Varanini and P Nannipieri, eds *The rhizosphere. Biochemistry and organic substances at the soil–plant interface.* Marcel Dekker, New York.

Srivastava DS (1999). Using local-regional richness plots to test for species saturation: pitfalls and potentials. *Journal of Animal Ecology,* **68,** 1–16.

Stachowicz JJ, Whitlatch RB and Osman RW (1999). Species diversity and invasion resistance in a marine ecosystem. *Science,* **286,** 1577–1579.

Stachowicz JJ *et al.* (in press). Reconciling pattern and process in marine bioinvasions: how important is diversity in determining community invisibility? *Ecology.*

Stadler J, Trefflich A, Klotz S and Brandl B (2000). Exotic plant species invade diversity hot spots: the alien flora of northwest Kenya. *Ecography,* **23,** 169–176.

Stahl PD and Smith WK (1984). Effects of different geographic isolates of *Glomus* on the water relations of *Agropyron smithii. Mycologia,* **76,** 261–267.

Staley JT (2001). A microbiological perspective of biodiversity. In JT Staley and A-L Reysenbach, eds *Biodiversity of microbial life: Foundation of Earth's biosphere.* John Wiley and Sons, New York.

Stanton NL (1988). The underground in grasslands. *Annual Review of Ecology and Systematics,* **19,** 573–589.

Stark S, Wardle DA, Ohtonen R, Helle T and Yeates GW (2000). The effect of reindeer grazing on decomposition, mineralization and soil biota in a dry oligotrophic Scots pine forest. *Oikos,* **90,** 301–310.

Steiner CF (2001). The effects of prey heterogeneity and consumer identity on the limitation of trophic-level biomass. *Ecology,* **82,** 2495–2506.

Steinmüller W and Bock E (1976). Growth of Nitrobacter in the presence of organic matter. I. Mixotrophic growth. *Archives of Microbiology,* **108,** 305–312.

Stephan A, Meyer AH and Schmid B (2000). Plant diversity affects culturable soil bacteria in experimental grassland communities. *Journal of Ecology,* **88,** 988–998.

Sterner RW (1995). Elemental stoichiometry of species in ecosystems. In CG Jones and JH Lawton, eds *Linking species and ecosystems,* pp. 240–252. Chapman and Hall, New York.

Sterner RW, Elser JJ, Chrzanwski TH, Schampel JH and George NB (1996). Biogeochemistry and trophic ecology: a new food web diagram. In GA Polis, KO Winemiller, eds *Food webs: integration of patterns and dynamics,* pp. 72–80. Chapman and Hall, New York.

Stevens MHH and Carson WP (1999). Plant density determines species richness along an experimental fertility gradient. *Ecology,* **80,** 455–465.

Stocker R, Körner C, Schmid B, Niklaus PA and Leadley PW (1999). A field study of the effects of elevated CO_2 and plant species diversity on ecosystem-level gas exchange in a planted calcareous grassland. *Global Change Biology*, **5**, 95–105.

Stockner JG and Porter KG (1988). Microbial food webs in freshwater planktonic ecosystems. In SR Carpenter, ed. *Complex interactions in lake communities*, pp. 70–83. Spriger-Verlag, New York.

Stohlgren TJ et al. (1999). Exotic plant species invade hot spots of native plant diversity. *Ecological Monographs*, **69**, 25–46.

Stoll P and Prati D (2001). Intraspecific aggregation alters competitive interactions in experimental plant communities. *Ecology*, **82**, 319–327.

Stork N (1997). Measuring global biodiversity and its decline. In ML Reaka-Kudla, DE Wilson and EO Wilson, eds *Biodiversity II*, pp. 41–68. Island Press, Washington DC.

Stork NE and Eggleton P (1992). Invertebrates as determinants and indicators of soil quality. *American Journal of Alternative Agriculture*, **7**, 1,2,38–47.

Streitwolf-Engel R, Boller T, Wiemken A and Sanders IR (1997). Clonal growth traits of two *Prunella* species are determined by co-occurring arbuscular mycorrhizal fungi from a calcareous grassland. *Journal of Ecology*, **85**, 181–191.

Streitwolf-Engel R, van der Heijden MGA, Wiemken A and Sanders IR (2001). The ecological significance of arbuscular mycorrhizal fungal effects on clonal plant growth. *Ecology*, **82**, 2846–2859.

Strong DR (1992). Are trophic cascades all wet? Differentiation and donor-control in speciose ecosystems. *Ecology*, **73**, 747–754.

Strong DR (1999). Predator control in terrestrial ecosystems: the underground food chain of bush lupine. In H Olff et al., eds *Herbivores between plants and predators*, pp. 577–602, Blackwell Science, Cambridge.

Sulkava P and Huhta V (1998). Habitat patchiness affects decomposition and faunal diversity: a microcosm experiment on forest floor. *Oecologia*, **116**, 390–396.

Sulkava P, Huhta V and Laakso J (1996). Impact of soil faunal structure on decomposition and N-mineralisation in relation to temperature and moisture in forest soil. *Pedobiologia*, **40**, 505–513.

Sundbäck K et al. (1991). Influence of sublittoral micro-phytobenthos on the oxygen and nutrient flux between sediment and water: a laboratory continuous-flow study. *Marine Ecology-Progress Series*, **74**, 263–279.

Suominen O (1999). Impact of cervid browsing and grazing on the terestrial gastropod fauna in the boreal forests of Fennoscandia. *Ecography*, **22**, 651–658.

Suominen O, Danell K and Bergström R (1999). Moose, trees and ground living invertebrates: indirect interactions in Swedish pine forest. *Oikos*, **84**, 215–226.

Sutton MA et al., eds (2000). *Terrestrial ecosystems research in Europe: successes, challenges and policy.* Office for the official publications of the European Communities, Luxembourg.

Swenson W, Wilson DS and Elias R (2000). Artificial ecosystem selection. *Proceedings of the National Academy of Sciences of the United States of America*, **97**, 9110–9114.

Swift DJ (1993). The macrobenthic infauna off Sellafield (North-Eastern Irish Sea) with special reference to bioturbation. *Journal of Marine Biological Association UK*, **73**, 143–162.

Swift MJ and Anderson JM (1993). Biodiversity and ecosystem function in agricultural systems. In ED Schulze and HA Mooney, eds *Biodiversity and ecosystem function*, pp. 15–41. Springer-Verlag, New York.

Swift MJ, Vandermeer JH, Ramakrishnan PS, Anderson JM, Ong C and Hawkins B (1996). Biodiversity and agroecosystem function. In HA Mooney, J Lubchenco, R Dirzo and OE Sala, eds *Biodiversity and ecosystem function. Global diversity assessment*, pp. 433–443. Cambridge University Press, Cambridge.

Swift MJ et al. (1998). Global change, soil biodiversity, and nitrogen cycling in terrestrial ecosystems: three case studies. *Global Change Biology*, **4**, 729–743.

Symstad AJ (2000). A test of the effects of functional group richness and composition on grassland invasibility. *Ecology*, **81**, 99–109.

Symstad A and Tilman D (2001). Diversity loss, recruitment limitation and ecosystem functioning: lessons learned from a removal experiment. *Oikos*, **92**, 424–435.

Symstad AJ, Tilman D, Willson J and Knops JMH (1998). Species loss and ecosystem functioning: effects of species identity and community composition. *Oikos*, **81**, 389–397.

Taylor AFS, Martin F and Read DJ (2000). Fungal diversity in ecto-mycorrhizal communities of Norway spruce (*Picea abies* [L.] Karst.) and Beech (*Fagus sylvatica* L.) along north-south transects in Europe. In Schulze, ed. *Carbon and nitrogen cycling in European forest ecosystems*, pp. 343–365. Springer-Verlag, Berlin.

Teuben A (1991). Nutrient availability and interactions between soil arthropods and microorganisms during decomposition of coniferous litter: a mesocosm study. *Biology and Fertility of Soils*, **10**, 256–266.

Teuben A and Roelofsma TAPJ (1990). Dynamic interactions between functional groups of soil arthropods and microorganisms during decomposition of coniferous litter in microcosm experiments. *Biology and Fertility of Soils*, **9**, 145–151.

Teyssier-Cuvelle S, Mougel C and Nesme X (1999). Direct conjugal transfers of Ti plasmid to soil microflora. *Molecular Ecology*, **8**, 1273–1284.

Thies C and Tscharntke T (1999). Landscape structure and biological control in agroecosystems. *Science*, **285**, 89–91.

Thomson JD (2001). Using pollination deficits to infer pollination declines: Can theory guide us? *Conservation Ecology*, **5(1)**, 6 [on line] URL: http://www.consecol.org/vol5/iss1/art6

Thompson JM (1996). Evolutionary ecology and the conservation of biodiversity. *Trends in Ecology and Evolution*, **11**, 300–303.

Thompson J *et al.* (2001). Frontiers in Ecology. *BioScience*, **51**, 15–24.

Thompson K, Green A and Jewels AM (1994). Seeds in soil and worm casts from a neutral grassland. *Functional Ecology*, **8**, 29–35.

Tiedje JM, Asuming-Brempong S, Nusslein K, Marsh TL and Flynn SJ (1999). Opening the black box of soil microbial diversity. *Applied Soil Ecology*, **13**, 109–122.

Tilman D (1982). *Resource competition and community structure*. Princeton University Press, New Jersey.

Tilman D (1987). Secondary succession and the pattern of plant dominance along experimental nitrogen gradients. *Ecological Monographs*, **57(3)**, 189–214.

Tilman D (1988). *Plant strategies and the dynamics and structure of plant communities*, pp. 360. Princeton University Press, Princeton.

Tilman D (1996). Biodiversity: Population versus ecosystem stability. *Ecology*, **77**, 350–363.

Tilman D (1997a). Distinguishing between the effects of species diversity and species composition. *Oikos*, **80**, 185.

Tilman D (1997b). Community invasibility, recruitment limitation, and grassland biodiversity. *Ecology*, **78(1)**, 81–92.

Tilman D (1999a). The ecological consequences of changes in biodiversity: a search for general principles. *Ecology*, **80**, 1455–1474.

Tilman D (1999b). Diversity and production in European grasslands. *Science*, **286**, 1099–1100.

Tilman D (2000). What *Issues in Ecology* is, and isn't. *Bulletin of the Ecological Society of America*, **81**, 240.

Tilman D (2001). Functional diversity. In SA Levin, ed. *Encyclopedia of biodiversity, Vol.3*, pp. 109–120. Academic Press.

Tilman D (2002, in press). *Ecology: Achievement and challenge*. In MC Press, NJ Huntley and SA Levin, eds Blackwell Science, Oxford.

Tilman D and Downing JA (1994). Biodiversity and stability in grasslands. *Nature*, **367**, 363–365.

Tilman D and Kareiva P (1997). *Spatial ecology. The role of space in population dynamics and interspecific interactions*. Princeton University Press, Princeton.

Tilman D, RM May, Lehman CL and Nowak MA (1994). Habitat destruction and the extinction debt. *Nature*, **371**, 65–66.

Tilman D, Wedin D and Knops J (1996). Productivity and sustainability influenced by biodiversity in grassland ecosystems. *Nature*, **379**, 718–720.

Tilman D, Knops J, Wedin D, Reich P, Ritchie M and Siemann E (1997a). The influence of functional diversity and composition on ecosystem processes. *Science*, **277**, 1300–1302.

Tilman D, Lehman CL and Thomson KT (1997b). Plant diversity and ecosystem productivity: theoretical considerations. *Proceedings of the National Academy of Sciences of the United States of America*, **94**, 1857–1861.

Tilman D, Lehman C and Bristow CE (1998). Diversity–stability relationships: statistical inevitability or ecological consequence? *American Naturalist*, **151**, 277–282.

Tilman D, Reich PB, Knops JMH, Wedin D, Mielke T and Lehman C (2001). Diversity and productivity in a long-term grassland experiment. *Science*, **294**, 843–845.

Tilman D, Knops J, Wedin D and Reich P (2002). Experimental and observational studies of diversity, productivity and stability. In A Kinzig, S Pacala and D Tilman, eds *Functional consequences of biodiversity: Experimental progress and theoretical extensions*, pp. 42–70. Princeton University Press, New Jersey.

Timmins SM and Williams PA (1991). Weed numbers in New Zealand's forest and scrub reserves. *New Zealand Journal of Ecology*, **15**, 153–162.

Torsvik V, Goksoyr J, Daae FL, Sorheim R, Michelsen J and Salte K (1994). Use of DNA analysis to determine the diversity of soil communities. In K Ritz, J Dighton and KE Giller, eds *Beyond the biomass: Compositional and functional analysis of soil microbial communities*, pp. 39–48. Wiley, Chichester.

Tracy BF and Frank DA (1998). Herbivore influence on soil microbial biomass and nitrogen mineralization in a northern grassland ecosystem: Yellowstone National Park. *Oecologia*, **114**, 556–562.

Tracy CR and George TL (1992). On the determinants of extinction. *American Naturalist*, **139**, 102–122.

Trappe JM (1987). Phylogenetic and ecological aspects of mycotrophy in the angiosperms from an evolutionary standpoint. In GR Safir, ed. *Ecophysiology of VA mycorrhizal plants*, pp. 5–25. CRC Press, Boca Raton, USA.

Trenbath BR (1974). Biomass productivity of mixtures. *Advances in Agronomy*, **26**, 177–210.

Trenbath BR (1976). Plant interactions in mixed crop communities. In RI Papendick, PA Sanchez and GB Triplett, eds *Multiple cropping*, pp. 129–170. American Society of Agronomy, Madison.

Trigueros JM and Orive E (2001). Seasonal variations of diatoms and dinoflagellates in a shallow, temperate estuary, with emphasis on neritic assemblages. *Hydrobiologia*, **444**, 119–133.

Troumbis AY and Memtsas D (2000). Observational evidence that diversity may increase productivity in Mediterranean shrublands. *Oecologia*, **125**, 101–108.

Troumbis AY, Dimitrakopoulos PG, Siamantziouras A-SD and Memtsas D (2000). Hidden diversity and productivity patterns in mixed Mediterranean grasslands. *Oikos*, **90**, 549–559.

Tscharntke T and Hawkins BA, eds. *Multitrophic level interactions*. Cambridge University Press, Cambridge.

Tunlid A (1999). Molecular biology: a linkage between microbial ecology, general ecology and organismal biology. *Oikos*, **85**, 177–189.

Turner MG, Baker WL, Peterson CJ and Peet RK (1998). Factors influencing succession: lessons from large, infrequent natural disturbances. *Ecosystems*, **1**, 511–523.

Ullmann I, Bannister P and Wilson JB (1995). The vegetation of roadside verges with respect to environmental gradients in southern New Zealand. *Journal of Vegetation Science*, **6**, 131–142.

UNEP (1995). *Global Biodiversity Assessment*. Cambridge University Press, Cambridge.

Van Breemen N (1993). Soils as biotic constructs favouring net primary productivity. *Geoderma*, **57**, 183–211.

Vance ED and Chapin FS (2001). Substrate limitations to microbial activity in taiga forest floors. *Soil Biology and Biochemistry*, **33(2)**, 173–188.

Van de Bund WJ, Goedkoop W and Johnson RK (1994). Effects of deposit-feeder activity on bacterial production and abundance in profundal lake sediment. *Journal of the North American Benthological Society*, **13**, 532–539.

Van der Heijden MGA (2002). Arbuscular mycorrhizal fungi as a determinant of plant diversity: in search for underlying mechanisms and general principles. In MGA van der Heijden and Sanders IR, eds *Mycorrhizal ecology*. Ecological Studies 157. Springer-Verlag, Heidelberg.

Van der Heijden MGA and Sanders IR, eds (2002). *Mycorrhizal ecology*. Ecological Studies 157. Springer-Verlag, Heidelberg.

Van der Heijden MGA, Boller T, Wiemken A and Sanders IR (1998a). Different arbuscular mycorrhizal fungal species are potential determinants of plant community structure. *Ecology*, **79**, 2082–2091.

Van der Heijden MGA *et al.* (1998b). Mycorrhizal fungal diversity determines plant biodiversity, ecosystem variability and productivity. *Nature*, **396**, 69–72.

Van der Heijden MGA *et al.* (1999). "Sampling effect" a problem in biodiversity manipulation? A reply to David A. Wardle. *Oikos*, **87**, 408–410.

Van der Krift TAJ and Berendse F (2001). The effect of plant species on soil nitrogen mineralization. *Journal of Ecology*, **89**, 555–561.

Van der Maesen LJG (1993). The Council of Europe and biodiversity. *Naturopa*, **73**, 6–7.

Vandermeer JH (1989). *The ecology of intercropping*. Cambridge University Press, Cambridge.

Vandermeer JH (1998). Maximizing crop yield in alley crops. *Agroforestry Systems*, **40**, 199–206.

Vandermeer JH and Andow DA (1986). Prophylactic and responsive components of an integrated pest management program. *Journal of Economic Entomology*, **79**, 299–302.

Vandermeer JH and Boucher D (1978). Varieties of mutualistic interaction in population models. *Journal of Theoretical Biology*, **74**, 549–558.

Vandermeer JH and Perfecto I (1997). The agroecosystem: a need for the conservation biologist's lens. *Conservation Biology*, **11**, 1–3.

Vandermeer JH and Schultz B (1990). Variability, stability, and risk in intercropping. In SR Gliessman, ed. *Agroecology: Researching the ecological basis for sustainable agriculture*, pp. 205–229. Springer-Verlag, New York.

Van der Putten WH, Van Dijk C and Peters BAM (1993). Plant-specific soil-borne diseases contribute to succession in foredune vegetation. *Nature*, **362**, 53–56.

Van der Putten WH *et al.* (2000). Plant species diversity as a driver of early succession in abandoned fields: a multi-site approach. *Oecologia*, **124**, 91–99.

Van der Putten WH, Vet LEM, Harvey JA and Wäckers FL (2001). Linking above- and belowground multitrophic interactions of plants, herbivores, pathogens and their antagonists. *Trends in Ecology and Evolution*, **16**, 547–554.

Van Emden HF and Williams GF (1974). Insect stability and diversity in agro-ecosystems. *Annual Review of Entomology*, **19**, 455–475.

Van Es FB (1982). Community metabolism of intertidal flats in the Ems-Dollard Estuary. *Marine Biology*, **66**, 95–108.

Vanni MJ (1996). Nutrient transport and recycling by consumers in lake food webs: implications for algal communities. In GA Polis and KO Winemiller, eds *Food webs: integration of patterns and dynamics*, pp 81–95. Chapman and Hall, New York.

Van Wensem J, Verhoef HA and Van Straalen NM (1993). Litter degradation stage as a prime factor for isopod interaction with mineralization processes. *Soil Biology and Biochemistry*, **25**, 1175–1183.

Varney PL, Scott TAJ, Cooke JS, Ryan PJ, Boatman ND and Nowakowski M (1995). *Clodinafop-propasgyl—a useful tool for management of conservation headlands*. Brighton Crop Protection Conference: Weeds. Proceedings of an International Conference. Brighton, UK, 20–23 Nov. 1995, Vol. 3, 697–972.

Vegter JJ (1983). Food and habitat specialization in coexisting springtails (Collembola, Entomobryidae). *Pedobiologia*, **25**, 253–262.

Vegter JJ, Joosse ENG and Ernsting G (1988). Community structure, distribution and population dynamics of entomobryidae (Collembola). *Journal of Animal Ecology*, **57**, 971–981.

Verhoef HA and Brussaard L (1990). Decomposition and nitrogen mineralization in natural and agroecosystems: the contribution of animals. *Biogeochemistry*, **11**, 175–211.

Verhoef HA and van Selm AJ (1983). Distribution and population dynamics of Collembola in relation to soil moisture. *Holarctic Ecology*, **6**, 387–394.

Verhoef HA and Witteven J (1980). Water balance in collembola and its relation to habitat selection; cuticular water loss and water uptake. *Journal of Insect Physiology*, **26**, 201–208.

Verhoef HA, Prast JE and Verweij RA (1988). Relative importance of fungi and algae in the diet of *Orchesella cincta* (L.) and *Tomocerus minor* (Lubbock). *Functional Ecology*, **2**, 195–201.

Verity PG (1998). Why is relating plankton community structure to pelagic production so problematic? *South African Journal of Marine Science*, **19**, 333–338.

Vitousek PM and Hooper DU (1993). Biological diversity and terrestrial ecosystem biogeochemistry. In ED Schulze and HA Mooney, eds *Biodiversity and ecosystem function*, pp. 3–14. Springer-Verlag, Berlin.

Vitousek PM and Walker LR (1989). Biological invasion by Myrica faya in Hawai'i: Plant demography, nitrogen fixation, ecosystem effects. *Ecological Monographs*, **59**, 247–265.

Vitousek PM, Weiners WA, Melillo JM, Grier CC and Gosz JR (1981). Nitrogen cycling and loss following forest perturbation: the components of response. In GW Barrett and R Rosenberg, eds *Stress effects on natural ecosystems*. John Wiley and Sons, New York.

Vitousek PM, Walker LR, Whiteaker LD, Muller-Dombois D and Matson PA (1987). Biological invasion by *Myrica faya* alters ecosystem development in Hawaii. *Science*, **238**, 802–804.

Vitousek PM *et al.* (1997a). Human alteration of the global nitrogen cycle: sources and consequences. *Ecological Aplications*, **7**, 737–750.

Vitousek PM, D'Antonio CM, Loope LL, Rejmanek M and Westbrooks R (1997b). Introduced species: a significant component of human-caused global change. *New Zealand Journal of Ecology*, **21**, 1–16.

von Euler F and Svensson S (2001). Taxonomic distinctness and species richness as measures of functional structure in bird assemblages. *Oecologia*, **129**, 304–311.

Waage JK (1991). Biodiversity as a resource for biological control. In DL Hawksworth, ed. *The biodiversity of microorganisms and invertebrates: its role in sustainable agriculture*, pp. 149–162. CAB International, Wallingford, UK.

Waage JK and Greathead DJ (1988). Biological control: challenges and opportunities. *Philosophical Tansactions of the Royal Society of London, Series B*, **318**, 111–128.

Waide RB *et al.* (1999). The relationship between productivity and species richness. *Annual Review of Ecology and Systematics*, **30**, 257–301.

Walker B (1992). Biological diversity and ecological redundancy. *Conservation Biology*, **6**, 18–23.

Walker B (1995). Conserving biological diversity through ecosystem resilience. *Conservation Biology*, **9**, 747–752.

Walker B, Kinzig A and Langridge J (1999). Plant attribute diversity, resilience, and ecosystem function: the nature and significance of dominant and minor species. *Ecosystems*, **2**, 95–113.

Wall DH and Moore JC (1999). Interactions underground. Soil biodiversity, mutualism, and ecosystem processes. *Bioscience*, **49**, 109–117.

Wall DH and Virginia RA (1999). Controls on soil biodiversity: insights from extreme environments. *Applied Soil Ecology*, **13**, 137–150.

Wall Freckman DH, Blackburn TH, Brussaard L, Hutchings P, Palmer MA and Snelgrove PR (1997). Linking biodiversity and ecosystem functioning of soils and sediments. *Ambio*, **26**, 556–562.

Wallace JB and Webster JR (1996). The role of macroinvertebrates in stream ecosystem function. *Annual Review of Entomology*, **41**, 115–139.

Wardle DA (1995). Impacts of disturbance on detritus food webs in agro-ecosystems of contrasting tillage and weed management practices. *Advances in Ecological Research*, **26**, 105–185.

Wardle DA (1998). A more reliable design for biodiversity study? *Nature*, **394**, 30.

Wardle DA (1999). Is 'sampling effect' a problem for experiments investigating biodiversity—ecosystem function relationships? *Oikos*, **87**, 403–407.

Wardle DA (2001). Experimental demonstration that plant diversity reduces invasibility—evidence of a biological mechanisms or a consequence of sampling effect? *Oikos*, **95**, 1–170.

Wardle DA (2002). *Communities and ecosystems: Linking the aboveground and belowground components.* Princeton University Press, Princeton.

Wardle DA and Giller KE (1997). The quest for a contemporary ecological dimension to soil biology. *Soil Biology and Biochemistry*, **28**, 1549–1554.

Wardle DA and Nicholson KS (1996). Synergistic effects of grassland plant species on soil microbial biomass and activity: implications for ecosystem-level effects of enriched plant diversity. *Functional Ecology*, **10**, 410–416.

Wardle DA, Nicholson KS and Rahman A (1995). Ecological effects of the invasive weed species *Senecio jacobaea* L. (ragwort) in a New Zealand pasture. *Agriculture, Ecosystems and Environment*, **56**, 19–28.

Wardle DA, Bonner KI and Nicholson KS (1997a). Biodiversity and plant litter: experimental evidence which does not support the view that enhanced species richness improves ecosystem function. *Oikos*, **79**, 247–258.

Wardle DA, Zackrisson O, Hörnberg G and Gallet C (1997b). The influence of island area on ecosystem properties. *Science*, **277**, 1296–1299.

Wardle DA, Zackrisson O, Hörnberg G and Gallet C (1997c). Biodiversity and ecosystem properties. *Science*, **278**, 1867–1869.

Wardle DA *et al.* (1999). Plant removals in perennial grassland: vegetation dynamics, decomposers, soil biodiversity, and ecosystem properties. *Ecological Monographs*, **69**, 535–568.

Wardle DA, Bonner KI and Barker GM (2000a). Stability of ecosystem properties in response to above-ground functional group richness and composition. *Oikos*, **89**, 11–23.

Wardle DA *et al.* (2000b). Biodiversity and Ecosystem Function: an Issue in Ecology. *Bulletin of the Ecological Society of America*, **81**, 235–239.

Wardle DA, Barker GM, Yeates GW, Bonner KI and Ghani A (2001). Impacts of introduced browsing mammals in New Zealand forests on decomposer communities, soil biodiversity and ecosystem properties. *Ecological Monographs*, **71**, 587–614.

Warren PH (1996). The effects of between-habitat dispersal rate on protist communities and metacommunities at two spatial scales. *Oecologia*, **105**, 132–140.

Warren MS *et al.* (2001). Rapid responses of British butterflies to opposing forces of climate and habitat change. *Nature*, **414**, 65–69.

Warwick RM and Clarke KR (1998). Taxonomic distinctness and environmental assessment. *Journal of Applied Ecology*, **35**, 532–543.

Weatherby AJ, Warren PH and Law R (1998). Coexistence and collapse: an experimental investigation of the persistent communities of a protist species pool. *Journal of Animal Ecology*, **67**, 554–566.

Webster JR, Waide JB and Patten BC (1974). Nutrient recycling and the stability of ecosystems. In FG Horwell, JB Gentry and MH Smith, eds *Mineral cycling in southeastern ecosystems.* National Technical Information Service, Springfield, Virginia.

Wedin D (1995). Species, nitrogen, and grassland dynamics: The constraints of stuff. In CG Jones and JH Lawton, eds *Linking species and ecosystems*, pp. 253–262. Chapman and Hall, New York.

Wedin DA and Tilman D (1996). Influence of nitrogen loading and species composition on the carbon balance of grasslands. *Science*, **274**, 1720–1723.

Weekers PHH, Bodelier PLE, Wijen JPH and Vogels GD (1993). Effects of grazing by the free-living soil amoebae *Acanthamoeba castellanii*, *Acanthamoeba polyphaga*, and *Hartmannella vermiformis* on various bacteria. *Applied and Environmental Microbiology*, **59**, 2317–2319.

Weiher E and Keddy P (1998). Assembly rules, null models, and trait dispersion: new questions from old patterns. *Oikos*, **74**, 159–164.

Weiher E, Clarke G and Keddy P (1995). Community assembly rules, morphological dispersion, and the coexistence of plant species. *Oikos*, **81**, 309–322.

Werner D (1992). *Symbiosis of plant and microbes.* Chapman and Hall, London.

Werner EE (1991). Nonlethal effects of a predator on competitive interactions betwen two anuran larvae. *Ecology*, **72**, 1709–1720.

Werner EE (1992). Individual behavior and higher-order species interactions. *American Naturalist*, **140**, S5–S32.

Werner EE (1998). Ecological experiments and a research program in community ecology. In WJJ Resetarits and J Bernardo, eds *Experimental ecology. Issues and perspectives*, pp. 3–26. Oxford University Press, Oxford.

Werner EE and Anholt BR (1996). Predator-induced behavioral indirect effects: Consequences to competitive interactions in anuran larvae, *Ecology*, **77**, 157–169.

Western D and Pearl MC, eds (1989). *Conservation for the twenty-first century.* Oxford University Press, New York.

Westoby M (1998). A leaf-height-seed (LHS) plant ecology strategy scheme. *Plant and Soil*, **199**, 213–227.

Whittaker RH (1975). *Communities and ecosystems*, 2nd edn. MacMillan, New York.

Wickham SA (1995). Trophic relations between cyclopid copepods and ciliated protists: complex interactions link the microbial and classic food webs. *Limnology and Oceanography*, **40**, 1173–1181.

Wicklum D and Davies RW (1995). Ecosystem health and integrity? *Canadian Journal of Botany*, **73**, 997–1000.

Willey RW (1979). Intercropping—its importance and its research needs. Part II. Agronomic relationships. *Field Crop Abstracts*, **32**, 73–85.

Williams BL and Griffiths BS (1989). Enhanced nutrient mineralization and leaching from decomposing Sitka spruce litter by enchytraeid worms. *Soil Biology and Biochemistry*, **21**, 183–188.

Williams RJ and Martinez ND (2000). Simple rules yield complex food webs. *Nature*, **404**, 180–183.

Wilsey BJ and Potvin C (2000). Biodiversity and ecosystem functioning: importance of species evenness in an old field. *Ecology*, **81**, 887–892.

Wilson DS (1992). Complex interactions in metacommunities, with implications for biodiversity and higher levels of selection. *Ecology*, **73**, 1984–2000.

Wilson D and Faeth SH (2001). Do fungal endophytes result in selection for leafminer ovipositional preference? *Ecology*, **82**, 1097–1111.

Wilson EO (1988). The current state of biological diversity. In EO Wilson, ed. *Biodiversity*, pp. 3–18. National Academy Press, Washington DC.

Wilson EO, ed. (1998). *Biodiversity*. National Academy Press, Washington DC.

Wilson GWT and Hartnett DC (1997). Effects of mycorrhizae on plant growth and dynamics in experimental tallgrass prairie microcosms. *American journal of Botany*, **84**, 478–482.

Wilson JB (1999). Guilds, functional types and ecological groups. *Oikos*, **86**, 507–522.

Wiser SK, Allen RB, Clinton PW and Platt KH (1998). Community structure and forest invasion by an exotic herb over 23 years. *Ecology*, **79**, 2071–2081.

Wolfe MS (1985). The current status and prospects of multiline cultivars and variety mixtures for disease resistance. *Annual Review of Phytopathology*, **23**, 251–273.

Wolter C and Scheu S (1999) Changes in bacterial numbers and hyphal lengths during the gut passage through *Lumbricus terrestris* (Lumbricidae, Oligochaeta). *Pedobiologia* **43**: 891–900.

Wolters V (1988). Effects of *Mesenchytraeus glandulosus* (Oligochaeta, Enchytraeidae) on decomposition processes. *Pedobiologia*, **32**, 387–398.

Wolters V (1997). *Functional implications of biodiversity in soil*. Office for official publications of the European Community.

Wolters V *et al.* (2000). Effects of global changes on above- and belowground biodiversity: implications for ecosystem functioning. *Bioscience*, **50**, 1–10.

Woods KD (1993). Effects on invasion by *Lonicera tatarica* L. on herbs and tree seedlings in four New England forests. *American Midland Naturalist*, **130**, 62–74.

Woods LE, Cole CV, Elliott ET, Anderson RV and Coleman DC (1982). Nitrogen transformations in soil as affected by bacterial-microfaunal interactions. *Soil Biology and Biochemistry*, **14**, 93–98.

Wooten JT (1995). Effects of birds on sea urchins and algae: a lower intertidal trophic cascade. *EcoScience*, **2**, 321–328.

Workneh F and van Bruggen AHC (1994). Microbial density, composition, and diversity i organically and conventionally managed rhizosphere soil in relation to suppression of corky root of tomatoes. *Applied Soil Ecology*, **1**, 219–230.

Wright DH and Coleman DC (1993). Patterns of survival and extinction of nematodes in isolated soil. *Oikos*, **67**, 563–572.

Yachi S and M Loreau (1999). Biodiversity and ecosystem productivity in a fluctuating environment: The insurance hypothesis. *Proceedings of the National Academy of Sciences of the United States of America*, **96**, 1463–1468.

Yamamura N, Yachi S and Higashi M (2001). An ecosystem organization model explaining diversity at ecosystem level: coevolution of primary producer and decomposer. *Ecological Research*, **16**, 975–982.

Yang YH, Yao J, Hu S and Qi Y (2000). Effects of agricultural chemicals on DNA sequence diversity of soil microbial community: a study with RAPD marker. *Microbial Ecology*, **39**, 72–79.

Yates F (1934). The analysis of multiple classifications with unequal numbers in the different classes. *Journal of the American Statistical Association*, **29**, 51–66.

Yeates GW and King KL (1997). Soil nematodes as indicators of the effect of management on grasslands in the New England Tablelands (NSW): comparison of native and improved grasslands. *Pedobiologia*, **41**, 526–536.

Yeates GW, Watson RN and Steele KW (1985). Complementary distribution of *Meloidogyne*, *Heterodera*, and *Pratylenchus* (Nematoda: Tylenchida) in roots of white clover. Proceedings of the 4th Australian Conference on Grassland Invertebrate Ecology, pp. 71–79.

Yeates GW, Bardgett RD, Cook R, Hobbs PJ, Bowling PJ and Potter JF (1997). Faunal and microbial diversity in three Welsh grassland soils under conventional and organic management regimes. *Journal of Applied Ecology*, **34**, 453–470.

Yodzis P (1981). The stability of real ecosystems. *Nature*, **289**, 674–676.

Yodzis P (1988). The indeterminacy of ecological interactions as perceived through perturbation experiments. *Ecology*, **69**, 508–515.

Yu DW and Wilson HB (2001). The competition-colonization trade-off is dead; long live the competition-colonization trade-off. *The American Naturalist*, **158**, 49–63.

Zak DR *et al.* (1994). Plant production and soil microorganisms in late-successional ecosystems: a continental-scale study. *Ecology*, **75**, 2333–2347.

Zeller V, Bardgett RD and Tappeiner U (2001). Site and management effects on soil microbial properties of subalpine meadows: a study of land abandonment along a north-south gradient in the European Alps. *Soil Biology and Biochemistry*, **33**, 639–649.

Zheng DW, Bengtsson J and Agren GI (1997). Soil food webs and ecosystem processes: decomposition in donor-control and Lotka-Volterra systems. *American Naturalist*, **149**, 125–148.

Zhu Y Chen H *et al.* (2000). Genetic diversity and disease control in rice. *Nature*, **406**, 718–722.

Zobel M (1992). Plant species coexistence—the role of historical, evolutionary and ecological factors. *Oikos*, **65**, 314–320.

Zwart KB *et al.* (1994). Population dynamics in the belowground food webs in two different agricultural systems. *Agriculture Ecosystems and Environment*, **51**, 187–198.

Index

Printed in the United States
By Bookmasters